Drought and Drought Mitigation in Europe

Advances in Natural and Technological Hazards Research

VOLUME 14

The titles published in this series are listed at the end of this volume.

Drought and Drought Mitigation in Europe

Edited by

JÜRGEN V. VOGT
Space Applications Institute, JRC, Ispra, Italy

and

FRANCESCA SOMMA
Department of Agricultural Engineering,
University of Catania, Italy

SPRINGER-SCIENCE+BUSINESS MEDIA, B.V.

A C.I.P. Catalogue record for this book is available from the Library of Congress.

ISBN 978-90-481-5568-2 ISBN 978-94-015-9472-1 (eBook)
DOI 10.1007/978-94-015-9472-1
EUR 19513 EN

Printed on acid-free paper

Contents

Part 4: Drought Monitoring

Part 5: Drought Impact

Part 6: Drought Mitigation Strategies

Part 7: Conclusions and Recommendations

Preface

Drought is one of the major weather related disasters. Persisting over months or years, it can affect large areas and may have serious environmental, social and economic impacts. These impacts depend on the duration, severity and spatial extent of the precipitation deficit, but also and to a large extent on the environmental and socio-economic vulnerability of affected regions. The development and implementation of adequate mitigation strategies, however, gives the opportunity for a significant decrease of such vulnerability.

Recent European droughts have emphasised that the impact on European economies can be significant. While Europe as a whole has abundant freshwater resources, there exists a strong regional imbalance across the continent. Water shortage, therefore, is an important problem in many European regions, specifically in the semi-arid and continental climatic zones.

In response to the growing concern about this natural hazard, the Joint Research Centre (JRC) of the European Commission initiated a research and development project on drought monitoring and mitigation in Europe in 1997. The aim of this project is to perform and promote applied research in this field and to stimulate co-operation in the European context. In the frame of these activities an international workshop on 'Drought and Drought Mitigation in Europe' was organised in March 1999. The workshop, which was held at the Space Applications Institute (SAI) of the JRC in Ispra (Varese), Italy, was attended by about 30 experts from Europe, the United States and Israel with professional backgrounds in meteorology, hydrology, geography, agronomy, economics and politics. The intention of the workshop was to analyse the state-of-the-art in drought research, drought planning and drought mitigation strategies in Europe and to elaborate recommendations for future action. The papers presented in this volume result from this workshop. All papers have been discussed at the workshop and have been reviewed and revised in the following months. The material has been organised in a way to reflect the various areas of concern as related to the drought phenomenon. It covers aspects of drought definition, drought risk assessment, drought monitoring, and drought mitigation strategies.

During the workshop a lively and often controversial discussion emerged over the issues at hand. This controversy of opinions is also reflected in the papers and hopefully reflects the variety of opinions in both the scientific and political communities. In spite of this controversy, the participants agreed to a list of generally accepted conclusions and common recommendations for future action. The importance of a better link between scientific and political programmes and of the necessity to raise the awareness of the political community for the need to implement a pro-active approach to the problem are certainly among the most important ones. As a major outcome, an organisational and thematical structure for the set-up of a European Network on Drought Research and Mitigation has been proposed. This network should work towards a comprehensive understanding of the problem and should provide the necessary link between science, policy and management.

The problem of drought is a very complex one. It involves many aspects from science, socio-economics, politics and management, and in this sense is truly interdisciplinary. Yet we hope to offer some useful insights into the various problems related to drought, paying particular attention to the situation in Europe. With this regional focus the book is to be seen as complementary to other books published on the subject, covering both theoretical considerations and examples of drought management and policy in the United States and world-wide[i]*.*

We believe that the book will be of benefit to both the scientific community interested in the various aspects of drought detection, drought monitoring and drought management as well as to managers and politicians involved in problems related to water management, risk assessment and spatial planning at the European, national and regional levels. Finally, students in Earth sciences, especially in geography, climatology, hydrology and agriculture will find useful material in this collection of papers.

We wish to thank the Space Applications Institute, JRC Ispra, and the UK National Co-ordination Committee for the International Decade for Natural Disaster Reduction (IDNDR) for their financial and logistic support for the meeting as well as for the continuous encouragement of our work.

We hope that the papers as well as the conclusions and recommendations will help to raise the awareness of this important problem and to stimulate future co-operation and co-ordination in this field. A lot of progress is still to be made towards a comprehensive Drought Mitigation Framework for the European continent, including both adequate monitoring facilities and the implementation of pro-active mitigation strategies.

Ispra, May 2000

Jürgen Vogt
Francesca Somma

---

[i] For example:

Wilhite, D.A. and Easterling, W.E. (Eds.) (1987) *Planning for Drought. Toward a Reduction of Societal Vulnerability*, Westview Press, Boulder and London.

Wilhite, D.A. (Ed.) (1993) *Drought Assessment, Management and Planning: Theory and Case Studies*, Kluwer Academic Publishers, Boston, Dordrecht and London.

White, D.H. (Ed.) (1998) Drought Policy, Assessment and Declaration, Special Issue of *Agricultural Systems*, 57(3).

Wilhite, D.A. (Ed.) (2000) *Droughts. A Global Assessment*, Routledge, London and New York.

Part 1
Introduction

INTRODUCTION

J.V. VOGT and F. SOMMA

Space Applications Institute
EC Joint Research Centre
21020 Ispra (Va), Italy

During recent years, natural disasters gained increasing attention in the minds of both the general public and the politicians. In general, concern is rising about the frequency and impact of weather-related disasters such as droughts, floods and storms. All of these events originate from extremes of the climate system such as persistent anticyclonic conditions or strong gradients in atmospheric pressure and temperature. Climate change scenarios predict a likely increase in water-related stresses in Europe, especially in those areas which are already sensitive in terms of their hydrology: the Mediterranean region, the Alps, northern Scandinavia, coastal zones and Eastern Europe (IPCC, 1997).

Out of the weather-related disasters, drought is certainly the most complex one and both its causes and multifaceted impacts are not well understood. Persisting over months or years, droughts affect extended areas and large populations. From a climatic point of view, drought is a recurring phenomenon. It results from a shortfall in precipitation over an extended period of time, as well as from the inadequate timing or the ineffectiveness of the precipitation. Its environmental and socio-economic impacts, however, stem not only from the duration, severity and spatial extent of the precipitation deficit, but also and to a large extent from the environmental, social and economic vulnerability of affected regions. Key factors in this context are inadequate land use practices, unsustainable management of water resources and inadequate risk management.

Perceptions of what is to be called a drought and of its impact, however, vary to a large extent. Unlike a flood or a storm, a drought does not result in a sudden impact. It is rather a creeping phenomenon that establishes over a longer time period. Drought impacts are less obvious and spread over larger areas than damages resulting from other natural hazards (Wilhite, 1993). These spatial and temporal aspects and the complex interactions between environment and society make it difficult to evaluate the damage caused by a drought. Negative effects range from reduced agricultural production and shortfalls in public water supply to reduced hydropower generation and long-term environmental degradation. Questions already arise with respect to the start and the end of a drought and different groups will perceive the situation in completely different ways. While the general public and the tourist may enjoy the sunny weather, farmers and managers of water-related industries will be concerned with the effect on crop yields and production. As a consequence, there is no universal definition of a drought and

J.V. Vogt and F. Somma (eds.), Drought and Drought Mitigation in Europe, 3–5.

drought definitions must necessarily be region and application specific in order to be of use in practical circumstances.

Droughts frequently occur in many regions of Europe and their impact results in significant economic, social and environmental costs. The severe drought experienced throughout the Mediterranean in the early nineties, for example, caused an estimated damage of several billion Euros in Spain alone (Garrido and Goméz-Ramos, 2000). The current winter (1999-2000) has again seen strong precipitation deficits both in the Mediterranean and in Central-Eastern Europe with serious effects on cereal production. For a more detailed listing of European droughts see the following chapter by Bradford.

Differences in the quantity and the use of the abstracted water across Europe influence the potential impact of drought events. In the humid climates of Northern and Western Europe, the impact of prolonged dry spells is mainly seen in the ecological consequences for wetlands and in problems to provide sufficient drinking water of high quality. In Southern Europe, on the other hand, where 60% to 70% of the abstracted water is used for irrigation, the impact on agriculture is of major concern. In the Mediterranean, where prolonged dry periods are a characteristic of the climate, negative effects of drought have been aggravated due to constantly increasing demands for water, leading to an overexploitation of the available water resources in many areas (EEA, 1995; EEA 1998; EEA 1999; EEA and UNEP, 1997; Estrela *et al.*, 1997).

A significant decrease in agricultural production, an increased likelihood of forest fires and problems to ensure a municipal water supply of adequate quality and quantity are among the immediate and obvious effects of drought. Mean and long-term effects such as land degradation and desertification are, however, most severe consequences of repeated drought events. In Europe, this is especially true for the Mediterranean regions but also for parts of Central and Eastern Europe.

Growing concern about the quality and sustainable management of Europe's water resources and predictions of climate change impacts have led to numerous political initiatives over the last decade. Examples are the "Environmental Programme for Europe", the "Water Resources Framework Directive", the "Groundwater Action Programme" or the "Task Force on Environment-Water" (EEA and UNEP, 1997; EC, 1998). Any of these initiatives as well as the environmental reports of the European Environment Agency (EEA, 1995; EEA 1998; EEA 1999) or political frameworks such as the European Spatial Development Perspective (ESDP) highlight the problem and promote the development of adequate management strategies.

Despite the increasing awareness of the drought hazard, tools and strategies for risk assessment, prediction, monitoring and mitigation of droughts are, however, not well defined. There is no European drought policy as such and institutional frameworks to cope with drought situations are only weakly developed or not existing in most European countries. In general, strategies are more reactive than pro-active and standardisation of methods, as well as co-ordination and co-operation at a European level are weak. Consequently, there is a need to develop effective monitoring tools and mitigation strategies as well as to pursue research on the immediate and long-term impact of drought events. The simultaneous set-up of a policy and an institutional framework on the regional, national and European levels is certainly a most important issue.

This book is dedicated to further our understanding of the drought problem in Europe and to discuss policy and management options to mitigate its impacts. Representing a compilation of contributions from a meeting of experts from the various disciplines involved, it gives insights from different perspectives and concludes with a series of recommendations for further action. It has been structured in a way to reflect the various aspects of the drought phenomenon. After the introductory overview on the drought problems in Europe (part one), the difficulties related to the definition of drought are highlighted from different hydrological and agricultural perspectives (part two). Questions of drought risk are then discussed from climatological, hydrological and pedological points of view, including issues such as vulnerability and climate change. Here we include an overview on the development of a comprehensive drought policy in the United States, which could serve as a model for Europe (part three). Conventional and more advanced approaches to drought monitoring are then presented and recent progress in the United States is described (part four). The quantification of drought impact in socio-economic, hydrologic and agricultural terms and different views of how to mitigate such impacts are central themes of part five and six. Part seven gives the conclusions and recommendations formulated by the plenum of experts during the workshop.

REFERENCES

Estrela T., Marcuello C., Iglesias A. (1997) *Water Resources Problems in Southern Europe*, EEA Topic Report 15 - 1996 - Inland Waters, Office for Official Publications of the European Communities, Luxembourg.

EC - European Commission – Environment Water Task Force (1998) *Freshwater. A Challenge for Research and Innovation. A Concerted European Response*, European Communities, July 1998, EUR 18098 EN.

EEA - European Environment Agency (1995) *Europe's Environment. The Dobrís Assessment*, edited by D. Stanners and Ph. Bourdeau, Office for Official Publications of the European Communities, Luxembourg.

EEA - European Environment Agency (1998) *Europe's Environment. The Second Assessment,* Office for Official Publications of the European Communities, Luxembourg.

EEA - European Environment Agency (1999) *Environment in the European Union at the Turn of the Century*, Office for Official Publications of the European Communities, Luxembourg.

EEA and UNEP - European Environment Agency and United Nations Environment Programme (1997) *Water Stress in Europe – Can the Challenge be Met?* Copenhagen and Chatelaine.

Garrido, A. and Gómez-Ramos, A. (2000) Socio-Economic Aspects of Droughts, in J.V. Vogt and F. Somma (eds.) *Drought and Drought Mitigation in Europe*, Kluwer Academic Publishers, Drodrecht, 197-207 (this volume)

IPCC - Intergovernmental Panel on Climate Change (1997) *The Regional Impacts of Climate Change. An Assessment of Vulnerability*, edited by R.T. Watson, M.C. Zinyowera, R.H. Moss, Cambridge University Press, Cambridge.

Wilhite, D.A. (Ed.) (1993) The Enigma of Drought, in: D.A. Wilhite (ed.) *Drought Assessment, Management and Planning: Theory and Case Studies*, Kluwer Academic Publishers, Dordrecht, 3-15.

DROUGHT EVENTS IN EUROPE

R.B. BRADFORD
Institute of Hydrology
Wallingford, Oxon, OX10 8BB, UK

Abstract. This paper presents a short summary of a report on '*Drought Events in Europe*' (Bradford, 1999) commissioned by the Space Applications Institute (SAI) of the EC Joint Research Centre to identify particular drought-related topics for discussion during the workshop on 'Drought and Drought Mitigation in Europe'. This report was based on a review of recent literature on drought occurrence and extent, regional aspects of European droughts, and drought mitigation and planning, with an emphasis on hydrological droughts over the past decade. It included a brief description of common drought indices, impacts, events and mitigation measures illustrated by examples of studies on drought-related issues in Europe.

1. INTRODUCTION

Droughts are a complex, natural hazard that, to a varying degree, affect some part of Europe almost every year. Several severe and prolonged droughts have occurred in Europe in recent years, whilst current climate change scenarios suggest that the frequency and impacts of such events could increase. This has raised the awareness of the potential vulnerability of Europe to drought hazards in both the short and longer-term, particularly with increasing demands on available water resources, and the corresponding need for more research on their origins, potential impacts, appropriate mitigation strategies and more effective monitoring tools.

2. DEFINING DROUGHT OCCURRENCE AND EXTENT

No two droughts have the same extent, duration, intensity or impacts. Furthermore, they may be perceived differently by different water users and consequently drought definitions need to be related to a particular problem (Bidwell, 1973). The main types of drought are meteorological, agricultural, hydrological and water resources. These represent a gradation of frequency, severity and persistence.

As yet no universal and objective definition of a drought has been agreed that would satisfy the different perspectives of scientists, water planners and policy makers or represent all the different variables and spatial or temporal factors involved. The reliability of each method tends to depend on the length and reliability of the data series available, although, in general, there is a lack of suitable regional long-term data. A consistent definition may require a pooled analysis to define an independent sequence of droughts taking into account possible non-stationarity (Tallaksen, 2000).

J.V. Vogt and F. Somma (eds.), Drought and Drought Mitigation in Europe, 7–20.

A wide variety of different indices and methods have been developed to define and monitor droughts. Common methods include: Palmer Drought Severity Index (PDSI), Crop Moisture Index (CMI), Standard Precipitation Index (SPI) and Drought Magnitude (DM), Rainfall Index (RI), Surface Water Supply Index (SWSI), Threshold (Truncation) method, Concept of Runs and Cumulative Rainfall Departure (CRD). A summary of different drought indices is given in Tate and Gustard (2000).

A deficit in precipitation is commonly used to measure drought severity, although the timing of the deficit is important. The threshold approach is being used increasingly to express a drought in terms of its duration and severity since this approach can be applied to hydro-meteorological variables as well as sector criteria (Dracup *et al.*, 1980). Low flow measures, such as flow duration curves or the regionalisation of recession parameters (Demuth *et al.*, 2000), have a range of applications in drought studies and for strategic and operational mitigation measures.

Byun and Wilhite (1999) have identified four main weaknesses of conventional drought indices:

- Most lack sufficient precision to detect the onset and end of drought conditions or the accumulated effects.
- The aggravating effects of runoff and evapotranspiration that accumulate with time are not effectively taken into account.
- They are often based on a monthly time step that limits their usefulness.
- Most do not distinguish the different effects of drought on surface and groundwater supplies.

They have proposed a more objective, practical approach based on effective precipitation (EP), the sum of daily precipitation with a time-dependent reduction function. This is complemented by three indices: MEP (mean daily EP), to define the climatological characteristics of precipitation as a water resource for a station or area; DEP (deviation of EP from MEP); and SEP (standardised value of DEP). Consecutive days of negative SEP are used to define the onset, end and duration of a water deficit period. A further four indices are then used to show drought severity:

- Accumulation of negative SEP to define the combined duration and severity of a precipitation deficit.
- Accumulated precipitation deficit to show the extent to which precipitation departs from normal during the deficit period.
- Precipitation amount necessary for the return to normal.
- Effective drought index, a standardised index that can be applied to assess drought worldwide.

Increasingly remote sensing data are applied to develop spatially and temporally consistent drought indices to identify, assess and monitor drought conditions at different scales. These are based on a combination of remote sensing data such as from the Advanced Very High Resolution Radiometer (AVHRR) aboard the US NOAA satellites, digital terrain models and land use/land cover maps together with long-term meteorological data (Vogt *et al.*, 2000).

Gustard *et al.* (1997) have used gauged and ungauged flow data from the FRIEND European Water Archive to calculate the flow duration percentile for each day from February to August 1976, which was then mapped at the European scale. This technique could be used to show the spatial coherence of hydrological droughts at the European scale and their growth and decay. However, further research is required for practical applications in real-time, to accommodate more localised controls on catchment response (e.g. hydrogeological properties) and to map the progression of droughts in ungauged areas.

3. REGIONAL ASPECTS OF EUROPEAN DROUGHTS.

The spatial variability in the occurrence, duration and severity of droughts across Europe is influenced by the general climatic conditions prevailing within a particular region and by local influences, such as altitude and basin size. Some of the more severe and widespread droughts that have affected European countries recently are listed in table 1. These include Greece (1987/8 to 1992/3), Italy (1988-90), Spain (1990-1996), and UK (1988-92 and 1995/6).

The spatial and temporal patterns of drought in Europe are related to patterns of anomalous climatic behaviour: the more severe droughts are often a regional phenomenon resulting from large-scale anomalies. Persistent anticyclonic conditions are a common feature of European droughts, although the predominance of anticyclonic (or cyclonic) conditions within one region may be compensated by the opposite anomalous conditions in another region.

The frequency and duration of extreme events such as droughts is expected to increase with climate change. In general, multi-objective systems and water resources systems that are already highly stressed will be more sensitive to climate change. The greatest sensitivity to change is in the drier parts of southern and eastern Europe, where it has been suggested that droughts are becoming more frequent and of longer duration (Arnell, 1997). By the 2050s there could be a decrease in annual average runoff of over 30% in some areas of southern Europe, where the rate of soil degradation is also expected to increase leading to reduction in soil water storage capacity. Low flows could become more extreme across most of Europe whilst a substantial reduction in snowfall will alter the current temporal distribution of river flows. There has been limited work on the impact of climate change on groundwater recharge.

Drought impacts can be aggravated by anthropogenic activities, such as abstraction for water supplies or changes in land use and land drainage. Such activities, particularly over the past 25-30 years, have altered the hydrological regimes of many rivers and lakes in terms of their water quality, availability and water balances. In general, catchments with increased reservoir storage and/or river regulation are likely to be less vulnerable to drought.

Table 1: Recent drought events in Europe.

Date	Region	Characteristics
1968	Hungary	Rainfall (Feb-July) 10% of normal.
1969	Sweden, UK, Denmark	Very dry summer.
1971	Most of Europe	Exceptionally dry year. Lowest winter rainfall in Spain for 30 years; intensive summer drought in Poland; Rhine lowest since 1818.
1972	USSR	Lowest river levels for 50-80 years.
1973	North and East Europe	Very dry spring in eastern UK; low winter rain/snowfall in Austria, Germany and Czechoslovakia.
1974	Scandinavia, France, Holland, Austria	Dry spring in Norway (rainless April in some areas), Denmark, Holland, Austria; 9 week spring drought in Sweden; low rainfall April-August in France.
1975	North and East Europe	Dry winter in Eastern Europe, low river levels; February-August rainfall in Ireland, UK lowest this century; summer rainfall in Sweden lowest on record; October rainfall in Belgium lowest on record.
1976	Northern Europe (Scandinavia to France)	Severe drought, especially in SE England, N France. Hot dry summer following a dry winter. Record rainfall deficits. Surface water and groundwater deficits.
	Effects also spread to Eastern Europe	UK: 16 month duration, unprecedented intensity. Very dry summer in Scotland, impact worst in SE England with supply restrictions. March-September rainfall in Netherlands, Denmark, Norway, Sweden, Scotland and N France lowest on record. Very low soil moisture deficits in Ireland.
1977	UK	Dry summer from May-August. Scotland (mid) – driest summer since 1868 N Ireland – seventh successive summer with below average rainfall.
1978	South-East UK and W France	SE England – driest autumn since 1752. W France – driest Oct/Nov since records began.
1984	North and West UK	Very dry spring and summer. Affected surface water.
1988-1992	Most of Europe	Prolonged abnormal circulation pattern causing rainfall deficiency over wide geographic area interspersed with short wet periods, high summer temperatures and above average winter temperatures with reduced rainfall. Timing and severity of maximum runoff deficit varied widely. Water resource problems across Europe exacerbated by increased demand, e.g. 3000 rivers dried up in France in late 1990, irrigation restrictions. Severe summer drought in NE Germany in 1992, with crop production reduced by 22%. Above average rainfall in Norway and Scotland.
1990-1995	Spain, Portugal	Prolonged drought across all of Spain except North coast. Most intense between September 1994 and August 1995. HEP stations shut down.

Table 1: Recent drought events in Europe (continued).

Date	Region	Characteristics
1992-1993	Bulgaria, Hungary	Very hot dry summer 1992. Continued with below average rainfall to October 193. Severe loss of agricultural production in Bulgaria. Worst drought in USSR for 10 years.
1995	Ireland and UK Norway, Sweden	Hot dry summer and autumn. Dry soil. Impact on surface water sources, not ground water. Low temperature, little winter snow in Nordic countries.
1996	Bulgaria	Hot dry summer across whole country.
1997	France, UK, Ireland, Portugal, Germany	Very low rainfall.

Socio-economic factors that have led to increased water demands have made even the wetter parts of Europe susceptible to drought, especially in areas of high population density. It has been estimated that some 84 million people living in cities with more than 100,000 inhabitants already experience occasional limitations in water supply (European Commission, 1994). The vulnerability to drought and the frequency of water shortages will increase as demands converge with the availability of water supplies. Rising water demands also reduce the flexibility to meet all demands during drought events and have led to increased and more intensive exploitation of water resources, such that some areas have limited scope for meeting demands during droughts. Many countries have accepted the need to reduce water demand through improved water use, re-cycling, etc. Measures taken during a drought to use water more efficiently can bring longer-term benefits, such as postponing major investments in new sources.

The Mediterranean region relies heavily on irrigation for agricultural production. Agriculture accounts for 80% of total water demand in Greece, more than 50% in Italy, 68% in Spain and 52% in Portugal, compared with under 10% in northern Europe. The demand for irrigation water continues to increase placing greater stress on already limited water resources during the summer months. Consequently, less severe droughts in future could have a greater or more frequent impact as water demands for irrigation increase. There is also a shift of population from agriculture to industry, particularly in southern Europe. Coastal regions with a seasonal tourist industry can be particularly affected by water shortages.

Uncontrolled groundwater development has led to a reduction in baseflows, wetland degradation and adverse water quality impacts. These effects may be difficult to distinguish from natural variations, but even so tend to aggravate the onset and impact of a drought and possibly delay the recovery of ecosystems following a drought.

The type of land cover directly influences storage conditions and the hydrological response of a catchment. Changes in land use, such as afforestation, land drainage and agricultural intensification, can alter the vulnerability of an area to drought. Poor land use practices, particularly in more arid areas, and vegetation loss lead to soil erosion which reduces the capacity for infiltration and aggravates the vulnerability of an area to droughts and drought impacts.

The impacts of a drought are varied and affect many aspects of the environment, economy and society. They may be direct or indirect, singular or cumulative, and immediate or delayed. The magnitude of drought impacts is often related to the timing of the onset of the shortage in precipitation and the antecedent conditions. Often the area affected by a severe drought spreads gradually and hence droughts differ in terms of their spatial characteristics. Stored water resources are often used for multiple purposes and this complicates the sequence and quantification of impacts, whilst competition for these resources also increases during droughts. The complex interactions that take place between physical and socio-economic systems during droughts are difficult to quantify.

In general, the agricultural sector is often the first to be affected by drought due to its dependence on soil moisture storage. Changes in agriculture, primarily for economic objectives, have had a detrimental effect on water resources in many areas. Such areas have consequently become more susceptible to the impacts of drought.

Drought impacts may affect other sectors for months or even years depending on the use and availability of stored surface and groundwater supplies. The length of a recovery period is a function of water use, water supply source(s), drought intensity and duration, and the amount and timing of the precipitation as the drought episode finishes.

The economic impacts of drought are numerous, ranging from direct losses, such as in the food, energy and industrial production sectors, to less obvious indirect losses, such as in the recreation and insurance sectors (e.g. land subsidence). A drought can affect the economy of whole regions and increase conflicts between water users. Social impacts include public safety, public health and water use conflicts. However, quantifying the socio-economic impacts of a drought can be difficult due to the range and number of groups that can be affected and the geographical extent of a drought-affected area.

Environmental impacts associated with droughts include damage to habitats, forest fires, soil erosion, degradation of landscape and water quality, raised water temperatures, fish mortality, algal blooms, weed growth and poor air quality. Poor dilution of effluents can seriously affect aquatic ecosystems. Environmental impacts, however, are often difficult to quantify. Natural systems tend to adapt to dry periods and, whilst such impacts may persist, they may not necessarily be permanent. New EU environmental regulations to preserve water quality and ecosystems are now an important consideration for water managers during droughts.

4. DROUGHT MITIGATION AND PLANNING IN EUROPE

Drought perception is a function of the sensitivity of an activity to water availability, the time of year and the magnitude of the deficit (Bonacci, 1993). In general, droughts are not perceived as natural disasters to the same extent as floods, because droughts are less frequent, develop slowly and extend over a longer period. This usually allows strategic mitigation measures to be put in place or to be introduced operationally as a drought develops. However, for similar reasons the organisational activities for drought alleviation and loss compensation are often less advanced than for flood events. Perceptions are beginning to change as a result of several severe and widespread droughts in recent years, increasing stress on available resources and the environment, the prospect that droughts may become a greater hazard in future, and from increased public awareness.

Drought impact mitigation actions can be categorised in various ways, but all are aimed at alleviating water shortages by increasing water availability and/or reducing water demands. They can be considered as pro-active (strategic) or reactive (emergency) measures, each of which includes both technological (structural) and non-technological (non-structural) actions (Rocha and Santos, 1996). Short-term drought response measures, such as water supply operations, may need to be integrated with longer-term drought mitigation measures, such as inter-basin transfers, for which information on the likely drought severity recurrence interval needs to be assessed.

Operational measures will normally be undertaken at the local, regional and national level, whereas more strategic measures can be promoted at the national, international and EU level. The emphasis on drought-related actions differs from country to country according to the sectors most affected, local priorities, level of investment, previous experience, pre-planning and the type, duration or severity of the drought event. Priority will normally be given to maintaining water supplies for potable use during a drought, although emergency measures may have uncertain longer-term effects. An effective response to drought situations requires an appropriate institutional framework, whilst a multi-criteria approach can assist the decision-making process (Rossi, 2000). An international approach is required to manage drought impacts where resources are shared by different countries.

Monitoring of monthly hydro-meteorological variables (e.g. temperature, rainfall, soil moisture, cumulative recharge, gauged and naturalised flows, groundwater levels) usually concentrates on a comparison of current conditions with long-term averages to indicate potential or cumulative drought-induced stress. Europe has a relatively dense network of hydrological and meteorological stations. Although few (if any) stations are used specifically for drought monitoring or form part of a drought monitoring network, most countries have adopted some form of 'index' to gauge the severity of a drought. There may also be institutional division of responsibilities. For example, responsibilities for water supply may be separated from water management or data collection. In addition, despite the regional and widespread nature of droughts, neighbouring countries seldom appear to exchange data in order to monitor droughts.

Satellite-based techniques combined with GIS offer a means of monitoring drought stress in agriculture and natural ecosystems. Such techniques can provide spatially and temporally consistent indices to detect and monitor droughts at different scales (Vogt *et al.*, 2000, Viau and Vogt, 2000). A number of international data sets are available or are being established in Europe, such as FRIEND European Water Archive, EUCOS, MED-HYCOS, EURO-WATERNET (EEA) and EIONET. However, in general these are geared towards research, weather prediction, compliance with EC directives, and analysis of historical data rather than for strategic or operational water resources management. In several countries hydrological and other relevant data are used to produce regular summaries on the status of water resources.

Drought management plans have been established in a number of countries with the use of models and decision support systems (DSS). A recent example includes the application of DSS to analyse drought policy option and potential drought responses for the metropolitan area of Athens (Karavitis, 1999). The development of integrated, comprehensive and effective pro-active drought management strategies for the wide range of conditions in Europe is a major challenge.

Numerous European, national and regional initiatives are underway to address the wider problem of water stress (water quantity and quality) in Europe, improve the way in which water resources are used and managed, and to protect the environment. Many of these have a bearing on drought mitigation and planning. For example, the forthcoming EU Water Framework Directive will require member states to establish monitoring networks to assess water status using a consistent classification scheme with the river basin considered as the management unit. Other innovative approaches to support policy development include a shift in focus from Facilitating Infrastructure Supply (FIS – dams, pipelines, etc) to Demand Side Management (DSM – water use efficiency) and new types of policy instruments to cope with the diversity of conditions throughout Europe and the complexity of water-related problems.

An evaluation of present policies and directions for the future sustainability of strategic water resources in Mediterranean countries (France, Greece, Italy, Spain, Portugal, Cyprus and Malta) has recently been completed (JRC, 1998; see also EEA, 1996). One of the policy options relating to the European Spatial Development Perspective (ESDP) is the improvement of the balance between water supply and demand in areas prone to drought. Co-operation between different Mediterranean countries is being promoted through such projects as ECOWAT (European Co-operation Network on Water Supply and Management) and HYDRE (Monitoring of Water Resources to the benefit of Regional Policies, Agriculture and Environment).

5. RECOMMENDATIONS

Based on the foregoing analysis of the European situation the following issues are recommended for further discussion and as a basis for the development of a European drought mitigation strategy.

Formation of a European Network for Drought Research and Mitigation (ENDRM).

The proposed ENDRM would complement other activities on natural hazards (INDR Hazards Forum) and on desertification (Desertification Information Network), sustainable development (Local Agenda 21), climate change (UNEP/WMO, IPCC), land use (IGBP/IHDP) and water resources (EEA, EURAQUA). It would also provide a basis for developing related research or advice on the formulation and implementation of EU legislation.

Every country in Europe would appear to have its own mechanisms or procedures for monitoring and managing a drought. A possible objective for an ENDRM would be to review these different approaches in Europe as well as elsewhere in order to identify the best practices for drought management appropriate to European conditions. As part of its remit, the ENDRM could address the viability of a pan-European network of real-time drought index stations and the usefulness of an early-warning system for droughts. Any such networks would still need to be based on existing national monitoring programmes.

Analysis of the applicability and selection of appropriate drought indices and monitoring techniques to detect, monitor and assess drought phenomena at different levels (EU, national, regional, local).

A large number of drought indices have been developed for each of the main types of drought and for different European countries. Further research into the development of multiple indices that integrate the different types of drought and their particular characteristics is required. There is also the need to develop indices at a range of different scales, since this is central to monitoring and impact assessment.

Analysis of the extent of drought risk and preparedness in various regions of Europe, and raising awareness at the political and public level.

There is usually more opportunity to manage and mitigate the impacts of a drought compared to short-term events, such as floods, but it is by no means clear whether this is the case throughout Europe, particularly with changing water demands associated with higher living standards and greater economic activity. Furthermore, it seems likely that climatic variability and other human influences could increase the potential socio-economic impacts of drought, sensitivity to less severe droughts and the frequency of drought events. It would seem timely to review existing national and regional drought plans.

Formation of an international drought monitoring network and database.

Hydrological and other monitoring data are essential for developing local and regional drought management plans and guidelines, including the development of decision support systems. The collection and evaluation of data related to droughts will become increasingly valuable in understanding interactions between drought variables, to provide an early warning of changes or to mitigate impacts, to trigger emergency relief actions and for the development and testing of predictive models.

Development of strategic drought mitigation measures.

Developing new sources of supply to meet water demands, as in the past, has become a less viable option. There is a need to identify and quantify the consequences of a drought and to determine the potential magnitude and costs, as well as the probability of impacts, the perception of risk by different sectors, and the willingness to pay for or implement strategic drought mitigation measures. However, there appears to be relatively limited information available on the social, economic and environmental impacts of drought.

Analysis of consequences resulting from current climate change scenarios.

Precipitation patterns may have changed over the past decade or so with a trend towards unusually warm weather and periods of intense drought interspersed with periods of higher precipitation. This increased variability may itself be superimposed upon longer-term fluctuations and trends. Further research into the historical causes of the different types of droughts and into climate driving mechanisms will assist in more reliable predictions of potential changes in the extent, frequency and severity of future drought events. The analysis of climate change scenarios should focus on those that suggest a reduction in precipitation or changes in monthly rainfall patterns.

ACKNOWLEDGEMENTS

The review of 'Drought Events in Europe' was funded by the Space Applications Institute (SAI) of the Joint Research Centre, Ispra, Italy, and the UK International Decade for Natural Disaster Reduction (IDNDR). Technical contributions from Dr Jürgen Vogt of the SAI and from Dr Alan Gustard, Dr Gwyn Rees, Emma Tate and Gwyneth Cole of the Institute of Hydrology are gratefully acknowledged.

REFERENCES

For reasons of clarity, it was not practical to include all the sources of information used in the original review or the tables and figures in this summary. A full list of these references is given below.

Main Reference:

Bradford, R.B. (1999) *Drought Events in Europe*, Institute of Hydrology, UK, for Space Applications Institute, DG Joint Research Centre, European Commission, Ispra, Italy, 41p.

Cited References:

Arnell, N.W. (1997) The Impact of Climate Change on Hydrological Regimes and Water Resources in Europe, Final Report to EC-DGXII.

Bidwell, V.J. (1973) A Methodology for Analysing Agricultural Drought, in E.F. Schulz, V.A. Koelzer, and Khalid Mahmood (eds.), *Floods and Drought, Proceedings of 2nd International Symposium in Hydrology*, Fort Collins, Colorado, USA.

Bonacci, O. (1993) Hydrological Definition of Drought, *Hydrol. Processes* 7: 249-262.

Byun, H.R. and Wilhite, D.A. (1999) Objective Quantification of Drought Severity and Duration, *J. Climatology* 1: 2747-2756.

Demuth, S., Lehner, B., and Stahl, K. (2000) Assessment of the Vulnerability of a River System to Drought, in J.V. Vogt and F. Somma (eds.), *Drought and Drought Mitigation in Europe*, Kluwer Academic Publishers, Dordrecht, 209-219.

Dracup, S., Lee, K.S., and Paulson, E.G. (1980) On the Definition of Droughts, *Wat. Resour. Res.* 16(2): 297-302.

EEA (European Environment Agency) (1996) *Water Resources Problems in Southern Europe*, European Topic Centre for Inland Waters Topic Rep. 15.

European Commission (1994) Europe 2000+: Co-operation for European Territorial Development, EC Regional Policies, Brussels, Luxembourg.

Gustard, A., Rees, H.G., Croker, K.M., and Dixon, J.M. (1997) Using Regional Hydrology for Assessing European Water Resources, in A. Gustard (ed.), *FRIEND'97 - Regional Hydrology: Concepts and Models for Sustainable Water Management*, Proceedings of 3rd International Conference on FRIEND, Postojna, Slovenia, IAHS 246: 107-115.

JRC (Joint Research Centre) (1998) *Towards a Sustainable/Strategic Management of Water Resources: Evaluation of Present Policies and Orientations for the Future*, European Commission, DG XVI (Regional Policy and Cohesion).

Karavitis, C.A. (1999) Decision Support Systems for Drought Management Strategies in Metropolitan Areas, *Water International* 24: 10-21.

Rocha, J.S. and Santos, M.A. (1996) Country Paper of Portugal, in Federal Institute of Hydrology (Bundesanstalt für Gewasserkunde) (ed.), *Management and Prevention of Crisis Situations: Floods, Droughts and Institutional Aspects*, Proceedings Euraqua, Rome, pp. 151-169.

Rossi, G. (2000) Drought Mitigation Measures: a Comprehensive Framework, in J.V. Vogt and F. Somma (eds.), *Drought and Drought Mitigation in Europe*, Kluwer Academic Publishers, Dordrecht, 233-246.

Tallaksen, L.M. (2000) Streamflow Drought Frequency Analysis, in J.V. Vogt and F. Somma (eds.), *Drought and Drought Mitigation in Europe*, Kluwer Academic Publishers, Dordrecht, 103-117.

Viau, A.A. and Vogt, J.V. (2000) Scale Issues in Drought Monitoring, in J.V. Vogt and F. Somma (eds.), *Drought and Drought Mitigation in Europe*, Kluwer Academic Publishers, Dordrecht, 185-193.

Vogt, J.V., Niemeyer, S., Somma, F., Beaudin, I., and Viau, A.A. (2000) Drought Monitoring from Space, in J.V. Vogt and F. Somma (eds.), *Drought and Drought Mitigation in Europe*, Kluwer Academic Publishers, Dordrecht, 167-183.

Further Reading:

Alley, W.M. (1984) The PDSI: Limitations and Assumptions, *J. of Clim. and App. Met.* 23: 1100-1109.

Batini, G., Benedini, M., Passino, R., and Ubertini, L. (1996) Country Paper of Italy, in Federal Institute of Hydrology (Bundesanstalt für Gewasserkunde) (ed.), *Management and Prevention of Crisis Situations: Floods, Droughts and Institutional Aspects*, Proceedings Euraqua, Rome, pp. 97-122.

Ben-Zvi, A. (1987) Indices of Hydrological Drought in Israel, *J. Hydrol.* 92: 179-191.

Beran, M.A. and Rodier, J.A. (eds.) (1985) Hydrological Aspects of Droughts, Studies and Reports in Hydrology 39, UNESCO-WMO.

Bredenkamp, D.B., Botha, L.J., van Tonder, G.J., and van Rensberg, H.J. (1995) Manual on Quantitative Estimation of Groundwater Recharge and Aquifer Storativity, Water Research Commission of South Africa Report TT 73/95.

Briffa, K.R., Jones, P.D., and Hulme, M. (1994) Summer Moisture Variability Across Europe, 1892-1991: An Analysis Based on the Palmer Drought Severity Index, *Int. J. Clim.* 14: 475-506.

Bryant, S.J., Arnell, N.W., and Law, F.M. (1994) The Long-Term Context for the Current (UK) Hydrological Drought, *J. Inst. Water Environ. Mgmt.* 8: 39-51.

Bussay, A. and Szinell, C. (1996) Drought Continues in Hungary in 1995, *Drought Network News* 8(1).

Chang, T.J. and Stenson, J.R. (1990) Is it Realistic to Define a 100-Year Drought for Water Management? *Wat. Res. Bull.* 26(5): 823-829.

Czamara, W., Jakubowski, W., and Radzuk, L. (1997) Probabilistic Analysis of Extreme Low Flows in Selected Catchments in Poland, in A. Gustard (ed.), *FRIEND'97 – Regional Hydrology: Concepts and Models for Sustainable Water Management*, Proceedings of 3rd International Conference on FRIEND Postojna, Slovenia, IAHS 246:159-168.

Demuth, D. and Heinrich, B. (1997) Temporal and Spatial Behaviour of Drought in South Germany, in A. Gustard (ed.), *FRIEND'97 - Regional Hydrology: Concepts and Models for Sustainable Water Management*, Proceedings of 3rd International Conference on FRIEND, Postojna, Slovenia, IAHS 246: 151-157.

EEA and UNEP - European Environment Agency and United Nations Environment Programme (1997): Water Stress in Europe – Can the Challenge be Met? Copenhagen and Chatelaine.

Europe's Environment (1995) Statistical Compendium for the Dobris Assessment. (Compiled jointly by Eurostat, European Commission and others) 450 p.

Farat, R., Kepinska-Kasprzak, M., Kowalczak, P., and Mager, P. (1998) Droughts in Poland, *Drought Network News* 10(1).

Gommes, R. and Petrassi, F. (1994) *Rainfall Variability and Drought in sub-Saharan Africa since 1960*, FAO Agromet, Series WP 9, Rome, Italy.

Gottschalk, L., Jensen, J.L., Lundquist, D., Solantie, R., and Tollan, A. (1979) Hydrological Regions in Nordic Countries, *Nordic Hydrol.* 10: 273-286.

Institute of Hydrology (in press), *Sustainable Water Use in Europe, Part 3: Extreme Hydrological Events – Droughts*, European Topic Centre for Inland Waters.

Institute of Hydrology (1980) *Low Flow Studies*, Research Report 1.

Jones, P.D., Hulme, M., Briffa, K.R., Jones, C.G., Mitchell, J.F.B., and Murphy, J.M. (1996) Summer Moisture Availability over Europe in the Hadley Centre General Circulation Model based on the Palmer Drought Severity Index, *Int. J. Climatol.* 16: 155-172.

Jones, P.D. and Lister, D.H. (1998) Riverflow Reconstruction for 15 catchments over England and Wales and an Assessment of Hydrologic Drought since 1865, *Int. J. Climatol.* 18: 999-1013.

Karl, T.R. and Knight, R.W. (1985) Atlas of Monthly PDSI (1931-1983) for the Contiguous US., *Hist. Clim. Series 3-7*, National Climate Data Center, Asheville, North Carolina, USA.

Kasparek, L. and Novicky, O. (1997) Application of a Physically-Based Model to Identify Factors Causing Hydrological Droughts in Western and Central European Basins, in A. Gustard (ed.), *FRIEND'97 - Regional Hydrology: Concepts and Models for Sustainable Water Management* Proceedings of 3rd International Conference on FRIEND, Postojna, Slovenia, IAHS 246: 197-204.

Koleva, E. (1995) Drought in the Lower Danube Basin, *Drought Network News* 7(1).

Krasovskaia, I., Arnell, N.W., and Gottschalk, L. (1994) Flow Regimes in Northern and Western Europe: Development and Application of Procedures for Classifying Flow Regimes, in *FRIEND: Flow Regimes from International Experimental and Network Data*, Proceedings of the Braunschweig Conference, October 1993, IAHS Publ. No. 221, 1994.

Lambin, E.F. and Ehrlich, D. (1998) Land Cover Changes in sub-Saharan Africa (1982-91): Application of a Change Index Based on Remotely Sensed Surface Temperatures and Vegetation Indices at a Continental Scale, *Remote Sens. Environ.* 61: 181-200.

Le Houerou, H.N., Popov, G.F., and See, L. (1993) *Agro-Bioclimatic Classification of Africa*, FAO Agromet, Series WP 6, Rome, Italy.

MacCarthaig, M. (1996) An Assessment of the 1995 Drought, Environmental Protection Agency, Dublin.

Marsh, T. J. and Lees M.L. (1985) The 1984 (UK) Drought, *Hydrological Data*, UK Series.

Marsh, T.J., Monkhouse, R.A., Arnell, N.W., Lees, M.L., and Reynard, N.S. (1994) The 1988-92 Drought, *Hydrological Data*, UK Series.

Mawdsley, J.A., Petts, G.E., and Walker, S. (1994) Assessment of Drought Severity, British Hydrological Society Occasional Paper No. 3.

Mayes, J.M. (1994) Recent Changes in the Monthly Distribution of Regional Weather Types in the British Isles, *Weather* 49(5): 149-168.

McKee, T.B., Doesken, N.J., and Kleist, J. (1993) The Relationship of Drought Frequency and Duration to Time scales, *Proceedings 8th Conference on Applied Climatology*, Anaheim , California, pp. 179-184.

McKee, T.B., Doesken, N.J., and Kleist, J. (1995) Drought Monitoring with Multiple Timescales, *Proceedings 9th Conference on Applied Climatology*, Dallas, Texas, pp. 233-236

Moll, J.R., Parmet, B.W., and Sprong, T.A. (1996) Country Paper of the Netherlands, in Federal Institute of Hydrology (Bundesanstalt für Gewasserkunde) (ed.), *Management and Prevention of Crisis Situations: Floods, Droughts and Institutional Aspects*, Proceedings Euraqua, Rome, pp. 123-141.

Moore, R.J. (1996) Country Paper of the United Kingdom, in Federal Institute of Hydrology (Bundesanstalt für Gewasserkunde) (ed.), *Management and Prevention of Crisis Situations: Floods, Droughts and Institutional Aspects*, Proceedings Euraqua, Rome, pp. 197-225.

Niemeyer, S. and Vogt, J.V. (1998) Towards Monitoring Drought Conditions in Sicily Using an Energy Balance Approach, *Proceedings Int. Conf. on Comp. Tech. in Agric. (ICCTA)*, Florence, Italy.

Ozga-Zielinska, M. (1989) Droughts and Floods - Their Definition and Modelling, in M.L. Kavvas (ed.), *New Directions for Surface Water Modelling*, IAHS 181: 313-322.

Palmer, W.C. (1965) Meteorological Drought, US Dept. of Commerce Weather Bureau, Washington Res. Paper 45.

Palmer, W.C. (1968) Keeping Track of Crop Moisture Conditions Nationwide: the New Crop Moisture Index, *Weatherwise*, 21: 156-161.

Querner, E.P., Tallaksen, L.M., Kasparek, L., and van Lanen, H.A.J. (1997) Impact of Land Use, Climate Change and Groundwater Abstraction on Streamflow Droughts Using Physically Based Models, in A. Gustard (ed.), *FRIEND'97 - Regional Hydrology: Concepts and Models for Sustainable Water Management*, Proceedings of 3rd International Conference on FRIEND, Postojna, Slovenia, IAHS 246: 171-179.

Rossi, G., Benedini, M., Tsakaris, G., and Giakoumakis, S. (1992) On Regional Drought Estimation and Analysis, *Water Resources Management* 6: 249-277.

Saalthun, N.R. and Tollan, A. (1996) Country Paper of Norway, in Federal Institute of Hydrology (Bundesanstalt für Gewasserkunde) (ed.), *Management and Prevention of Crisis Situations: Floods, Droughts and Institutional Aspects*, Proceedings Euraqua, Rome, pp. 143-150.

Sharma, T.C. (1997) A Drought Frequency Formula, *Hydrol. Sci.* 42(6): 803-814.

Shiklomanov, I.A. (1997) *Comprehensive Assessment of the Freshwater Resources of the World*, World Meteorological Organisation.

Shorthouse, C.A. and Arnell, N.W. (1997) Spatial and Temporal Variability in European River Flows and the North Atlantic Oscillation, in A. Gustard (ed.), *FRIEND'97 - Regional Hydrology: Concepts and Models for Sustainable Water Management,* Proceedings of 3rd International Conference on FRIEND, Postojna, Slovenia, IAHS 246: 77-96.

Smith, D.I., Hutchinson, M.F., and MacArthur, R.J. (1993) Australian Climatic and Agricultural Drought: Payments and Policy, *Drought Network News* 5(3).

Szentimrey, T., Farago, T., and Szalai, M. (1992) Window Technique for Climate Trend Analysis, *Climate Dynamics* 6: 127-134.

Tallaksen, L.M., Madsen, H., and Clausen, B. (1997) On the Definition and Modelling of Streamflow Drought Duration and Deficit Volume, *Hydrol. Sci.* 42(1): 15-33.

Tallaksen, L.M. and Hisdal, H. (1997) Regional Analyis of Extreme Streamflow Drought Duration and Deficit Volume, in A. Gustard (ed.), *FRIEND'97 - Regional Hydrology: Concepts and Models for Sustainable Water Management*, Proceedings of 3rd International Conference on FRIEND, Postojna, Slovenia, IAHS 246: 141-150.

Tase, N. (1976) Area-Deficit-Intensity Characteristics of Droughts, *Hydrological Paper* 87, Colorado University.

Tate, E.L. and Gustard, A. (2000) Drought Definition: a Hydrological Perspective, in J.V. Vogt and F. Somma (eds.), *Drought and Drought Mitigation in Europe*, Kluwer Academic Publishers, Dordrecht, 23-48.

Turkes, M. (1996) Meteorological Drought in Turkey: A Historical Perspective, 1930-1993, *Drought Network News* 8(3).

UK Climate Impacts Programme (1998) Climate Change Scenarios for the United Kingdom, Tech. Rep. 1.

Ungani, L.S. and Kogan, F.N. (1998) Drought Monitoring and Corn Yield Estimation in Southern Africa from AVHRR Data, *Remote Sens.Environ.* 63: 219-232.

Van Lanen, H.A.J., Tallaksen, L.M., Kasparak, L., and Querner, E.P. (1997) Hydrological Drought Analysis in the Hupsel Basin Using Different Physically-Based Models, in A. Gustard (ed.), *FRIEND'97 - Regional Hydrology: Concepts and Models for Sustainable Water Management*, Proceedings of 3rd International Conference on FRIEND, Postojna, Slovenia, IAHS, 246: 189-196.

Wilby, R.L., O'Hare, G. and Barnsley, N. (1997) The North Atlantic Oscillation and British Isles Climate variability, 1865-1996, *Weather* 52(9): 266-276.

Wilhite, D.A. (ed.) (1993) *Drought Assessment, Management and Planning: Theory and Case Studies*, Kluwer Academic Publishers.

Willeke, G.J., Hosking, J.R., Wallis, J.R., and Guttman, N.B. (1994) The National Drought Atlas. Inst. for Wat. Resources 94-NDS-4, US Army Corps of Engineers.

World Meteorological Organisation (1994) Guide to Hydrological Practices: Data Acquisition and Processing, Analysis, Forecasting and Other Applications, Pub. No.168.

World Meteorological Organisation (1997) Climate and Water in Europe: Water Quality and Aquatic Ecosystems.

Yevjevich, V. (1967) An Objective Approach to Definitions and Investigations of Continental Hydrologic Droughts, Hydr. Paper 23, Colorado University.

Zelenhazi, E. and Salvai, A. (1987) A Method of Streamflow Drought Analysis, *Wat. Resources Res.* 23/1: 156-168.

Part 2
Drought Definition

DROUGHT DEFINITION: A HYDROLOGICAL PERSPECTIVE

E.L. TATE and A. GUSTARD
Institute of Hydrology
Wallingford, Oxon, OX10 8BB, UK

Abstract. This paper summarises different techniques for defining a drought. Hydrologists as distinct from economists or social scientists define drought according to water deficits in some component of the hydrological cycle (precipitation, soil moisture, river flow and groundwater) or the impact on the level of service provided to public water supply, irrigation or hydropower demands. Generic features of droughts are their severity, frequency, duration and spatial extent. For operational purposes it is essential that a wide range of analytical procedures are used according to the response characteristics of the resource system. However, for comparisons of drought severity over time and between countries, there may be merit in applying simple procedures for drought frequency assessment.

1. INTRODUCTION

There is no universally accepted definition of drought. Ideally, definitions should be related to the region under study and be specific to the particular application. Drought indicators can be either qualitative (usually descriptive, linguistic definitions of drought severity) or quantitative (requiring the use of statistical analysis). McMahon and Diaz Arenas (1982) define drought as a period of abnormally dry weather sufficiently prolonged for the lack of precipitation to cause a serious hydrological imbalance, carrying connotations of a moisture deficiency with respect to man's usage of water. Several general definitions have been proposed by hydrological authors, perhaps most succinctly by Beran and Rodier (1985):

> *'The chief characteristic of a drought is a decrease of water availability in a particular period and over a particular area.'*

A comprehensive classification of drought considers its type, intensity, duration and spatial extent in conjunction with its perceived impact, since only through such a broad description can a fully objective picture be obtained.

Moreover, drought definitions are continually being updated, particularly to take account of impacts on, and complex interactions with, the environment and society; for example, the relationship between rainfall variability and impacts can depend upon the specifics of a particular agro-ecological zone or the economy. An issue for investigation is, therefore, an economically useful definition of drought (Benson and Clay, 1994). Furthermore, hydrological definitions require further development to include the impact

J.V. Vogt and F. Somma (eds.), Drought and Drought Mitigation in Europe, 23–48.

of drought on ecosystems; for example, the resilience of in-stream ecology to river flow deficiencies. Despite the difficulties with finding a suitable definition of drought for a particular application, it provides a practical way to assess a drought situation and to give a relative picture of its severity.

The following categories of drought are reviewed from a hydrological perspective in this paper:

1) Climatological drought
2) Agro-meteorological drought
3) River flow drought
4) Ground-water drought
5) Operational drought

The aim is to provide a concise presentation of each drought type, as a quick reference for hydrologists and those interested in assessing drought from a hydrological viewpoint. It is not intended to be comprehensive, rather to act as a summary, with key definitions and indices referenced for the benefits of further research. A simple description of each drought category is given here, followed by a summary of the most common drought indices and descriptors falling into each category.

2. CLIMATOLOGICAL DROUGHT

Climatological drought can be simply expressed in terms of a rainfall deficit, in relation to some average amount, and the duration of the dry period. One of rainfall's main merits for assessing drought severity is that long-term records are often available. However, rainfall has limitations as a measure of water availability in droughts: measurements over short periods take no account of initial ground conditions nor seasonal evaporation variations. Effective rainfall may be a more useful drought indicator, but requires estimates of actual evapotranspiration. The indirect nature of evapotranspiration and lack of readily available data on the subject mean it is not in widespread use as a drought severity indicator (DSI).

The most widely accepted definition of drought, amplifying the notion of a prolonged and abnormal moisture deficiency, is Palmer's (1965): '*an interval of time, generally of the order of months or years in duration, during which the actual moisture supply at a given place rather consistently falls short of the climatically expected or climatically appropriate moisture supply.*'

However, Marsh and Lees (1985) comment that simple definitions requiring a fixed number of dry days in sequence, although still enshrined in certain official definitions, have generally been abandoned in favour of methods depending on accumulated departures from the mean value linked to some criterion for determining that the drought has ended. This approach was used by Foley (1957) and the procedure has been used by the Meteorological Office as a guide to rainfall deficiency in the UK. Two specific problems attend the use of indices of this type. First, average rainfall itself is not a constant value; it is subject to random and systematic changes, which may reflect climatic trends or oscillations. Second, the identification of a drought is seldom clear cut. A summary of climatological drought definitions is given in table 1 (after WMO, 1975).

2.1. Palmer Drought Severity Index

The Palmer Drought Severity Index (PDSI) was developed by W. C. Palmer, in 1965, in the USA and has been used in various other countries; for example, since 1982 it has been used in South Africa and a data bank of digital maps has been produced. These maps can be used to identify the presence of drought and to monitor its progress. The underlying concept of the PDSI method is that the amount of precipitation required for the near-normal operation of the established economy of an area, during some stated period, is dependent on the average climate of the area and on the prevailing meteorological conditions both during and preceding the period under consideration. Drought severity is related to the accumulated weighted differences between actual precipitation and the precipitation requirement. It takes into account the current weather and soil conditions and the normal climate of the area.

The PDSI is widely used in the USA. It is routinely calculated by the NOAA (National Oceanic and Atmospheric Administration) for all 344 climate divisions in the country. Many sectors of government and private industry use this index in real-time drought management and planning (Karl, 1991). However, du Pisani (1990) in the SARCCUS (Southern African Regional Commission for the Conservation and Utilisation of the Soil) workshop on drought claims that the South African experience with the PDSI has been somewhat less than satisfying. It has been refined as far as possible with regard to the calculation and mapping, but still does not satisfy the needs of policy makers and does not always yield index values which correlate well with actual on-site evaluations. Bruwer (1990) says a limitation is that it is not generally indicative of short-term (*i.e.* periods of several weeks) changes in moisture status affecting crops and farming operations. Also, there is a lack of meteorological data in the arid areas of South Africa, which makes it difficult to apply. The PDSI is relatively complex as it requires a great number of primarily meteorological and hydrological data.

Working on the contiguous United States, with data from 1895-1981, Diaz (1983) considered the following limits for PDSI values:

- -2 to +2 was normal,
- -2 to -3 was indicative of 'moderate' drought,
- -3 to -4 was 'severe', and
- < -4 was considered indicative of 'extreme' drought conditions.

In the analysis, a drought sequence was taken to be the occurrence of three or more consecutive months with a PDSI value ≤ -2.0, whilst a series of six or more months was a major drought event. The end of a drought sequence was taken as the last month in the run where the PDSI retained a value ≤ -2.0.

2.2. Drought Indices Developed from the Palmer Model

Four indices which have been developed from the popular Palmer model are:

- the PDSI, as described above;
- the Palmer Hydrologic Drought Index (PHDI);
- the monthly moisture anomaly index, or Z index (ZINX);
- the Colorado Palmer Drought Index (CPDI).

The indices have different rates of response to changes in supply and demand for moisture and therefore represent different types of drought. Despite being derived from basic water balance variables, the calculation procedures for the indices are complex. In terms of drought types, the ZINX is used most appropriately as a measure of agricultural drought intensity. The PDSI is usually presented as a measure of meteorological drought intensity. It tends to respond too slowly to be used as a measure of agricultural drought and too rapidly to assess conditions of hydrological drought (*i.e.* river flow and ground-water levels) accurately. Thus its primary use is as a measure of drought severity. By modifying the PDSI calculation procedures to produce the PHDI, an index is developed that should better assess moisture anomalies that impact river flow, ground-water availability, and lake or reservoir levels.

The Colorado Palmer Drought Index (CPDI) is a version of the index developed by Palmer, but uses data for smaller areas. It is useful at lower elevations particularly for soil moisture applications for agriculture.

Guttman *et al.* (1992) stated that water resource managers and planners should define an extreme drought, or any other category of drought, in terms of an index value that corresponds to the expected quantile (return period) of the event rather than using Palmer's arbitrary thresholds.

The approach taken by Palmer for defining drought has two important implications. One is that the imbalance between water supply and water demand is the key to defining droughts. The other is that the severity of a drought depends not only on the area's water deficit but also the region's drought resistance characteristics.

2.3. Surface Water Supply Index

Several American states have adopted their own index for monitoring water conditions. The Surface Water Supply Index (SWSI) (Dezman *et al.*, 1982), developed in Colorado, integrates historical data with current figures of reservoir storage, river flow and precipitation at high elevations into a single index number. The SWSI is intended to act as a complement to the PDSI. It is useful for evaluating water availability in snow-pack, reservoirs, and river flows.

The PDSI does not account for snow accumulation and melt. The purpose of the SWSI is 'to be an indicator of basin-wide water availability for the [mountain-water dependent] sector, be predictive, and permit comparison of water supply conditions between basins to assess relative drought severity.' The original SWSI was formulated as a rescaled sum of non-exceedance probabilities of four hydrologic components: snow-pack, precipitation, river flow, and reservoir storage. The mathematical formulation of the SWSI is as follows:

$$SWSI = \frac{aP_{snow} + bP_{prec} + cP_{strm} + dP_{resv} - 50}{12} \tag{1}$$

where a, b, c, and d are weights for each hydrologic component; $a + b + c + d = 1$; P_i = the probability of non-exceedance (in %) for component i; and *snow*, *prec*, *strm*, and *resv* = the snow-pack, precipitation, river flow, and reservoir storage hydrologic components, respectively. Subtracting 50 and dividing by 12 are a centring and compressing procedure designed to make the value have a similar magnitude to the PDSI (Garen, 1993).

2.4. Rainfall Anomalies

Many definitions based solely upon rainfall and short period 'droughts' or 'dry spells' appear to be tied to climatic experience in the British Isles and Europe or north-eastern USA. In these regions, rainfall is normally received at fairly frequent intervals and crop and animal husbandry and water-storage operations are not geared to the long spells of rainless weather, which are seasonally normal in some semi-arid regions. Definitions which do not specifically include duration have a wide application and have been used with success in the drier continents of the world (WMO, 1975). Descriptions of examples of the use of percent anomalies in various countries follow, along with examples of indices and a mapping procedure based on similar theories.

Southern African rainfall trends from 1900/01 to 1995/96 were plotted having been expressed as percent anomalies from the 1961-90 average. The 1991/92 drought seems to have been linked to the El Niño/Southern Oscillation (ENSO) problem (Hulme, 1996).

Droughts in South Africa have been broadly defined as less than 70% of normal precipitation. This becomes a *disaster* or *severe* drought when two consecutive seasons experience 70% or less rainfall. However, it should be summarised for several time periods (annual, water year, crop growing season) (Bruwer, 1990).

Drought in Wales (1975-76) was gauged as April-September 1975 rainfall as a percentage of average 1941 to 1970 April-September rainfall (plotted as contour maps). Durations of 1, 3, 6, 9, 12 and 18 months were also studied to rank the drought events with reference to the 1941 to 1970 period (Welsh National Water Development Authority, 1976).

2.4.1. *Foley Drought Index*

To examine the temporal pattern of wet and dry periods, the excesses or deficiencies of monthly (or other period) rainfall compared with the respective long-term average may be integrated to produce a graph of cumulative departures (sometimes called a 'residual mass curve'). A succession of deficiencies may amount to a considerable total deficiency, thereby identifying a 'drought' period. The graph has the advantage that it clearly shows both the duration and amount of the rainfall deficiency. It has the disadvantage that all deficiencies carry the same weight irrespective of the water requirements of vegetation. In the work of Foley (1957), deficiencies are expressed in thousandths of the annual rainfall, called 'units' for brevity. Units are cumulated as a method of assessing a drought period and may also be divided by the number of months in a period to give an index of drought severity. This method has been used successfully in Australia in long-term drought studies (WMO, 1975).

2.4.2. *Rainfall Deficiency Indices*

This type of index has been commonly used in the UK. Such drought indices readily identify periods of rainfall deficiency but they require cautious interpretation since the apparent magnitude of each low rainfall event can be sensitive to the criteria used to define the end of each drought. In the case of the 1988-92 drought in the UK, the drought was considered terminated when a period of six months registered above average rainfall (Marsh *et al.*, 1994).

Herbst *et al.* (1966) describe a method, based on monthly rainfall data, whereby it is possible to determine the duration and intensity of droughts and their months of onset and termination. A drought index is also calculated which enables the intensity of droughts to be compared irrespective of their seasonal occurrence. It is suggested that this method is more applicable for South Africa than British and Australian methods owing to the acute, prolonged and widespread nature of the drought prevailing in South Africa at the time the paper was written. The method recognises that the benefit to vegetation of above-average rainfall persists for a period of time after the rain actually fell, and that, similarly, vegetation takes some time to recover after the end of a drought event. Thus, carry-over of the effects of rainfall are considered so that a deficit or excess for a particular month is obtained and this value is used to derive the 'effective rainfall'. A mean monthly rainfall deficit can then be calculated from the effective rainfall relative to the mean monthly rainfall. The drought index is calculated from average monthly rainfall deficiency divided by mean monthly deficits for the same period, multiplied by the period of the drought.

2.4.3. *Isoline Maps*

Rossi *et al.* (1992) show how an isoline map for the description of past droughts in a given time interval may be drawn by using various descriptors. For instance, the rainfall depth at a specified time interval is usually expressed as a percentage of the mean for the same interval, over a certain number of years. A second descriptor may be the deviation from normal rainfall in a past drought period of given duration, expressed as absolute value or ratio or percentage of the difference, with reference to the mean rainfall, *i.e.* 'rainfall deficit'. Another drought descriptor is the standardised deficit obtained as the ratio, for a given interval, of the precipitation deviation from the mean over standard deviation. The analysis of historical drought events can also be carried out by mapping the contours of a given probability (of return periods) of drought events of a given time interval.

A more compact description of the areal extent of an historical drought can be drawn from the relationship between the percent of the mean rainfall (or other drought indices) and the areal coverage of the drought.

Regional drought characteristics can be estimated through stochastic models. Two methods can be adopted for separating the deterministic and stochastic components: parametric and non-parametric methods. In both cases, the stationary stochastic component series for a given station is obtained from the standardised series of monthly precipitation. A general mathematical model of the time structure of the monthly precipitation may be obtained through Fourier series analysis for the description of the periodic variability of the parameters (mean and standard deviation). Once the mathematical model of the basic hydrologic process is completed, the selected drought indices can be computed and analysed. A truncation level has to be chosen. A simple approximate way is to assume that a water demand equals the mean or median of the basic variable (Rossi *et al.*, 1992). Tase (1976) used the following indices in an investigation of areal drought characteristics on a unit square grid basis:

- deficit area, defined for a time interval as the number of unit areas presenting a deficit for a given truncation level of the standardised variable;
- total areal deficit defined as the sum of the deficits over the region in the time interval;
- maximum deficit in the time interval.

Rossi used similar indices to Tase but the grid system was given by a Thiessen polygon network.

2.5. Percentile Method

According to empirical, cumulative distributions, or the calculated theoretical distribution, each value of a meteorological or hydrological element can be assigned a cumulative frequency, expressed in percentiles. From the percentile value it is possible to draw conclusions about the severity of a given value of the meteorological or hydrological element in the site under study. This method is apparently simple to use and its results can be used for preliminary analyses (Bonacci, 1993).

2.6. Deciles of Precipitation

This concept has been applied to classify drought. Monthly and annual precipitation totals are ranked, highest to lowest, and decile ranges are determined from the cumulative frequency distribution. The first decile represents precipitation values in the lowest 10% of the distribution; the second decile between 10% and 20%, and so forth. Deciles give a reasonably complete picture of a particular rainfall distribution.

2.7. Poisson Process

The Poisson Process involves distribution of maximum dry periods (drought durations) between rainfall events at a point within a season. The distribution of the number of precipitation events within a season is assumed to follow a homogeneous Poisson process. Based on the Poisson assumption, the distribution functions of the two random variables, namely the maximum dry period within a season and the waiting time to experience the first critical dry period, are given. The statistical moments for both cases can then be computed (Gupta, 1972).

2.8. Palfai Aridity Index

The Palfai Aridity Index (PAI) is the ratio of the mean temperature during the period from April to August and the rainfall from October to August. In work carried out in Hungary, it correlates well with the PDSI and the author (Palfai, 1990) chose a threshold value PAI=6.5 to define 'severe' droughts. The index is focused on the main crop growing season. However, Pereira (1990) declares that further research is necessary before using the index outside of Hungarian climatic conditions.

Table 1: Definition of Climatological Drought (after WMO, 1975).

Author	Definition of drought or associated concepts	Region and comments
Abu-Zeid and Abdel-Dayem (1990)	Use the five years' moving average of annual runoff.	
Baldwin-Wiseman (1941)	Engineers' drought in Australia is 3 or more consecutive months with deficit of 50% from mean rainfall.	Australia
Banerji and Chabra (1964)	A seasonal rainfall deficit of more than 50% is considered a severe drought.	Andhra Pradesh, India
Bates (1935)	Annual precipitation is 75% of normal, or monthly precipitation is 60% of normal.	USA
Bazza and Stockton (1990)	Use the coefficients of variation of annual and monthly precipitation.	
Blumenstock (1942)	A period with precipitation less than some small amount such as 0.10 inches in 48 hours.	
British Rainfall Organisation (1936)	*Absolute drought*: at least 15 consecutive days, none of which receive as much as 0.25 mm. *Partial drought*: at least 29 days during which mean rainfall does not exceed 0.25 mm per day. *Dry spell*: 15 consecutive days, none of which has received as much as 1 mm.	Britain; inapt in normally drier regions.
Brounov (early 20th cent.)	Ten days with rainfall not exceeding 5 mm.	Ref. Tannehill (1947)
Bryant *et al.*, (1992)	Define the DSI (drought severity indicator) as: 'the accumulated monthly deficit relative to the mean for a standard period - in this case the 1951-1989 mean.'	
Clarke (1991)	Annual rainfall < 75% of the average is regarded as a drought.	*i.e.* 13.2% of India suffers a drought more than 1 in 3 years.
Cole (1933)	15 days with no rain.	USA
Condra (1944)	A period of strong wind, low precipitation, high temperature, and unusually low relative humidity.	
Emberger (1955)	$I = \frac{100p}{(M-m)(M+m)}$ where M is the mean max. temperature in the hottest month and m is the mean min. temp. in the coldest month; p in mm and M and m in °C.	France. Based on de Martonne's index, (M-m) is an index of continentality.

Table 1: Definition of Climatological Drought (after WMO, 1975) continued.

Author	Definition of drought or associated concepts	Region and comments
Fitzpatrick (1953)	Period terminated by at least 6.4 mm during any 48 hours.	Australia. Evaluated probability that dry spells of any length would occur at any time throughout the year.
Henry (1906)	21 days or more when rainfall is 30% or less of average for the time and place. Extreme drought when rainfall fails to reach 10% of normal for 21 days or more.	USA
Hoyt (1936)	Any amount of rainfall < 85% of normal.	USA
Kulshreshtha and Klein (1990)	Use a comparison of annual time series of precipitation and wheat yields.	
de Martonne (1926)	Index of aridity: $I = \frac{P}{t+10}$ where P is monthly precipitation (mm) and t is mean monthly temperature (°C). Monthly index of I is approximate indicator of aridity. Index modified to: $I = \frac{n*\bar{p}}{t+10}$ where n is number of days during a certain period from a few days to a year and $\bar{p}$ is daily mean precipitation in the period.	Used to define climatic limits of deserts, prairies and forests. Not good in cool places where t + 10 approaches zero. Used extensively by geographers and biologists to compute aridity.
Popov (1948)	Index of aridity: $P = \frac{\Sigma g}{2.4(t-t')r}$ where P is index of aridity; Σg is annual amount of effective precipitation; t-t' is annual mean wetbulb depression °C; r is factor depending on day length; and g is that part of precipitation which is available for plants.	
Ramdas (1950)	When actual rainfall for a week is half of normal or less.	India

Table 1: Definition of Climatological Drought (after WMO, 1975) continued.

Author	Definition of drought or associated concepts	Region and comments
Selyaninov (1930)	Index given by: $k = \frac{\Sigma p}{\Sigma t / 10}$ where p is sum of rainfall (mm) during those months when mean temp. above 10°C and t is the sum of daily mean temperatures above 0°C for the same period.	USSR. Author suggested that a period be considered as a dry spell when k<1 and as a drought when k<0.5.
Tannehill (1947)	A period of 21 days when rainfall is 30% less than normal for the place and time.	
Tardieu and Plus (1990)	Use the three years moving average of annual rainfall.	
Tennessee Valley Authority (1944)	When no interval of 21 consecutive days received precipitation $>^1/_3$ of normal.	
Thornthwaite (1931)	Precipitation effectiveness as a function of mean temp. $P/E = 1.65(\frac{P}{T+12.2})^{(10/9)}$ in mm and °C where P/E is the precipitation/evaporation ratio, P is monthly precipitation, and T is mean temperature in °C.	USA

3. AGRO-METEOROLOGICAL DROUGHT

Agro-meteorological drought links various categories of meteorological and hydrological drought to agricultural impacts, focussing primarily on soil water deficits and differences between actual and potential evapotranspiration (Tate *et al.*, 2000). A drought exists when root-zone soil moisture reserves are insufficient to sustain crops and pasture between rainfalls.

Soil drought is the most common form of drought, entailing water deficiency in the root zone. It may be expressed by the water content of the soil profile per metre depth. Atmospheric drought is present when humidity is so low that it cannot be substituted by the actual soil moisture content. Thus, humidity becomes a limiting factor. Physiological drought is a limiting water deficiency in plant cells when the soil and air contain water in sufficient quantities. Most frequently it is caused by an abrupt and marked increase in air temperature while soil temperature remains low (Petrasovits, 1990).

A summary of agro-meteorological drought definitions is given in table 2 (after WMO, 1975).

3.1. FAO Water Satisfaction Index

This index was developed by Frere and Popov (1979) for the purpose of agro-meteorological crop monitoring and yield prediction and has been used extensively in both Botswana and Lesotho. The method calculates the soil moisture balance for 10-day periods (decads) and ultimately calculates an index, the cumulative or crop water satisfaction index, that is determined by the degree to which the crop water requirement is met. The following weaknesses in the FAO model have called for the establishment of an alternative model:

- The index does not consider crop- or stage-specific sensitivity to drought stress;
- The index only takes account of cumulative moisture deficits but not of the effects of consecutive deficits on potential crop yield;
- It can only partly accommodate actual evapotranspiration;
- The model assumes equal soil moisture availability over the entire range between field capacity and wilting point.

The model has been modified to subdivide the crop cycle into phenological stages, and a moisture budget model is used which contains a root growth function as well as a moisture depletion function. It then starts emulating a crop growth model (du Pisani, 1990).

3.2. Agrohydropotential (AHP)

The AHP is an index which can express the actual water requirement covering ability of a given field for any crop. It is the ratio between the actual water consumption and the optimal water requirement of a crop, and is crop specific. The AHP values can be calculated daily and per pentad according to the following formula:

$$AHP = \frac{W_c}{W_r} \tag{2}$$

where W_c is the actual water consumption; the actual evapotranspiration limited by the water balance, and W_r is the water requirement; quasi optimal, unlimited evapotranspiration.

The AHP index may vary between 1 and 0. The obtained AHP shows the degree of drought. Using the AHP method it is possible to prepare comparative drought-prognoses for any crop and also to estimate the degree of drought susceptibility by the frequency of occurrence (Petrasovits, 1990). However, Pereira (1990) claims that it should be considered as a crop water deficit index, not as a drought index. This index is a more basic form of the modified FAO Water Satisfaction Index, and works on a different time-step.

3.3. Index of Moisture Adequacy

The Index of Moisture Adequacy (IMA), a ratio of actual to potential evapotranspiration during the crop growth period, was developed by Sastri *et al.* (1981). It was based on studies of pearl millet and pulses for the arid region of western India. An integrated drought classification scheme was developed by Ramana Rao *et al.* (1981) which considered values of IMA during each crop growth stage (Sastri, 1993).

3.4. Dry-day Sequences

Hershfield *et al.* (1972) propose the frequency of dry-day sequences as a measure of agricultural drought. However, the occurrence and severity of the drought depends upon the length of the sequence in combination with the water content and water holding capacity of the soil, and the water-use pattern of the crop. Brutsaert (1972) also takes dry-day sequences as a measure of agricultural drought, implying that it is the most important factor from an agricultural point of view. This is based on the observation that agricultural drought exists as a result of the depletion of soil moisture in the root zone, impeding plant growth. Dry-day sequences are sequences of days with less than a certain threshold precipitation, *e.g.* 0.25 inches.

3.5. Generalised Hydrologic Model

This is a conceptual system for modelling the headwater portion of the hydrological cycle, which has been used to simulate the relationship between hydrological drought and crop yields, and is a useful tool for managing agricultural yields. It is based on a visualisation of percolation, soil moisture storage, drainage, and evapotranspiration characteristics, which are intended to represent the significant hydrological process in a rational manner. It is consistent with observations of the soil moisture profile made by experimental studies. Application of the system to prolonged precipitation records can result in a history of computed soil moisture inventories, which may be useful in providing keys to defining drought. By establishing the values of the model parameters, which were associated with past drought conditions, it can be possible to evaluate future conditions. It can also be possible to forecast the number of rainless days which must occur before an area could be considered as undergoing a drought (Burnash and Ferral, 1972).

Bidwell (1972) says that a fairly accurate estimate of the probability of drought occurrence is required. This can be solved by finding the joint probability distribution of the set of hydrological variables related to droughts and the functional relationship between the hydrological variables and drought damage. A simulation by mathematical modelling is described and techniques are introduced for constructing and analysing models to simulate the relationship between hydrological drought and crop yields.

3.6. Thornthwaite Method

The Thornthwaite model of the water balance, which may be used on a daily or long-period basis, has been applied to the solution of numerous soil-water problems, including the study of agricultural drought. It has been used to assess the feasibility of irrigation as it provides information on the total volume of water needed at any time and gives a definitive measure of drought. When compared with the water surplus in other seasons it shows whether there is enough water present during the year to permit irrigation (WMO, 1975).

3.7. Crop Moisture Index

The Crop Moisture Index (CMI) is a complement to the PDSI. It is more responsive to short-term changes in moisture conditions and therefore is more useful in the interpretation of agricultural drought. The CMI was developed by Palmer (1968) to indicate the status of agricultural drought or wetness. It measures the degree to which moisture requirements of growing crops are met and responds rapidly to changes in rainfall. However, it is not applicable to germinating and shallow-rooted or cool-season crops. Timing of rainfall is also important in using the CMI since the response of a specific crop varies according to growth stage (Heddinghaus, 1991).

3.8. Moisture Availability Index

The Moisture Availability Index (MAI), the difference of weekly normalised precipitation and normalised potential evapotranspiration, provides a relative measure of moisture available for a given crop (Heddinghaus, 1991).

Table 2: Definition of Agro-meteorological drought (after WMO, 1975).

Author	Definition of drought or associated concepts	Region and comments
Bova (1941)	A drought index K: $K = \frac{10(H+Q)}{\Sigma t}$ where H is productive soil moisture in mm in the top 100 cm of soil at beginning of spring; Q is precipitation in mm accumulated daily from start of spring; Σt is the temperature (°C) sum counted from the day of the passage of mean daily temperature through zero.	USSR. When K ≤ 1.5 start of drought damage to plants is indicated.
Fitzpatrick (1965)	A water-use model with range of available soil moisture 0-10 cm and evapotranspiration losses (E_t) computed from Australia sunken evaporimeter (E_A). $Et = 0.8\ E_A$ when soil moisture > 64 mm; $Et = 0.4\ E_A$ when soil moisture ≤ 64 mm	Australia. Has been used in climatic studies and land-use surveys.
Palmer (1965)	A water-balance model involving rainfall, a coefficient of evapotranspiration, runoff and available soil water.	USA. Method based on the Thornthwaite concept of PE.
Palmer (1968)	Severity of agricultural drought is defined in terms of the magnitude of the computed abnormal ET deficit and expressed as a crop moisture index.	USA. Inputs are weekly values of temperatures and rainfall.
Rickard (1966)	Agricultural drought exists when the soil water in the root zone is at or below the permanent wilting percentage. The condition continues until rain falls in excess of daily ET.	New Zealand. Drought relief would not occur with say one day of excess rainfall, *e.g.* 2.5 – 5 mm.
Ross (1990)	When the crop moisture demand exceeds soil moisture and the system ability to supply.	Lethbridge Northern Irrigation District, Alberta
Van Bavel and Verlinden (1956)	A day on which the available soil moisture was depleted to some small percentage of available capacity.	

4. RIVER FLOW DROUGHT

River flow droughts are related primarily to the effects of periods of rain shortfall on surface and subsurface water supply. They usually lag behind climatic and agro-meteorological droughts, and as a result, the impacts are usually out of phase with those in other economic sectors.

The analysis of river flow drought suffers from the effects of artificial influences in many cases (*e.g.* urbanisation, effluents, abstractions) which makes drought severity calculation difficult. However, drought impacts often need to be assessed within the context of contemporary flow conditions.

The month is the usual time unit for river flow drought studies. As a time unit, the month is long enough to eliminate all less significant, extreme events, and is short enough to allow the monitoring of drought effects in agriculture, water supply and ground-water levels.

4.1. Flow Frequency Curves

Flow frequency curves, which show the average interval in years (return period) between which the flow falls below a given discharge, can be used to identify low flow periods. The plots may be derived for the lowest daily discharge in each year or from flows averaged over longer durations. The procedure for producing such curves is described fully in the Low Flow Studies Report (Institute of Hydrology, 1980). However, the use of flow frequency curves provides no idea of the sequence of flows nor the duration of low flow events.

The fact that many river flow records are too short to permit an extensive analysis has led to the development of methods of synthesising data. One simple method is to establish relations between the short-period station and one that has been in existence for much longer. The relationship may be graphical or mathematical and is assumed to apply to low flows, their durations and probabilities. In contrast to this very simple type of approach, there are methods of river flow simulation which treat discharge as a stochastic process (WMO, 1975).

Ben-Zvi (1987) defines drought as a severe shortage in the appearance of natural waters with respect to normal for place and time. The term 'severe shortage' means a deep, continuous and widely extended shortage. For reasons of generality, the normal and the shortage are described by the mean and the standard deviation of the recorded data. Some other definitions are:

- A deep shortage - when annual runoff is lower than the mean, by at least one standard deviation;

- A continuous shortage - when annual volumes are lower than the mean, during at least four consecutive years;
- An extended shortage - when a deep or continuous shortage extends over the entire region under consideration.

4.2. Gumbel's Extreme Events

Gumbel treated drought, a hydrologic extreme event, as an annual minimum flow value. He recognised that, since a drought can be defined as an extreme value of a statistical realisation of discharge, the drought can be analysed by the statistical theory of extreme values. This enables a recurrence interval to be assigned to a drought that corresponds to an annual minimum flow (Chang and Stenson, 1990).

4.3. Discrete Markov Process

Sen (1976; 1980) stated that the drought characteristics of any phenomenon are dependent on the underlying generating mechanism and can be modelled by a suitable stochastic process such as an independent, first- and second-order Markov process. The most important information which can be obtained by this method is that each deficit is connected to the previous surpluses and deficits. Bonacci (1993) states that this process should be used for the identification of extreme droughts.

4.4. Truncation Levels

River flow droughts have been analysed by Zelenhasić and Salvai (1987). River flow droughts are related to river flow deficit measured from a certain reference discharge (or 'truncation level') Q_r. This method uses instantaneous or daily data instead of mean annual flows as used by Sen in similar work. The method deals with droughts lasting less than a year, and is a method of analysing and describing the stochastic process of river flow droughts. Important components of river flow droughts are considered: deficit, duration, time of occurrence, number of river flow droughts in a given interval, largest river flow drought deficit, and the largest river flow drought duration in a given time interval. The truncation method is now applied in drought analysis in the majority of studies. Drought can be defined as a period when the analysed characteristic is either equal to or lower than the truncation level of drought. Dracup *et al.* (1980) defined the truncation level by the following expression:

$$X_0 = X_m - e * S_D \qquad (3)$$

where X_m is the series mean, S_D is the series standard deviation and is multiplied by e, the elective scaling factor. The truncation level is chosen to be a given measure of the central tendency of the drought sample. This results in approximately half the events being classified as 'high' and half as 'low' (Bonacci, 1993). For regional drought, two truncation levels are required: one for the hydrological variable of interest, the other for the critical area.

4.5. Rainfall and Runoff Deficiency Indices

Drought indices presented in Marsh and Lees (1985), for studies made on British droughts, are based upon a measure of the degree of deficiency in either rainfall or river flow from average conditions. Both the rainfall and runoff indices are derived from monthly data. A dry period is defined by the departure of the monthly totals below the average value. Where a sequence of monthly values is less than the mean, the percentage deficiency is accumulated. Thus, as the dry period progresses, the cumulative index steadily increases. A period of rainfall deficiency is considered to be terminated if:

- one month exceeds 200% of the mean, or
- two months together exceed 250% of the mean, or
- three months together exceed 300% of the mean.

The rainfall index may fluctuate as the drought progresses and the most intense period of the drought may not necessarily coincide with the final month. In this study, Marsh and Lees (1985) used rainfall data averaged over the 1941-70 period for the UK.

Runoff data can be used instead of rainfall data to produce a similar drought index giving a more direct estimate of a drought's magnitude in hydrological terms. Unfortunately, runoff records tend to be rather short and so largely preclude historical comparisons. A runoff deficit is considered to start when discharge falls below the monthly average and to be terminated when flows return to 90% of the monthly average.

Cumulative runoff diagrams can be used as a drought indicator, by plotting the period of interest alongside the long-term average cumulative runoff curve (Marsh and Lees, 1985).

4.6. Theory of Hydrologic Runs

In Yevjevich's opinion, runs, as statistical properties of sequences in both space and time, represent the best basic concept for an objective definition of drought (Yevjevich, 1967). A run is a succession of the same kind of observation preceded and succeeded by one or more observations of a different kind. The method involves plotting a time series and setting a threshold value, c_0. The sequence of consecutive negative deviations is called a negative run, and may be associated with a drought, and vice versa. Note that c_0 can be a fixed number, a deterministic or stochastic time series, or both, according to circumstances. In the study of water deficits, the quantity c_0 is represented by the demand series (Millan, 1972, also described in Millan and Yevjevich, 1971). The number of consecutive intervals where the variable has lower values than the truncated level (length of deficit run) indicates the drought duration. The sum of deviations between the truncation level and the variable values along the deficit run represents the deficit amount. If the deficit is divided by the duration, the deficit intensity is obtained.

Dracup *et al.* (1980) suggest a systematic approach to drought definition, incorporating the use of the theory of runs. Alternatives are discussed, as are the implications of selecting particular components.

The main advantage of using runs as the elements of drought definitions is the possibility of determining their properties (distribution, time dependence, *etc.*) analytically or by data generation methods once the characteristics of the basic variable are known. It should be possible to determine the probability of a given run-sum of a given run-length of monthly values at a precipitation station, once that series is available and properly described mathematically (Yevjevich, 1967). However, Takeuchi (1974) comments that Yevjevich's runs theory has limited application since it is limited to time series involving river flow with a constant seasonal truncation level. Nevertheless, it is a realistic approach for measuring the severity of droughts for areas which depend mainly upon river flow for their water supply.

The WMO Guide to Hydrological Practices (WMO, 1994) says that another method to characterise a drought period is to investigate, by multi-variate statistical analysis, the variables T_1, T_2,..., T_N, which are defined as the maximum continuous time periods during a drought when discharge remains less than Q_1, Q_2,..., Q_N, respectively (where N is the number of years of record).

5. GROUND-WATER DROUGHT

Ground-water drought is a highly subjective phenomenon. Essentially, a ground-water drought is outlined by lower than average annual recharge for more than one year. Smaller, shallower aquifers are more susceptible and sensitive to ground-water drought since they have what can be considered as a finite storage. Larger, for example sedimentary, aquifers are less affected by low annual recharge volumes since they have enormous storage reserves. Ground-water *storage* is considered an unsuitable drought indicator because data are not readily available. Ground-water *levels* are a better indicator in an aquifer area.

Ground-water drought can also be assessed by looking at its secondary effects. If there is a climatological/hydrological drought and runoff is lower than average, less flow will reach rivers. Rivers which have a low base flow index (*i.e.* those rivers which receive only a small proportion of their flow from stored sources, such as ground-water aquifers) will suffer more than those with a high base flow index which receive a greater proportion of their flow from surface water sources. The level of storage in an aquifer declines naturally due to supply to springs, seepage lines, base flow in rivers, and the sea. On top of the natural decline is superimposed the artificial decline caused by pumping, for local water supply and other applications.

An engineering drought occurs when ground-water levels get so low that measures such as extra pumping, and possibly pumping to greater depths, are required. An economic drought occurs when levels get so low that it is prohibitively expensive to draw any more water. The biggest droughts occur when recharge volumes are low, when normal water supply sources are depleted and in a state of emergency the ground-water stocks are turned to. Long-term average ground-water levels can be plotted along with monthly average and extreme levels for the preceding record. On top of this can be plotted the levels for the period concerned, and abnormally low levels can be easily identified. Whether or not these low levels constitute a ground-water drought depends upon other engineering and economic factors as described above.

In practice, the severity of a ground-water drought is normally measured in terms of recharge received by an aquifer and the minimum, or trough level, of the water-table during the drought years. However, the difficulties attending the precise measurement of hydrological variables are especially significant in the ground-water context (Marsh *et al.*, 1994).

6. OPERATIONAL DROUGHT

An operational drought is not necessarily a naturally-induced event: it is a complex combination of a natural shortage of precipitation, the development of major water resources systems, and recent operational management decisions (Mawdsley *et al.*, 1994).

A moderate water supply drought in the UK is defined as one where demand restrictions are necessary to reduce the total consumption by the public consumers by 10%. Serious and severe droughts are linked to the risk of standpipes (vertical pipes connecting taps to the main water distribution system) being imposed in the near future. The difference between them relates to the risk of this happening: a serious drought occurs when the risk is greater than 20%, and a severe drought when the risk is greater than 50%. For a 'short' class of drought, the risk of standpipes refers to the risk within the current year; for a 'long' class of drought, the risk is highest for standpipes to be imposed in subsequent years (Mawdsley *et al.*, 1994).

Takeuchi (1974) states that whenever the amount of water, which has been expected and relied upon for use in any of man's activities, cannot be met for some reason, a drought condition is established. A water shortage is defined as a situation in which a water surplus does not exist on the average and a potential water demand exists.

Operational drought is essentially an infrastructure problem. Reservoir and groundwater stocks may be full, but water supply companies in Britain, for example, have had problems meeting surges of demand. On hot, dry summer days, at peak demand times of the day, surges in demand can reach several times the normal demand. It is impossible for this amount to be supplied at the usual pressure. Indeed, it would be uneconomical to design water supply networks to cope with this size of demand for very short periods of the year. Marsh *et al.* (1994) state that surface water reservoir stocks may often be indicative of a drought severity which can appear inconsistent with the contemporary hydrological drought intensity.

6.1. Storage Yield Analysis

Reservoir storage is a most useful drought severity indicator, due to data availability often on a daily or weekly basis. Although absolute reservoir storage is useful, these data are strongly influenced by how the reservoir has been operated. Yet when reservoir storage is combined with predicted inflows, it is quite a powerful indicator of drought severity.

The severity of a drought in terms of reservoir storage can be estimated by carrying out a computer simulation of how the reservoir would have behaved when subjected to a historical sequence of inflows. By expressing the yield as a percentage of the mean discharge and the storage as a percentage of the mean annual runoff, storage-yield relationships derived from different flow records may be compared. Furthermore, the return period of each storage requirement can be estimated from the storage-yield graph (Marsh and Lees, 1985).

In a study by Tase (1976), a drought is defined by using the differences between the water supply and water demand series. The water supply series are defined by the stochastic component in the monthly precipitation, which is presumed to be a time independent stationary stochastic process. The water demand series are given either by the mean of the stochastic component or by its probability quantiles as:

$$q = F(\xi_0) = P(\xi \leq \xi_0) \tag{4}$$

where ξ is a stationary stochastic component series for a given station, and ξ_0 is the truncation level.

In a study by Guerrero-Salazar and Yevjevich (1975), a drought is similarly defined using the differences between the processes of water supply and water demand. The supply processes or supply time series may be precipitation over an area, river flow at a point in a river, moisture in the soil, storage of water in an aquifer or reservoir, and similar hydrologic variables. The demand process or time series may be a single-purpose water use, such as that used for agriculture, for continuous irrigation, hydropower, or for water supply. When demand exceeds supply, the water shortage occurs, and the start of the drought is noted.

The 'percentage shortage index' (PSI) acts as an index of the severity of a water shortage:

$$PSI = (D\text{-}V)*100/D \qquad (5)$$

The demand D and the Normal supply V are then divided by the system's safe yield level SY, which is determined from the most severe water shortage experienced in the region's recorded history.

6.2. Minimum Flow Approach

This critical period approach uses low flow sequences only, and was proposed by Waitt (1945). From the historical river flow record, the lowest sub-sequences of flows of various durations are selected and plotted as flow volume versus duration. Constant draughts plot as straight lines on this so-called 'drought-curve'. The critical storage is given by the maximum intercept between the draught line and the drought curve; the corresponding duration at this position is the critical drought period for the reservoir.

6.3. Alexander's Method

The minimum flow approach does not provide an estimate of the storage corresponding to a given probability of failure. Alexander (1962) extended the earlier approach by developing a series of drought curves for different probabilities of occurrence and from these derived generalised storage-regulation-probability curves. The method uses a fundamental property of the Gamma distribution, *i.e.* the sum of n independent random variables from a Gamma distribution with parameters α and β is itself Gamma distributed (McMahon and Mein, 1986).

Since reservoirs have a fixed amount which they can store, the date of onset of drawdown of stocks is important. Even if there has been a very wet winter, if the spring season is dry then drawdown begins early and the stocks will not last through to the end of a dry summer. Enough water may have fallen during the winter to fill the reservoirs and aquifers several times, but it is impossible to store that amount of water.

In Southern Africa, the wet and dry seasons are rather clearly defined, and once the wet season has started, it is predictable that it will be wet for some time. In Britain and other temperate zones, this is much more difficult.

6.4. The Threshold Approach

From an economic viewpoint, drought may be considered as a time period when, for example, water shortages do not allow satisfaction of water requirements of all water users. In this case, drought threshold discharge, Q_d, using economic criteria, denoted by e, may be taken as:

$$Q_d^e = Q_m + \sum_{j=1}^{k} Q_{u,j} \tag{6}$$

where Q_m is the minimum acceptable flow, $Q_{u,j}$ the water requirements of jth user and k the number of water users. Ozga-Zielinska (1989) defines various thresholds according to criteria such as navigation.

7. CONCLUSIONS

This paper has reviewed a number of definitions of droughts from a hydrological perspective. Their application is dependent on the availability of data, the particular aspect of a drought to be characterised and the preferences and prejudices of the user. Whilst there may be merit in identifying one or two simple definitions of droughts for comparison between countries, it is essential that droughts are analysed according to the requirements of a specific local, regional or international investigation.

It is concluded that despite the difficulties with finding a suitable definition of drought for a particular application, such definitions can provide a practical way to assess a drought situation and to give a relative picture of its severity.

REFERENCES

Abu Zeid, M. and Abdel-Dayem, S. (1990) The Nile, the Aswan High Dam and the 1979-1988 Drought, in *Transactions of 14th Congress on Irrigation and Drainage*, Rio de Janeiro, Vol. 1-C, International Commission on Irrigation and Drainage (ICID), 319-334.

Alexander, G. N. (1962) The Use of the Gamma Distribution in Estimating Regulated Output from Storages, *Civil Engineering Transactions*, The Institution of Civil Engineers, Australia, CE 4(1): 29-34.

Baldwin-Wiseman, W.R. (1941) Further Cartographic Studies of Drought, *Quart. J. Roy. Met. Soc.* 67(288): 39-47, quoted from WMO (1975).

Banerji, S. and Chabra, B.M. (1964) Drought Characteristics and Estimating Probabilities of Their Occurrence, in *Surface Waters Symposium*, WMO/IASH, Belgium, Publication No. 63, 189-192.

Bates, C.G. (1935) Climatic Characteristics of the Plains Region, in M. Silcox, *et al.* (eds.), *Possibilities of Shelterbelt Planting in the Plains Region*, Washington D.C., quoted from WMO (1975).

Bazza, M. and Stockton, C.W. (1990) Variabilité spatiale et chronologique du climat Marocain et importance de l'irrigation dans l'amélioration de ses potentialités pour les cultures pluviales, in *Transactions of 14th Congress on Irrigation and Drainage*, Rio de Janeiro, Vol. 1-C, International Commission on Irrigation and Drainage (ICID) 233-249.

Benson, C. and Clay, E. (1994) *The Impact of Drought on Sub-Saharan African Economies: A Preliminary Examination*, Overseas Development Institute (ODI) Working Paper 77, London, ISBN 0 85003 212 1.

Ben-Zvi, A. (1987) Indices of Hydrological Drought in Israel, *J. Hydrol.* 92: 179-191.

Beran, M. and Rodier, J.A. (1985) Hydrological Aspects of Drought, *Studies and Reports in Hydrology 39*, UNESCO-WMO, Paris.

Bidwell, V.J. (1972) A Methodology for Analysing Agricultural Drought, in *Floods and Droughts. Proceedings of the Second International Symposium in Hydrology*, Sept. 11-13, 1972, Colorado, USA, 515-522.

Blumenstock, G. (1942) *Drought in the United States Analysed by Means of the Theory of Probability*, USDA Technical Bulletin 819, 5 p, quoted from WMO (1975).

Bonacci, O. (1993) Hydrological Identification of Drought, *Hydrological Processes* 7: 249-262.

Bova, N.V. (1941) Agrometeorological Indices in the South-Eastern USSR, *Socialistic Grain Farming*, No. 3, quoted from WMO (1975).

British Rainfall Organisation (1936) *British Rainfall*, Air Ministry, Meteorological Office, London, quoted from WMO (1975).

Brutsaert, W. (1972) Moisture Availability as it Relates to Droughts, in *Floods and Droughts, Proceedings of the Second International Symposium in Hydrology*, Sept. 11-13, 1972, Colorado, USA.

Bruwer, J.J. (1990) Drought Policy in the Republic of South Africa, in *Proceedings of the SARCCUS Workshop on Drought*, June 1989, ISBN 0 949986 24 0.

Bryant, S.J., Arnell, N.W., and Law, F.M. (1992) The Long-term Context for the Current Hydrological Drought, *Proceedings of the IWEM Conference on the Management of Scarce Water Resources*.

Burnash, R.J.C. and Ferral, R.L. (1972) Generalized Hydrologic Modelling, a Key to Drought Analysis, in *Floods and Droughts, Proceedings of the Second International Symposium in Hydrology*, Sept. 11-13, 1972, Colorado, USA, 503-514.

Chang, T.J. and Stenson, J.R. (1990) Is it Realistic to Define a 100-year Drought for Water Management? *Water Resources Bulletin* 26(5): 823-829.

Clarke, R. (1991) *Water: the International Crisis*, Earthscan Publications with Swedish Red Cross, London, ISBN 1853831050.

Cole, H.S. (1933) Droughts in Arkansas, *Monthly Weather Review* 61: 129-140, quoted from WMO (1975).

Condra, G.E. (1944) Drought, Its Effects and Measures of Control in Nebraska, *Nebraska University, Conservation and Survey Division, Conservation Bulletin* 25, quoted from WMO (1975).

Dezman, L.E., Shafer, B.A., Simpson, H.D., and Danielson, J.A. (1982) Development of a Surface Water Supply Index – A Drought Severity Indicator for Colorado, in *Proceedings Int. Symp. on Hydrometeorology*, American Water Resources Association (AWRA), June 13-17, 1982, Colorado, USA, 337-341.

Diaz, H.F. (1983) Drought in the United States. Some Aspects of Major Dry and Wet Periods in the Contiguous United States, 1895-1981, *Journal of Climate and Applied Meteorology* 22(1): 3-16.

Dracup, J.A., Seong Lee, K., and Paulson Jr, E.G. (1980) On the Definition of Droughts, *Water Resources Research* 16: 297-302.

Emberger, L. (1955) *Une classification biogéographique des climats*, Recueil des travaux, Fac. Sc. de l'Univ. de Montpellier, Fasc. 7, quoted from WMO (1975).

Fitzpatrick, E.A. (1953) *Probability Analysis of Rainfall Factors in Drought in New South Wales*, Rural Bank of N. S. Wales and N. S. Wales Dept. of Lands, Australia, 54 p, quoted from WMO (1975).

Fitzpatrick, E.A. (1965) *Climate of the Tipperary Area*, CSIRO, Melbourne, Land Research Series No. 13, quoted from WMO (1975)

Foley, J. C. (1957) *Drought in Australia*, Bureau of Meteorology, Bulletin No. 43, 281 p.

Frere, M. and Popov, G.F. (1979) *Agrometeorological Crop Monitoring and Forecasting*, Plant Production and Protection Paper 17, FAO, Rome.

Garen, D.C. (1993) Revised Surface-Water Supply Index for the Western United States, *Journal of Water Resources Planning and Management*, 119(4): 437-454.

Guerrero-Salazar, P. and Yevjevich, V. (1975) *Analysis of Drought Characteristics by the Theory of Runs*, Hydrology Paper 80, Colorado State University, Colorado.

Gupta, V.K. (1972) Droughts as Extreme Distributions from a Point Rainfall Process, in *Floods and Droughts. Proceedings of the Second International Symposium in Hydrology*, Sept. 11-13, 1972, Colorado, USA.

Guttman, N.B., Wallis, J.R. and Hosking, J.R.M. (1992) Spatial Comparability of the Palmer Drought Severity Index, *Water Resources Bulletin* 28(6): 1111-1119.

Heddinghaus, T.R. (1991) Monitoring and Dissemination of Drought Conditions at the Joint Agricultural Weather Facility, in D. A. Wilhite, D. A. Wood, and P. A. Kay (eds.), *Proceedings of the Seminar and*

Workshop on Drought Management and Planning, Institute of Agriculture and Natural Resources, University of Nebraska-Lincoln, 67-72.

Henry, A.J. (1906) Drought, in *Climatology of the United States,* U.S. Weath. Rev. Bull. Q. Washington, WB Report No. 361: 51-58, quoted from WMO (1975).

Herbst, P.H., Bredenkamp, D.B., and Barker, H.M.G. (1966) A Technique for the Evaluation of Drought from Rainfall Data, *Journal of Hydrology* 4: 264-272.

Hershfield, D.M., Brakensiek, D.L., and Comer, G.H. (1972) Some Measures of Agricultural Drought, in *Floods and Droughts, Proceedings of the Second International Symposium in Hydrology*, Sept. 11-13, 1972, Colorado, USA, 491-502.

Hoyt, J.C. (1936) *Droughts of 1930-1934*, USGS Water Supply Paper No. 680, 106 p, quoted from WMO (1975).

Hulme, M. (ed.) (1996) *Climate Change and Southern Africa: an Exploration of Some Potential Impacts and Implications for the SADCC region*, Report commissioned by WWF International and co-ordinated by Climatic Research Unit, University of East Anglia, Norwich, UK.

Institute of Hydrology (1980) *Low Flow Studies*, Institute of Hydrology, Wallingford, U.K.

Karl, T.R. (1991) Climatological Drought Prediction in D.A. Wilhite, D.A. Wood, and P.A. Kay (eds.), *Proceedings of the Seminar and Workshop on Drought Management and Planning*, Institute of Agriculture and Natural Resources, University of Nebraska-Lincoln, 9-16.

Kulshreshtha, S.N. and Klein, K. (1990) Drought Mitigation Through Large-scale Intensive Irrigation Projects - A Case Study of the South Saskatchewan River Project, in *Transactions of 14th Congress on Irrigation and Drainage*, Rio de Janeiro, Vol. 1-C, International Commission on Irrigation and Drainage (ICID), 13-28.

Marsh, T.J. and Lees, M.L. (1985) *The 1984 Drought*, Hydrological Data: UK Series, Institute of Hydrology/British Geological Survey.

Marsh, T.J., Monkhouse, R.A., Arnell, N.W., Lees, M.L., and Reynard, N.S. (1994) *The 1988-92 Drought*, Hydrological Data: UK Series, Institute of Hydrology/British Geological Survey.

de Martonne, E. (1926) Une nouvelle fonction climatologique: l'indice d'aridité, *La Météorologie* 2: 449-458, quoted from WMO (1975).

Mawdsley, J., Petts, G., and Walker, S. (1994) *Assessment of Drought Severity*, British Hydrological Society (BHS) Occasional Paper No. 3.

McMahon, T.A. and Diaz Arenas, A. (1982) *Methods of Computation of Low Streamflow*, Studies and Reports in Hydrology 36 UNESCO Paris 95.

McMahon, T.A. and Mein, R.G. (1986) *River and Reservoir Yield.* Water Resources Publications, USA, ISBN 0-918334-61-6.

Millan, J. (1972) Statistical Properties of Runs as Applied to Hydrologic Droughts, in *Floods and Droughts. Proceedings of the Second International Symposium in Hydrology*, Sept. 11-13, 1972, Colorado, USA, 534-550.

Millan, J. and Yevjevich, V. (1971) *Probabilities of Observed Droughts*, Hydrology Paper 50, Colorado State University, Colorado.

Ozga-Zielinska, M. (1989) Droughts and Floods - Their Definition and Modeling, in *New Directions for Surface Water Modeling*, IAHS Publication No. 181, 313-322.

Palfai, I. (1990) Description and Forecasting of Droughts in Hungary, in *Transactions of 14th Congress on Irrigation and Drainage*, Rio de Janeiro, Vol. 1-C. International Commission on Irrigation and Drainage (ICID), 151-158.

Palmer, W.C. (1965) *Meteorological Drought*, U. S. Weather Bureau, Washington D.C., Research Paper No. 45, quoted from WMO (1975).

Palmer, W.C. (1968) Keeping Track of Crop Moisture Conditions Nationwide: The New Crop Moisture Index, *Weatherwise* 21:156-161, quoted from WMO (1975).

Pereira, L.S. (1990) The Role of Irrigation in Mitigating the Effects of Drought, in *Transactions of 14th Congress on Irrigation and Drainage*, Rio de Janeiro, Vol. 1-F. International Commission on Irrigation and Drainage (ICID), G43.1 – G43.27.

Petrasovits, I. (1990) General Review on Drought Strategies, in *Transactions of 14th Congress on Irrigation and Drainage*, Rio de Janeiro, Vol. 1-C, International Commission on Irrigation and Drainage (ICID), 1-12.

du Pisani, A.L. (1990) Drought Detection, Monitoring and Early Warning, *Proceedings SARCCUS Workshop on Drought*, June 1989, ISBN 0 949986 24 0, 6-11.

Popov, V.P. (1948) Moisture Balance in the Soil and the Dryness Indices of Climate in the Ukrainian S.S.R, *Sci. Rep. of the State Univ. of Kiev*, Vol. 7(1), quoted from WMO (1975).

Ramana Rao, B.V., Sastri, A.S.R.A.S., and Rama Krishna, Y.S. (1981) An Integrated Scheme of Drought Classification as Applicable to the Indian Arid Region, *IDOJARAS* 85:317-22, quoted from Sastri (1993).

Ramdas, D.A. (1950) Rainfall and Agriculture, *Ind. J. Met. and Geophys.* 1(4): 262-274, quoted from WMO (1975).

Rickard, D.S. (1966) The Occurrence of Agriculture Drought at Ashburton N.Z. *New Zealand Journal Agricultural Res.*, 431-441.

Ross, F A (1990) Irrigation District Operation on Water Shortage Conditions. The Alberta Experience, in *Transactions of 14th Congress on Irrigation and Drainage*, Rio de Janeiro, Vol. 1-C, International Commission on Irrigation and Drainage (ICID), 353-360.

Rossi, G., Benedini, M., Tsakiris, G., and Giakoumakis, S. (1992) On Regional Drought Estimation and Analysis, *Water Resources Management* 6: 249-277.

Sastri, A.S.R.A.S. (1993) Agricultural Drought Management Strategies to Alleviate Impacts: Examples from the Arid and Subhumid Regions of the Indian Subcontinent, in D.A. Wilhite (ed.), *Drought Assessment, Management, and Planning: Theory and Case Studies*, Kluwer Academic Publishers, Dordrecht, 65-86.

Sastri, A.S.R.A.S., Rama Krishna, Y.S., and Ramana Rao, B.V. (1981) A New Method for Classifying Agricultural Droughts, *Archiv für Meteorologie, Geophysik und Bioklimatologie Ser.* B 29: 283-287, quoted from Sastri (1993).

Selyaninov, G.T. (1930) Methods of Agricultural Climatology, in *Agricultural Meteorology,* No. 22 L, quoted from WMO (1975).

Sen, Z. (1976) Wet and Dry Periods of Annual Flow Series, *J. Hydraulics Division* 102 (HY10), American Society of Civil Engineers (ASCE) Proc. Paper 12457, 1503-1514.

Sen, Z. (1980) Statistical Analysis of Hydrologic Critical Droughts, *J. Hydraulics Division* 106 (HY1), American Society of Civil Engineers (ASCE) Proc. Paper 15134, 99-115.

Takeuchi, K. (1974) *Regional Water Exchange for Drought Alleviation*, Hydrology Paper 70, Colorado State University, Colorado.

Tannehill, I.R. (1947) *Drought, Its Causes and Effects*, Princeton University Press. 264 p, quoted from WMO (1975).

Tardieu, H. and Plus, B. (1990) Après la sécheresse des années 1984-1987 dans le bassin de la Garonne, France. Recommandations pour l'avenir, in *Transactions of 14th Congress on Irrigation and Drainage*, Rio de Janeiro, Vol. 1-C, International Commission on Irrigation and Drainage (ICID), 209-220.

Tase, N. (1976) *Area-deficit-intensity Characteristics of Droughts*, Hydrology Paper 87, Colorado State University, Fort Collins, Colorado.

Tate, E.L., Meigh, J.R., Prudhomme, C., and McCartney, M. (2000) *Drought Assessment in Southern Africa Using River Flow Data*, Final report to DFID for project R6573, Institute of Hydrology, Wallingford, UK.

Tennessee Valley Authority (1944) *Precipitation in the Tennessee River Basin*, TVA Hydraulic Data Division, quoted from WMO (1975).

Thornthwaite, C.W. (1931) The Climates of North America According to a New Classification, *Geogr. Rev.* 21: 633-655, quoted from WMO (1975).

Van Bavel, C.H.M. and Verlinden, F.J. (1956) *Agricultural Drought in North Carolina,* N.C. Agric. Experimental Station Tech. Bull. No. 122, quoted from WMO (1975).

Waitt, F.W.F. (1945) Studies of Droughts in the Sydney Catchment Areas, *J. Inst. Engineers, Australia,* 17(4-5): 90-97.

Welsh National Water Development Authority (WNWDA), Directorate of Resource Planning, (1976) *The 1975 Drought*, Brecon, Wales.

World Meteorological Organisation (1975) *Drought and Agriculture*, Technical Note No. 138, WMO No. 392, Geneva, Switzerland.

World Meteorological Organisation (1994) *Guide to Hydrological Practices. Data Acquisition and Processing, Analysis, Forecasting and Other Applications*, Publication No. 168.

Yevjevich, V. (1967) *An Objective Approach to Definitions and Investigations of Continental Hydrologic Droughts*, Hydrology Paper 23, Colorado State University.

Zelenhasić, E. and Salvai, A. (1987) A Method of Streamflow Drought Analysis, *Water Resources Research* 23: 156-168.

DEFINITION, EFFECTS AND ASSESSMENT OF GROUNDWATER DROUGHTS

H.A.J. VAN LANEN and E. PETERS
Sub-department of Water Resources
Wageningen University
6709 PA Wageningen, The Netherlands

Abstract. Natural groundwater droughts originate from reduced recharge over a prolonged period of time and these droughts are often enhanced by human activities (e.g. abstractions). Low groundwater heads and small groundwater gradients cause the droughts. Groundwater droughts are mainly associated with low well yields, which affect public water supply and irrigation practices and are related to other droughts (e.g. agricultural droughts). Groundwater drought monitoring and assessment methods are based upon an analysis of time-series of groundwater recharge or groundwater heads and require a threshold or probability occurrence, which can only be derived from a proper evaluation of the effects of groundwater droughts, such as lowering of well levels, reduction of springflow or reduction of capillary rise. Some of these effects are discussed to illustrate how eventually such threshold or probability of occurrence has to be determined.

1. INTRODUCTION

Under natural conditions groundwater droughts are recurrent phenomena which as other droughts originate from precipitation deficiency (including snowfall). Groundwater droughts are associated with falling groundwater levels resulting in low well yields, springflows and streamflows or even drying-up of wells, brooks and rivers. Eventually this can lead to degraded terrestrial and aquatic ecosystems. Intensive exploitation of groundwater resources, for irrigation in semi-arid regions and for industrial and domestic water supply in the more densely populated areas in the temperate regions, enhances or even creates groundwater droughts. Crops in normally irrigated areas may suffer from water stress and the public water supply might seriously be hampered because groundwater wells dry up. Over the last decade management of groundwater resources, including the development of drought mitigation measures, has emerged as a priority because of the natural variability of precipitation and because of human interference with the groundwater system in extended areas.

A drought, especially a meteorological drought, is generally well understood by the public but translating this awareness into a clear operational definition and into associated rigorous and objective procedures for quantifying the drought in terms of onset, duration, severity and spatial extent is far from straightforward (e.g. Marsh and Lees, 1985). However, transparent operational definitions and assessment procedures are required before reactive or pro-active drought mitigation measures can be identified and implemented.

J.V. Vogt and F. Somma (eds.), Drought and Drought Mitigation in Europe, 49–61.

A groundwater drought might be defined as a lack of groundwater recharge, or a lack of groundwater expressed in terms of storages or groundwater heads in a certain area and over a particular period of time. The lack is usually determined in comparison to some "normal" or average amount or level derived from frequency analysis using historical data. This definition looks simple and its application seems straightforward. The definition and the assessment procedures, however, are still not fully operational for the objective recognition of groundwater droughts and for the implementation of measures to mitigate these droughts. The objective of this paper is to elaborate the definition of groundwater droughts and to discuss associated effects (e.g. lower well yields, lower groundwater levels, reduction of spring flow, reduction of capillary rise) as well as a procedure to assess groundwater droughts.

2. SCOPE OF GROUNDWATER DROUGHTS

Different types of droughts are generally distinguished, namely meteorological, agricultural, groundwater and surface water droughts (e.g. Marsh and Lees, 1985; NDMC, 1995). Although all droughts have a precipitation shortfall as origin, it takes longer for groundwater droughts to show up in a region than for meteorological droughts. Not all meteorological droughts (low precipitation and possibly high evaporation rates) or all agricultural droughts (soil moisture deficiencies) lead to a groundwater drought. The relation between groundwater droughts and other droughts is presented in figure 1. In this diagram the effects of groundwater droughts are presented as well, because these effects are relevant for the definition and assessment procedure as discussed hereafter. Only natural causes for groundwater droughts are presented. Of course, man-induced causes could be added as well. For example, climatic change will influence the climate variability, land use change will have an effect on evapotranspiration, groundwater abstraction will lower the groundwater heads and land drainage will increase discharge to the surface water system, leaving less water in the aquifer to cover dry periods.

Under natural conditions groundwater droughts result from reduced groundwater recharge. A few days of below average groundwater recharge, however, are insufficient to cause a groundwater drought. A prolonged period (several months up to years) with reduced recharge is necessary before a groundwater drought may develop. A groundwater drought will only develop if reduced recharge coincides with low groundwater storage due to antecedent meteorological conditions.

A reduced recharge leads to lower groundwater heads and lower groundwater storage. Groundwater recharge, heads and/or storage could be used to define and to assess groundwater droughts. Groundwater heads are preferred because heads are readily measured. Observed levels give adequate information about the state of the aquifer and the groundwater gradients. Gradients are relevant because they determine the fluxes in a groundwater system. Both the levels and the gradients are essential to quantify the effects of groundwater droughts. In the context of groundwater drought recognition and assessment, groundwater storage is hardly used because it cannot be measured directly.

The often ill-defined bottom of the groundwater system and insufficient knowledge about the spatially variable storage coefficient hampers the computation of groundwater storage.

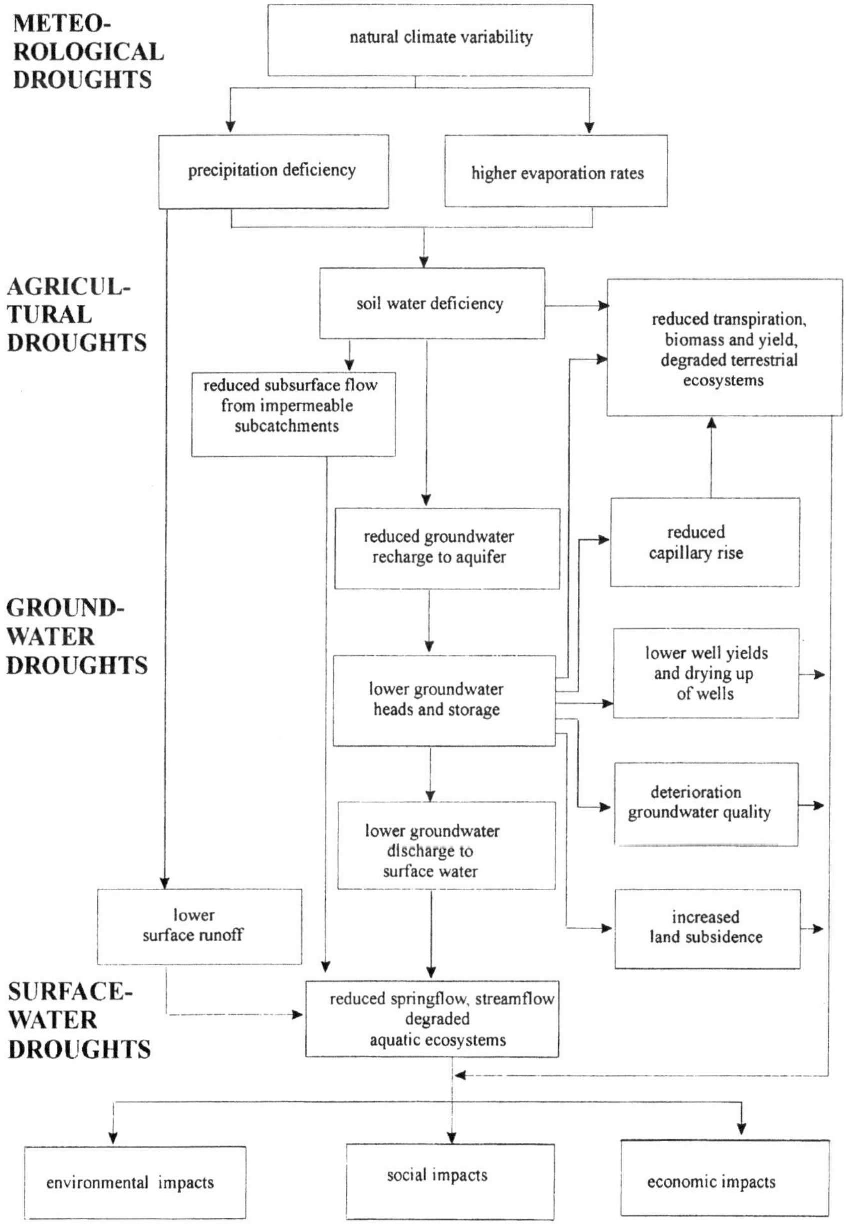

Figure 1: Schematic outline of the different droughts (adapted from NDMC, 1995).

Groundwater droughts can be recognized and evaluated in a strict and in a wide sense. Evaluation of groundwater droughts in the strict sense is restricted to the state of the aquifer itself. These droughts are characterized by low groundwater levels and associated lower groundwater pumping rates and in extreme situations even by drying-up of the boreholes (Figure 1). Lower extraction rates have an economic impact if more fuel has to be used to abstract the required amount of groundwater. If the fuel or the groundwater is unavailable (dry wells) then economic (e.g. loss of crop yield) and social consequences (lack of drinking water) may occur. Besides groundwater quantity issues, droughts can also induce a deterioration of the water quality in the aquifer itself. Well-known examples are the horizontal and/or vertical displacement of the fresh-salt water interface causing salt-water intrusion (e.g. Gupta and Gurunadha, 1998). This might lead to water supply and environmental problems. The quality issues especially emerge if aquifers are intensively exploited. Evaluation of groundwater droughts in the wider sense is not only restricted to the state and processes in the aquifer itself, but it also includes effects outside the boundaries of the aquifer, which are caused by the lower groundwater heads and smaller gradients (Figure 1). Examples are:

- lower water-tables having a negative effect on terrestrial ecosystems (e.g. oxygen concentration, nutrient fluxes);
- reduced capillary rise resulting in loss of agricultural crop yield and lower biomass production;
- lower groundwater discharge to the riparian area, springs and streams. This might have an impact on aquatic ecosystems and on inflow to surface water reservoirs;
- increased land subsidence leading to damage to the infrastructure (e.g. buildings, roads; Carrillo-Rivera 1998). In limestone areas sudden sinkhole formation can occur, especially if excessive pumping enhances the groundwater droughts.

In the next sections we will first discuss groundwater droughts in the strict sense and we will present an assessment method. Next some aspects of groundwater droughts in the wider sense will be discussed, which are needed for a proper application of the assessment method.

3. DROUGHTS AND GROUNDWATER HEADS

3.1. Lower Well Yields

Groundwater droughts in a region are primarily associated with groundwater heads below a certain level. Clear evidence of groundwater droughts are dry wells, which do not run dry under normal conditions. Groundwater droughts, however, may also occur without drying-up of wells. For instance, already when water levels fall below a certain level, this may negatively affect the community. Then reactive mitigation measures can be taken (e.g. lowering of submersible pumps) or particular actions can be imposed (e.g. temporal termination or reduction of abstractions). Abstraction reduction can restrict environmental impact, but it usually will lead to economic consequences, such as higher water supply costs or yield loss of irrigated crops. Lowering of pumps results into a higher lift and often lower extraction rates because permeability in many aquifers decreases with depth. For example, at a well field in the south of the Netherlands, groundwater is normally extracted from river gravel, which drain the underlying chalk. During (extremely) dry years the gravel desaturates and the pumps have to be lowered in the chalk. Abstraction rates then significantly decrease by about 40%. As a consequence, longer pumping and more fuel are required to fulfill the water demands.

3.2. Drought Monitoring and Assessment

Time-series of groundwater heads are used to monitor and assess groundwater droughts. Many different methods of varying complexity and suitability are available which can be applied at different scales. A common method to monitor groundwater droughts is to plot the currently observed groundwater level in a graph, which also shows the 95% probability of occurrence lines (Figure 2). These lines, which are identical for each year, are derived from statistical analysis of historical time-series. For each monitoring date the 95% probability (or any other probability) points are computed and subsequently connected to a line. Such monitoring of water levels can quickly reveal when a critical level is reached and how long the drought lasts. This procedure is already operational in some European countries (e.g. IH, 1999; NITG, 1998). Such monitoring can be used to start reactive drought mitigation actions.

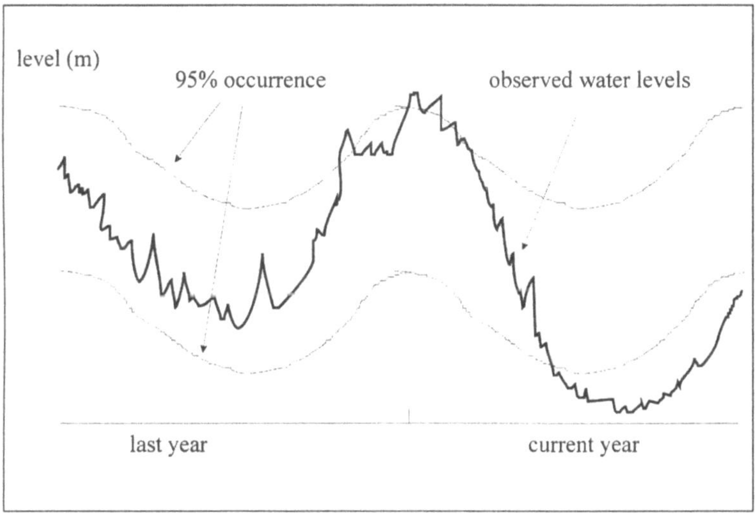

Figure 2: Monitoring of groundwater droughts using observed groundwater levels.

The most frequently applied procedure to assess groundwater droughts is based upon the determination of statistical drought characteristics (onset, duration, severity) derived from time-series of groundwater data. These characteristics can be used to implement pro-active drought mitigation measures (e.g. utilization of other resources, introduction of water saving policies). The analysis can include both observed data and simulated data. The latter data are needed to explore the negative impact of, for instance, groundwater abstraction or land drainage, but also to evaluate the positive impact of drought mitigation measures. Onset, duration and severity of droughts can be computed by considering situations where the groundwater head is below a certain threshold level (threshold approach; e.g. Tallaksen *et al.*, 1997). When all droughts are identified, a probability distribution can be calculated, if sufficient droughts show up in the time-series. A groundwater hydrograph of the Belgian-Dutch chalk region (Figure 3) has been used to illustrate the procedure.

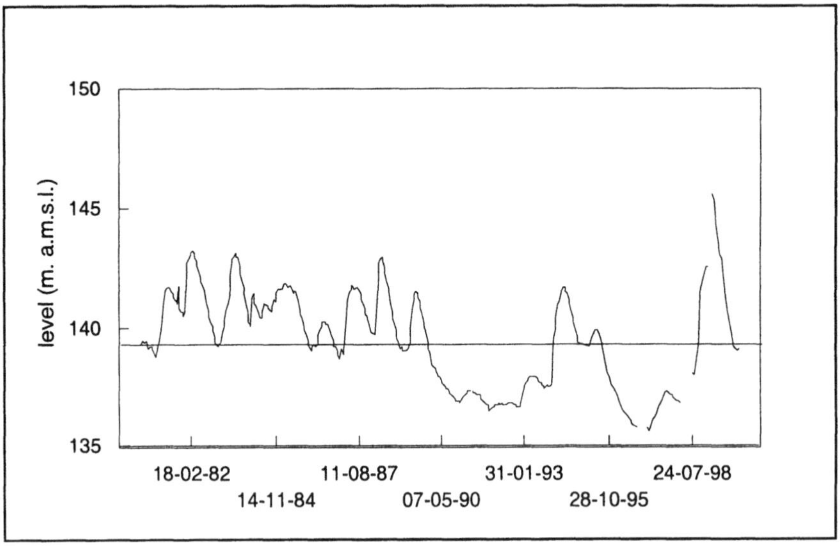

Figure 3: Example of a well (WP98) in the Noor catchment (The Netherlands) which shows drought characteristics.

All processing of the data was done with the software code EXDEV (⊂i ica and Novický, 1995). Onset and duration of groundwater droughts can be computed with the software code analogue to surface water droughts for which it has been developed. The severity of a groundwater drought can also be determined with EXDEV if it is defined as the cumulative departure from the threshold:

$$S(k) = \sum_{i}^{i=n} (\varphi(i) - \varphi_{th}) \tag{1}$$

where:

- $S(k)$: severity of drought k [m]
- $\varphi(i)$: groundwater head at day i [m]
- φ_{th}: threshold groundwater head [m]
- n: number of days that drought lasts [-]

The severity can also be normalized by dividing $S(k)$ by the number of days n.

In the well shown in figure 3, the average groundwater level over the years 1980-1997 is 139.3 meters. If we assume that groundwater droughts can only occur in situations when the groundwater level is below the average level, three minor droughts occur in the eighties and two major ones in the nineties. The major droughts have been observed in a lot of permeable catchments with deep groundwater levels in northwest Europe. These droughts are related to low groundwater recharge (Dijksma and van Lanen, 1998). In the given case, the number of droughts is too small to obtain reliable results from statistical analysis. However, modeling can be used to generate longer, synthetic time-series. For the relatively short time-series in figure 3 the number of groundwater droughts, and the onset and the duration of the most severe drought for different threshold values are given in table 1.

Table 1: Characteristics of groundwater droughts derived from groundwater data from well WP98 using different threshold values.

Threshold (m. a.m.s.l.)	Exceedance Probability (%)	Number of Droughts	Maximum Drought	
			Onset	Duration (days)
139	60	5	09.12.1989	1489
138	65	2	27.03.1990	1417
137	82	4	21.02.1996	528
136	97	1	06.08.1996	239

The results show that the chosen threshold has a large influence on the characteristics of the drought. For example, a threshold of 138 m picks out the two major droughts. The early-nineties drought is classified as the longest. The departure from the average groundwater level, however, is larger in the mid-nineties drought, which is picked out by using a threshold smaller than 138 m. A threshold of 137 m separates each of the two major droughts in the nineties in more, but shorter spells. It may be argued if this separation is relevant, which depends on the objective of the study. Of course, a low threshold value, such as 136 m only picks out the most severe event, i.e. the mid-nineties drought. The use of the departure from the threshold (eq. 1) as a criterion would also have given this result. The onset of the droughts (Table 1) shows that a groundwater drought in temperate regions does not have to coincide with the summer period where meteorological and agricultural droughts usually prevail. The droughts are out of phase, which may be an advantage because of a lower water demand during winter.

The above-mentioned threshold approach can also be used to show how groundwater drought will increase due to, for example, groundwater abstraction. A hypothetical drawdown is superimposed to illustrate the procedure for the observed groundwater levels of well WP98 (Figure 3)

The drawdown is supposed to vary over the year from 1.0 m (1 April) to 2.0 m (1 October) to simulate a higher groundwater abstraction during summer. The same thresholds as used for the situation without the drawdown (Table 1) are used to illustrate the procedure.

Table 2: Enhancement of groundwater droughts derived from groundwater data from well WP98, including a drawdown of 1.0-2.0 m using different threshold values.

Threshold (m. a.m.s.l.)	Number of Droughts	Maximum Drought	
		Onset	Duration (days)
139	9	28.08.1989	1625
138	8	21.10.1989	1544
137	2	20.02.1990	1413
136	4	20.12.1995	912

Lower groundwater tables, of course, result in more severe groundwater droughts (Table 2). Both the number of droughts and the duration of the most severe drought increase. The onset of the major drought will be several months earlier (thresholds 136, 138 and 139 m). A lowering of the water table could even lead to a shift of the most severe drought to another period in a decade (threshold 137 m).

Groundwater hydrographs from some permeable catchments in Mediterranean countries show more severe groundwater droughts than the ones experienced in the temperate part of Europe. Intensive exploitation of groundwater, primarily for irrigation, in combination with years with precipitation shortfall, has resulted in a very serious situation in the last decade. The groundwater level in the Upper-Guadiana basin (Spain), for example, has dropped by meters as shown in figure 4.

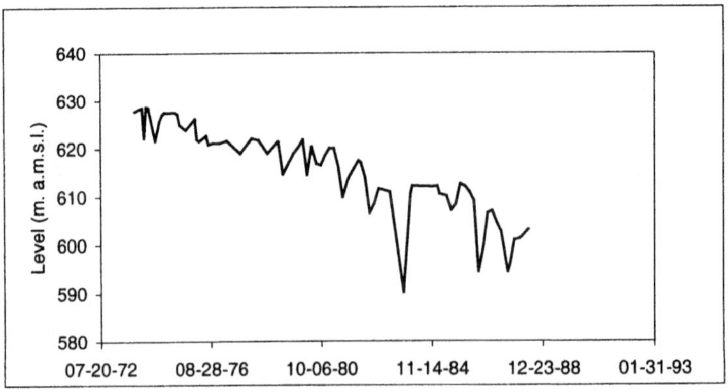

Figure 4: Groundwater hydrograph of well 2030-3002 (Upper-Guadiana basin, Spain).

Over the period 1973-1988 the drop is about 30 m in a particular borehole. In recent years the levels have recovered to some extent. Pulido Bosch (1991) and Rushton (1998) report similar or even larger groundwater declines in other semi-arid regions. In these situations the man-induced groundwater drought overshadows the natural groundwater droughts. In these cases application of the above-mentioned procedure usually results in one long-lasting drought.

Assessment of groundwater droughts by using groundwater heads in a catchment must not be based on the analysis from a single hydrograph (Figure 3, Tables 1 and 2), but hydrographs from several representative observation wells must be evaluated. In a catchment different aquifer units may occur, each with a different response to human activities, such as groundwater exploitation. Furthermore the meteorological conditions over a catchment can vary, which may give rise to differences in rainfall shortage and potential evapotranspiration rates. Especially for the definition and implementation of drought mitigation measures on a regional scale several representative wells have to be included in the evaluation. For example, the spatial extent of groundwater droughts in the United Kingdom can be derived from maps presented by IH (1999). On the maps the probability occurrence of groundwater levels (Figure 2) in key observations wells is presented. The map, which overlays a map with the major UK aquifers, is updated each month.

The above-mentioned drought monitoring and assessment procedure is a pure technical analysis, i.e. the results are presented in statistical terms. The severity of a drought, however, does not only depend on the statistics of the event, but primarily on how the associated effects affect economy, social life and environment (Figure 1). So, the classification of the severity will also depend on how the lower groundwater heads and smaller groundwater gradients will affect, for instance, crop productivity and streamflow. This implies an evaluation of the groundwater droughts in a wider sense, i.e. assessment of the effects outside the aquifer. A distinction has to be made between the effects of groundwater droughts on valley bottom hydrology (the riparian area, the springs and the streams), which prevail in all catchments, and the effects in lowland areas where additional consequences occur.

4. GROUNDWATER DROUGHTS AND LOWLAND HYDROLOGY

In lowland areas with shallow groundwater tables, e.g. deltas, groundwater droughts are associated with a decline of the water levels of the unconfined aquifer. These deeper water levels have a negative effect on the abiotic conditions in the root zone of wet and semi-wet vegetation types. The amount of soil moisture will decrease, and the oxygen concentration and the soil temperature will increase. Besides the direct effect of these changes on vegetation development, the nitrogen and phosphate cycle will substantially change. Usually more nutrients will be available. All these changes, which are triggered by small changes in the groundwater head, result into a shift from wet, nutrient-poor vegetation types to more common, drought-resistant, nutrient-rich types. The replacement of vegetation types implies a substantial loss of rare terrestrial ecosystems. For instance, in The Netherlands an evaluation of the national groundwater database has shown that human activities have resulted in a fall of the average groundwater level by 0.25-0.35 m and locally by more than 1.0 m over the last 40 years (Rolf, 1989). As The Netherlands is a country with shallow water tables, this decrease of the groundwater heads significantly enhanced groundwater droughts. About 560.000 hectares of nature land (16% of Dutch territory) were negatively affected in the mid-eighties due to groundwater droughts (Kors *et al.*, 1996). National policy involves a reduction of the affected area by 40% over the coming decades. So, ecological impact of groundwater droughts was the main reason to implement strategic drought mitigation measures in The Netherlands.

Groundwater droughts resulting in deeper groundwater levels can also affect the production of agricultural crops. In lowland areas, the soil moisture supply of agricultural crops partly depends on capillary rise. The land is reclaimed in such a way that slightly lower water tables can substantially decrease the upward flow from the water table. A lot of research has been carried out to define optimal groundwater levels (e.g. Bouma *et al.*, 1980; Bouma, 1989; van Wijk and Feddes, 1986) and to quantify the capillary rise to the root zone for different groundwater levels. The effect of a lower groundwater table on the capillary rise and therefore on the soil moisture deficit depends, among other factors, on the soil type (Figure 5).

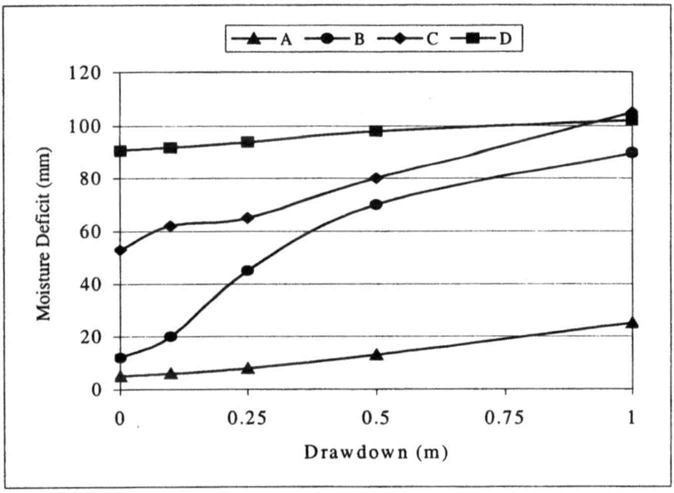

Figure 5: Relation between the soil moisture deficit and the groundwater drawdown for different soils (adapted from Bannink et al., 1987).

The moisture deficit of soil types A and D (sandy soils with deep water levels and a high, A, and low, D, soil moisture availability, respectively) is hardly susceptible to a drawdown of the water table compared to the average condition, whereas the deficits of other soil types show a substantial increase. Especially soil type B (sandy soil with a shallow water level and a soil moisture availability strongly dependent on the water level) is susceptible; the deficit increases from about 10 to 90 mm when the water table falls by 1.0 m. The larger soil moisture deficit, which is induced by a groundwater drought, enlarges the agricultural drought which originates from a precipitation deficit.

5. GROUNDWATER DROUGHTS AND VALLEY BOTTOM HYDROLOGY

Groundwater droughts decrease the groundwater gradient, which results in a lower groundwater discharge to the riparian area, the springs and the streams (Figure 1). Contrary to the processes dealt with in the previous section, the lower groundwater discharge occurs in all catchments. The groundwater levels in the riparian area typically show a small fluctuation. Differences between wet and dry years are small. Small changes in groundwater heads, which look insignificant, are crucial for the wetlands along the brook. More prolonged droughts or more severe droughts might kill the wetlands.

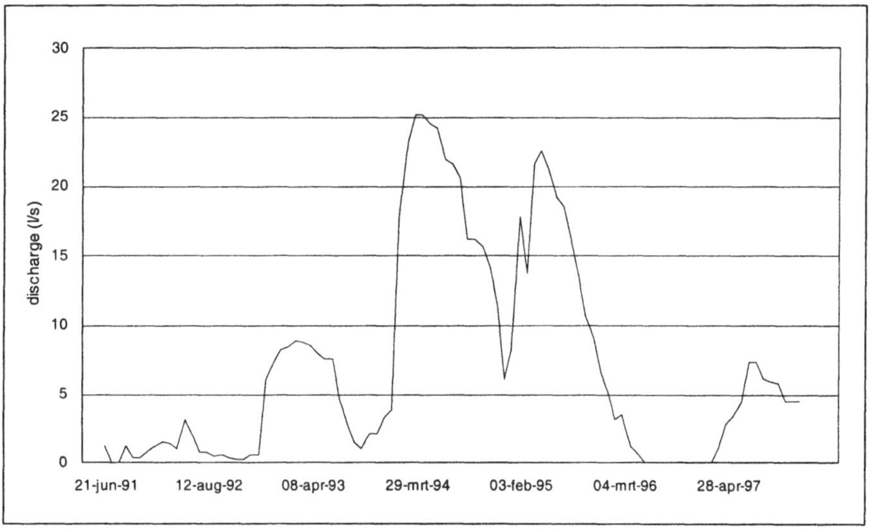

Figure 6: Discharge of the major spring in the Noor catchment (The Netherlands).

Groundwater droughts in these areas are usually better reflected in the spring discharge. Figure 6 shows the discharge for a major spring in the same region for which the groundwater levels are shown in figure 3. Low springflow (<1.0 ls^{-1}) or drying-up of the spring is clearly correlated with the two groundwater drought periods in the region (early and mid 1990's, Figure 3). So, there is a clear relation between the natural groundwater recharge, the groundwater heads and the springflow. Modeling has shown that the number of days that this spring would dry up due to more severe groundwater droughts will double if the average recharge would decrease by 20% (van Lanen and Dijksma, 1999). Land use or climatic change can cause such mutation in the natural recharge.

6. DISCUSSION AND CONCLUSIONS

A groundwater drought occurs if in an aquifer the groundwater heads have fallen below a critical level over a certain period of time, which results in adverse effects. Although the causes for the drought may be natural (rainfall shortage) and/or man-induced (abstractions, climatic change), no difference prevails between the type of effects. The basis for groundwater droughts are low groundwater levels and the associated smaller gradients. These cause effects in the aquifer itself, such as lower well yields and possibly changes in the chemical composition, but also effects outside the aquifer (e.g. lower crop yields, extinction of rare plant species, lower springflow and stream flow). These adverse effects have to be converted into an economic, social and environmental impact (Figure 1). As a result of the impact assessment and the identification of undesirable stresses, the critical value or threshold of the groundwater level can be determined. It is likely that a set of thresholds will be identified; each threshold is then related to a certain severity class. In this stage, and not earlier, the statistical analysis of time-series of groundwater levels can be performed using the appropriate thresholds. The results of the analysis (probability distributions of onset, duration and severity of droughts) can then be used to implement pro-active drought mitigation measures. Simulation modeling can be required to generate new groundwater hydrographs in order to investigate the impact of the drought mitigation measures in advance. Such statistical analysis can further be used to find the appropriate probabilities of drought occurrence, needed for groundwater drought monitoring (Figure 2). On the basis of the drought monitoring, reactive drought mitigation measures can then be issued.

It must be clear that the critical groundwater level below which adverse effects prevail cannot readily be assessed. It is not just a number to be derived from time-series of groundwater levels by using statistical analysis. Policy-makers, groundwater managers and scientists have to work together to make the definition and the assessment of groundwater droughts really operational.

ACKNOWLEDGEMENTS

The research presented in this paper was carried out as part of the program of the Wageningen Institute for Environment and Climate Research (WIMEK-SENSE) as a contribution to the ARIDE project, which is financed by the EU Climate and Environment Program (EVN4-CT97-0553). Some of the groundwater data were kindly provided by Centro de Estudios Hidrograficos, CEDEX (Madrid) and the N.V. Drinking Water Company (WML) in Maastricht.

REFERENCES

Bannink, M.H., Bles, B.J., and van Holst, A.F. (1987) Application of the Soil Map 1:50 000 to Support Provincial Water Management, *Cultuurtechnisch Tijdschrift* 27(3):175-189 (in Dutch).

Bouma, J. (1989) Using Soil Survey Data for Quantitative Land Evaluation, in B.A. Stewart (ed.), *Advances in Soil Science* 9:177-213.

Bouma, J., de Laat, P.J.M., Awater, R.H.C.M., van Heesen, H.C., van Holst, A.F., and van de Nes, Th. J. (1980) Use of Soil Moisture Data in a Model for Simulating Regional Soil Moisture Regimes, *Soil Science Society America Journal* 44: 797-802.

Carrillo-Rivera, J.J. (1998) Monitoring of Exploited Aquifers Resulting in Subsidence. Example: Mexico City, in H.A.J. van Lanen (ed.), *Monitoring for Groundwater Management in (Semi-) Arid Regions*, Studies and Reports in Hydrology 57, UNESCO, Paris, 151-165.

Dijksma, R. and van Lanen, H.A.J. (1998) Monitoring and Modelling of Springflow in the Noor Catchment (The Netherlands), in J. Bucek, M. Šir, and M, Tesa (eds.), *Catchment Hydrological and Biochemical Processes in a Changing Environment*, Proc. ERB Conference Liblice, Czech Republic, September 22-24, 1998, 21-24.

Gupta, C.P. and Gurunadha Rao, V.V.S. (1998) Monitoring Salt Water Intrusion: Some Imperatives and an Example: Kavaratti Island (India), in H.A.J. van Lanen (ed.), *Monitoring for Groundwater Management in (Semi-) Arid Regions*, Studies and Reports in Hydrology 57, UNESCO, Paris, 177-185.

IH (1999) *Water Watch*, Institute of Hydrology, Wallingford, OXON, UK, http:/www.nwl.ac.uk/ih/www/research/idroughtwatch.html.

Kors, A., Claessen, F., Vermulst, H., van Ek, R., Bos, H., Boven, E., de Lange, W., and Arnold, G. (1996) Water Systems Reconnaissance 1996, *Policy Analysis Water Systems Reconnaissance: Theme: Droughts and Groundwater*, RIZA Nota 97.041, Lelystad, 160 p (in Dutch).

Lanen, H.A.J. van and Dijksma, R. (1999) Water Flow and Nitrate Transport to a Groundwater-fed Stream in the Belgian-Dutch Chalk Region, *Hydrological Processes* 13: 295-307.

Marsh, T. and Lees, M. (1985) The 1984 Drought, *Hydrological Data UK*, Institute of Hydrology, Wallingford, OXON, UK 80 p.

NDMC (1995) Understanding and Defining Drought, National Drought Mitigation Center, University of Nebraska-Lincoln, http://enso.unl.edu/nmdc.

NITG (1998) Information on Groundwater Levels ('Op de (peil)hoogte'), Information Bulletin, December 1998, Netherlands Institute for Geosciences, Delft (in Dutch).

Pulido Bosch, A. (1991) The Overexploitation of Some Karstic Aquifers in the Province of Allicante (Spain), in L. Candela, M.B. Goméz, L. Puga, L.F. Rebollo, and F. Villarroya (eds.), *Aquifer Overexploitation.*, I.A.H. XXIII International Congress, Canary Islands, Spain, 557-561.

⊂i ica, J. and Novický, O. (1995) Experiments with Deficit Volumes (EXDEV). Description of the Program, Edition April 1995, Report Czech Hydrometeorological Institute, Prague.

Rolf, H.L.M. (1989) Fall of the Groundwater Levels in the Netherlands, Ministry of Transport, Public Works and Water Management, Lelystad (in Dutch).

Rushton, K.R. (1998) Monitoring an Over-exploited Aquifer in India, in H.A.J. van Lanen (ed.) *Monitoring for Groundwater Management in (Semi-) Arid Regions*, Studies and Reports in Hydrology 57, UNESCO, Paris, 141-149.

Tallaksen, L., Madsen, H. and Clausen, B. (1997) On the Definition and Modelling of Streamflow Drought Duration and Deficit Volume, *Hydrological Sciences Journal* 42(1): 15-33.

Wijk, A.L.M. van and Feddes, R.A. (1986) Simulating Effects of Soil Type and Drainage on Arable Crop Yield, in A.L.M. van Wijk, J. Weselling and A.A. Balkema (eds.), *Agricultural Water Management*, Rotterdam, 97-112.

AGRICULTURAL DROUGHT – A PRACTICAL APPROACH TO DEFINITION, ASSESSMENT AND MITIGATION STRATEGIES

G. MARACCHI
Department of Climatology and Agrometeorology
University of Florence
50144 Florence, Italy

Abstract. Despite the improvements in investigative and operative technology, drought remains one of the main problems in many countries, involving a large number of human activities. Over the past few years scientists have proposed specific formulas and procedures to detect drought phenomena or to give specific information to end-users or decision-makers. This paper illustrates some basic concepts on agricultural drought, with a collection of the main indices used to determine drought severity in this extremely sensitive economic sector. Some considerations on the actual knowledge of the problem are presented and some suggestions on possible improvements of the land and natural resource management capabilities are given.

1. INTRODUCTION

Aiming at a practical approach to the problem of agricultural drought in Europe, this paper has been written with the specific objective to answer some basic questions, more than to describe the general problem. The described approach could be useful to define general common goals of scientific community and to establish a methodology for improving the ability to manage drought, now and in the future, keeping in mind that drought represents only one aspect of the more general problem of water availability. Many initiatives have been taken in Europe about the topic of drought, which appears as a central issue in the Fifth Framework Programme of the European Union. A cooperative and well organised effort is necessary to analyse the problem and to monitor the situation in order to establish a correct policy of mitigation.

2. DEFINING AGRICULTURAL DROUGHT

In general, we should distinguish between two types of drought, a spatial drought and a temporal drought. The spatial drought concerns mainly geographers and is close to the concept of aridity. It is related to the scarcity of rainfall during the whole year and it is a climatic constant feature, caused by the general circulation pattern. In areas where this type of drought is common, the local civilisation has faced the problem over the centuries by adopting specific techniques and lifestyles. From an agricultural point of view we can assume that in all climatic situations, both in the past and nowadays, the local civilisation has fitted to these situations through appropriate farming techniques.

J.V. Vogt and F. Somma (eds.), Drought and Drought Mitigation in Europe, 63–75.

Keeping in mind these considerations, we should concentrate on the temporal variation of the water availability in the context of the climatic pattern of each region. This variation for each region is due to the change in the inter-annual variability, or to a trend in the average amount of rainfall, or to a different distribution of rainfall along the year (WMO, 1998). In addition to these causes of variation related to the climatic pattern, we have also to consider:

- the increase in the use of water and the increasing competition for water among various economic sectors such as agriculture, industry, and tourism;
- the change in land management and land use and the possible effects on the water balance;
- the worsening of the water quality and the lack of adequate agronomic technologies to optimise water use.

In any case, we can state that agricultural drought is the interaction between climatic conditions and some other factors (e.g. increasing water consumption, variation in land use or low efficiency in the use of water) that lead to a strong decrease in agricultural production or to a worsening of product quality. In this sense, agricultural drought is something broader than climatic drought, because the former can be due to different factors rather than just environmental conditions.

3. HOW CAN WE MEASURE AGRICULTURAL DROUGHT?

If we examine the indices developed during this century to assess drought (Tables 1 to 4), we can classify them into two groups according to the information needed for their calculation and the assumptions on which they are based:

1) indices requiring meteorological or agrometeorological parameters;
2) indices requiring information on local agricultural practices and climatic behaviour.

Specifically, within the first group we can identify three types of indices:

i) rainfall indices, developed to detect deviations in precipitation, used mainly by engineers and water managers;
ii) rainfall and temperature indices, used for climatic classifications on a large scale and adopted mainly by geographers and naturalists;
iii) rainfall, potential evapotranspiration (ETP) and soil moisture indices, concerning agricultural climate classifications and used by agronomists.

Because agricultural drought depends on many causes, we should develop a method that highlights these causes and is, at the same time, easy to adapt to a large range of situations. In order to do so we need more than the identification of one single index and the solution needs to follow a procedure based on two main components: (a) the characterisation of the regional agricultural practices and (b) the analysis of climatological trend (Maracchi, 1996).

Table 1: Drought indices with rainfall input (modified and updated from WMO, 1975).

Author	Definition of drought or associated concepts	Region and comments
Brounov, early 20th century *(Tannehill, 1947)*	Ten days with rainfall not exceeding 5 mm	
Henry, 1906	21 days or more when rainfall is 30% or less than average for the time and place. Extreme drought when rainfall fails to reach 10% of normal for 21 days or more	U.S.A.
Cole, 1933	15 days with no rain	U.S.A.
Bates, 1935	When annual precipitation is 75% of normal or when monthly precipitation is 60% of normal	U.S.A.
British Rainfall Organisation, 1936	*Absolute drought*: at least 15 consecutive days none of which received as much as 0.25 mm *Partial drought*: at least 29 days during which mean rainfall does not exceed 0.25 mm per day *Dry spell*: 15 consecutive days none of which has received as much as 1 mm	Britain
Hoyt, 1936	Any amount of rainfall less than 85% of normal	U.S.A.
Baldwin-Wiseman 1941	Engineers' drought in Australia for 3 or more consecutive months with deficit of 50% from mean rainfall	Australia
Blumenstock, 1942	Less than 2.5 mm in 48 hours	
Conrad, 1944 *(Conrad and Pollak., 1950)*	A period of 20 (or 30) consecutive days or more without 6.4 mm of precipitation in 24 hours during the season, from March to September included.	U.S.A.
Ramdas, 1950	When actual rainfall for a week is half of normal or less	India
Fitzpatrick, 1953	The dry period is considered concluded when at least 6.4 mm of rainfall are recorded during 48 hours.	Australia (based on Blumenstock). Evaluated probability that dry spell of any length would occur at any time throughout the year.
Foley, 1957	Computed residual mass curves of rainfall. Divided values by average annual rainfall to give "index of severity".	Australia. Dividing by annual average makes comparison between more reliable stations. Index is dimensionless.
Gibbs and Maher, 1967	State that rainfall is the best single index of drought and use rainfall deciles to demonstrate temporal and spatial distribution of drought. Areas where rainfall is included in the first decile (10% of normal precipitation of the period) coincides with drought areas.	Australia. Provides a useful presentation of spatial distribution of drought.
McKee *et al.*, 1993	Standardised Precipitation Index (SPI). The index is calculated by taking the differences of precipitation from the mean for a particular time scale, and then dividing by the standard deviation.	USA. An index based on the probability of precipitation for any time scale.

Table 1: Drought indices with rainfall input (continued).

Author	Definition of drought or associated concepts	Region and comments
Le Houérou, 1988 *(Le Houérou et al., 1993)*	Dependable Rains (DR). Based on the concept of dependable rains, defined as the amount of rainfall that occurs in four years of a five years period.	Africa. It is used to realise plans for agricultural production
Gommes and Petrassi, 1994	National Rainfall Index (RI). It is calculated for each country by taking a national average annual precipitation weighted according to the long term precipitation averages of all the individual stations.	Africa. The RI allows comparison to be made between years and between countries.
Phillips and McGregor, 1998	DSI3 - Three months Drought Severity Index . For this index the drought starts when the rainfall of a month is lower than the mean of the last three monthly mean. DSI6 - Six months Drought Severity Index. Based on the previous one to test the sensitivity / stability of DSI3.	Great Britain - Used to assess drought hazard in England.

Table 2: Drought indices with rainfall and mean temperature input (modified and updated from WMO, 1975).

Author	Definition of drought or associated concepts	Region and comments
Lang, 1915	Precipitation factor I = (Mean Annual Precipitation) / (Mean Annual Temperature)	Germany. Used for climatic classification of the soils. When I is lower than 40 the soil is arid.
De Martonne, 1926	Index of aridity I = (Mean Annual Precipitation) / (Mean Annual Temperature +10)	Used to define climatic limits of deserts, prairies and forests. Does not apply well in cool zones where T+10 approaches to zero. Used extensively by geographers and biologists to define aridity.
Koloskov, 1925	Ratio of annual precipitation and accumulated mean daily temperature during the vegetation period (divided by 100)	U.S.S.R. Ratio may be used as comparative agroclimatic index.
Selyaninov, 1930	Index of aridity	U.S.S.R. Author suggested that a period can be considered as a dry spell when the index is <1 and as a drought when the index is <0.5.
Köppen, 1931	Defines *"dry" climate* by: p<2t for regions of winter rain and p<2t+14 for regions of summer rain or no rainy season, where p is annual precipitation in cm and t is mean temperature in °C. *"Desert" climate* defined by: p<t for winter rain; p<t+14 for summer rain; p<t+7 for no rainy season.	Used extensively in classification of the dry climates of the world.

Table 2: Drought indices with rainfall and mean temperature input (continued).

Author	Definition of drought or associated concepts	Region and comments
Emberger, 1955	Index of aridity {I = (100 P) / [(M - m) (M + m)]} Where P is the sum of rainfall during those months when the mean temperature is above 10°C; M is the mean maximum temperature in the hottest month and m is the mean minimum temperature in the coldest month.	France. Based on de Martonne's index.
Knochenbauer, 1937	Daily maximum temperature and humidity, (both measured in the afternoon) are used to define a dry spell.	Germany.
Condra, 1944	Period of strong wind, low precipitation, high temperature and usually low relative humidity.	U.S.A. This anticipates the combination of low precipitation and high evapotranspiration
Henin and Ternisien, 1944	Computed evapotranspiration and drainage in function of temperature and precipitation.	France. Procedure improved by Turc (1954) incorporating additional factors.
Thornthwaite, 1931	Precipitation effectiveness.	U.S.A.
Gaussen, 1954	When total monthly precipitation in mm is less than twice the mean temperature in °C	An approximation to rainfall lower than evapotranspiration based on Köppen.
Budyko, 1970	Hydrothermal coefficient	U.S.S.R.
(Benincasa et al., 1991)		

Table 3: Drought indices with soil-water and crop parameters input (modified and updated from WMO, 1975).

Author	Definition of drought or associated concepts	Region and comments
Russell, 1896	A period of months or years during which little rain falls.	Australia. Author points out that the word drought is not used in Australia in the same sense as in England and elsewhere.
Bova, 1941	Drought index calculated as a function of rainfall, temperature and soil moisture.	U.S.S.R. When the index is lower than 1.5 beginning of drought damage to plants is indicated.
Barger and Thom, 1949	Evaluated precipitation climate from productive performance of crops.	U.S.A.
Van Bavel, 1953	Agricultural drought should be defined on the basis of soil-water status and resultant plant behaviour.	U.S.A.
Van Bavel and Verlinden, 1956	A condition in which there is insufficient soil water available to crops.	U.S.A.

Table 3: Drought indices with soil-water and crop parameters input (continued).

Author	Definition of drought or associated concepts	Region and comments
Thornthwaite and Mather, 1955	Used the water-balance concept with variable storage of soil water.	U.S.A. Use extended to other continents, some results of doubtful value.
White, 1955	Defined drought with respect to xerophilous species using comments on pasture conditions as guide.	Western N.S.W., Australia. Extrapolation to other areas not reliable.
Foley, 1957	Used report of conditions of crops and livestock published in official bulletins or newspapers together with rainfall analyses.	Australia. Necessarily rather qualitative but helpful to define the significance of rainfall deficiency.
Alpat'ev and Ivanova, 1958	Based definition of severity on the comparison of crop yields with long-term mean yields. Because not all yield decreases are a result of drought the author suggests that only years when yields decrease by 25% be classified as drought years.	Variations in yield due to different levels of agronomic practices are still greater than those due to droughts (Kulik, 1958).
Kulik, 1958	Used preceding meteorological conditions, soil characteristics and level of agronomic techniques in the region. *Semi-drought*: 10 days with soil water 20 mm in first 20 cm of soil. *Drought*: as above with 10 mm of water.	U.S.S.R.
Holmes, 1962	States that in the quantitative evaluation of drought for agricultural purpose, precise and regular soil-water observations are most essential.	Canada
Fitzpatrick, 1965	Developed a water-use model with range of available soil moisture 0-10 cm and evapotranspiration losses (Et).	Australia. It has been used in climatic studies and land-use surveys.
Palmer, 1965	Palmer Drought Severity Index (PDSI). A water-balance model, which involves rainfall, a coefficient of evapotranspiration, runoff and available soil water.	U.S.A. Method based on the Thornthwaite concept of potential evapotranspiration.
Rickard,1966	Agricultural drought exists when soil water in the root zone is at or below the permanent wilting percentage. The condition continues until rain falls in excess of daily evapotranspiration.	New Zealand. Drought relief would not occur if one day excess rainfall threshold.
Palmer,1968	Crop Moisture Index (CMI). Uses a meteorological approach to monitor week after week the crop conditions. Severity of agricultural drought is defined in terms of the magnitude of the computed abnormal evapotranspiration deficit	U.S.A. This index considers the weekly values of temperature and rainfall. A by-product of the 1965 drought work. Theoretically of universal applicability.
Shafer and Deszman, 1982	Surface Water Supply Index (SWSI). Developed to complement the Palmer Index for moisture condition across the state of Colorado. The index is designed to be an indicator of surface water condition and described as mountain water dependent, in which mountain snowpack is a major component.	Colorado (U.S.A.). It was modified and applied also in other western states as well: Oregon, Montana, Idaho and Utah.

Table 3: Drought indices with soil-water and crop parameters input (continued).

Author	Definition of drought or associated concepts	Region and comments
Karl and Knight, 1985	Palmer Hydrological Drought Severity Index (PHDSI). It can be considered an hydrological drought Index, based on precipitation, outflow and storage, does not take into account long term trends.	U.S.A. It is an evolution of the PDSI, able to operate in near real time.
Garrido, 1998	Normalised Precipitation Index (NPI). Obtained combining the Palmer Drought Severity Index (PDSI) and the Standardised Precipitation Index (SPI).	Spain (UE). A drought watch system for Southeast Spain..

Table 4: Drought indices with climatic indices and estimates of evapotranspiration as input (modified and updated from WMO, Tech.. Note No.138)

Author	Definition of drought or associated concepts	Region and comments
Vjsotskii, 1905	Index of aridity	U.S.S.R.
Ivanov, 1948	Indices of aridity	U.S.S.R.
Trumble,1937; Hounam, 1948; Prescott, 1949 *(Le Houérou et al., 1993)*	Waite index	Australia
Thornthwaite, 1948	Precipitation effectiveness index based on mean temperature gives water requirement for optimum growth. Water-balance model gives estimate of soil-moisture status.	Originally U.S.A. Use since extended to other continents. Reliability depends on climate and method of doubtful value in many areas.
Penman , 1948; 1961	Index of aridity	Originally U.K. and Europe. Use now extended to other continents with fair to good results. Of little value in dry areas where water supply to plants is limited.
Turc, 1954; 1955	Index of aridity	France
Baier and Robertson, 1966	Index of aridity	Canada
McIlroy, 1968	Introduces variable to consider leaf wetness and improves aerodynamic functions.	Australia. Used for conditions of limited soil moisture but measurement of many parameters restricts wide application.

Table 4: Drought indices with climatic indices and estimates of evapotranspiration as input (continued).

Author	Definition of drought or associated concepts	Region and comments
Sly, 1970	Climatic moisture index	Canada. Average seasonal values, indicating differences in water balance, used for soil-climate classification purposes.
Hargreaves 1974; 1988	This index compares rainfall with Moisture Availability Index (MAI)	South America and Africa. It defines a series of thresholds to classify water availability for crops.
Le Houérou and Popov, 1981	Use a ratio between rainfall and potential evapotranspiration.	Africa. Use a series of thresholds to classify the rainy season.

The agricultural characterisation may be done on the basis of (a) an administrative boundary, and (b) the classification of the main agricultural crops cultivated in the area of interest (Martin, 1996). The administrative boundary, on the one hand, is an important criterion, because all relevant decisions are taken on the basis of the administrative unit; on the other hand, the classification of the main crops takes into account possible differences from an agro-climatic point of view.

The climatological analysis may be done studying the meteorological situation during the growing seasons of the main crops of an area. If we consider that the variability of rainfall is larger than that of evapotranspiration, the anomalies of rainfall can be considered the main cause of drought and the changes in its amount and distribution the reason for water shortage. Based on these assumptions, the climatological analysis can be simplified using rainfall only. Thus the following procedure can be suggested for the characterisation of agricultural drought[1]:

1. Collection of daily historical meteorological data (at least 30 years).
2. Definition of the growing season for the main crops of each region.
3. Calculation of the ratio between the total amount of rainfall for the growing period of each year and its mean value for the period considered (e.g. 30 years).
4. Calculation of the ratio between the longest time period with daily rainfall < 5 mm for the growing period of each year and its mean value: $r_i = m_i/M$, where m_i is the maximum number of consecutive days in year i and M is the average of m_i over the period considered (e.g. 30 years).
5. Calculation of the ratio between the total number of days with rainfall < 5 mm for the growing period of each year and its mean value: $s_i = t_i/T$, where t_i is the total

[1] The calculations reported at points 2 to 6 must be repeated for each year.

number of days in year i and T is the average of t_i over the period considered (e.g. 30 years).

6. Calculation of the ratio between the total number of days in which Sum ETP > Sum Rainfall and its mean values for the period considered (e.g. 30 years). The ETP and Rainfall Sums for each year are relative to cumulated values starting at the beginning of the growing period.

These simple calculations are based on the following criteria :

- the necessity to relate the growing period to the main crops of the region
- the combination of the total amount of rainfall (3) with its distribution (4, 5)
- the water balance trend during the growing season
- the normalisation of all the values with respect to the mean value of the whole period considered (e.g. 30 years)

This analysis should underline the change in climatic patterns, from which it is possible to derive the effect on the productivity of each crop, in terms of quality and quantity. In many cases, according to the policy of Agenda 2000, quantity is more relevant than quality.

4. CONTRIBUTION OF MODERN TECHNOLOGIES

The procedure suggested can be implemented using the technologies developed over the last fifteen years:

- Geographic Information Systems (GISs) can be used for the characterisation of administrative districts and for the presentation and analysis of information on the main crops and their growing periods. GIS techniques allow an integrated analysis of the available information (Maracchi *et al.*, 2000).

- Data from satellites such as Meteosat or NOAA, together with Global Circulation Models (GCM) and Limited Area Models (LAM), can be used to regionalise rainfall and to identify precipitation patterns (Maracchi *et al.*, 1998).

- GCMs and LAMs can be applied to better understand climatic features characterising each region and the distribution of agro-climates.

- Commercial GIS packages can be used to organise and process such data and to apply the described procedures of agro-climatic analysis.

- Graphic software packages can be used to present the results in a synoptic way, easy to understand for policy makers and administrators.

What is important to agree on is a procedure that can be adjusted to the different natural and socio-economic situations in Europe, and to derive regionally comparable products.

5. THE INSTITUTIONAL FRAMEWORK

The context in which this activity can be seen includes various institutions that can give a contribution to the final outcome. For example:

- International agencies, such as WMO and FAO, can guarantee the links with international programmes such as the World Climate Programme, etc.
- The E.U. Directorates General of Research, Environment and Agriculture can guarantee the financial support and the general interest of all European countries.
- The National Ministries of Agriculture and Environment can contribute to fund the national activities.
- Public services, such as the meteorological, hydrological and agricultural services, can help providing some of the necessary data.
- Scientific institutions at national and European level can suggest, evaluate and validate the procedures for the agro-climatic analysis. The Space Applications Institute of the Joint Research Centre, for example, could guarantee the homogenisation of the results for all Europe. National institutions could apply the procedures taking into account the specific characteristics of their countries and regions.

6. HOW TO GUARANTEE A REAL TRANSFER OF THE RESULTS TO THE ECONOMIC ACTIVITIES?

The steps required to guarantee the transfer of the results to social and economic fields are the following:

A. Definition of the problem. The main activities that have to be carried out to achieve this result are the analysis of historical series of climatic data and the definition of actual agricultural drought risks in Europe. A reliable evaluation of existing trends and the understanding of the impact of human activities can be useful to determine the severity of future drought events on natural and agricultural ecosystems.

B. Identification of the users. The results of the preliminary analysis have to be transferred to local authorities (Ministries of Environment and Agriculture, Regional Governments, etc.) and European Institutions (Eurostat, Environmental Agency, etc.), in charge of land and resources management. Further activities should then be defined through a collaboration among the involved institutions, selecting the basic elements for the definition of the fundamental final information.

Maracchi, G., Battista, P., and Rapi B. (1998) *A Guide to Computer Science Application in Agrometeorology*, WMO - Regional Training Seminar for National Instructors of RA. I and RA. VI (Nairobi, Kenya, 20 April to 1 May 1998), 262 p.

Maracchi, G., Pérarnaud, V., and Kleschenko, A.D. (2000) *Applications of Geographical Information Systems and Remote Sensing in Agrometeorology*, Special Volume of Agricultural and Forest Meteorology (in press).

Martin, P. (1996) Regional Aspects of Climate Change, in *Climate Change Impact on Agriculture and Forestry*, Proceedings of the European School on Climatology and Natural Hazards, Volterra, Italy, 33-52.

McKee, T.B., Doesken, N.J., and Kleist, J. (1993) *The Relationship of Drought Frequency and Duration to Time Scales*, 8th Conference on Applied Climatology, Anaheim, CA, 17-22 Jan., 179-184.

McIlroy, I.C. (1968) *Evaporation and its Measurement. Part B.: Energy Balance and Combination Methods*, Proc. of WMO seminar on Agriculture Meteorology, Melbourne Australia, Pub. Australian Bureau of Meteorology 2: 409-431.

Palmer, W.C. (1965) *Meteorological Drought*, Research Paper 45, US Department of Commerce, Weather Bureau, Washington D.C.

Palmer, W.C. (1968) Keeping Track of Crop Moisture Conditions, Nationwide: the New Crop Moisture Index, *Weatherwise*, 21, 156-161.

Penman, H.L. (1948) *Natural evapotranspiration from open water, bare soil and grass*, Proc. Roy. Soc., London 193: 120-145.

Penman, H.L. (1961) Weather, Plant and Soil Factors in Hydrology, *Weather* 16 (7): 207-219

Phillips, I.D. and McGregor, G.R. (1998) The Utility of a Drought Index for Assessing the Drought Hazard in Devon and Cornwall, South West England, *Meteorol. Applications* 5: 359-372.

Ramdas, D.A. (1950) Rainfall and Agriculture, *Ind. J. Met. and Geophys.* 1(4): 262-274.

Rickard, D.S. (1966) The Occurrence of Agriculture Drought at Ashburton N.Z., *New Zealand Journal Agricultural Res.*, 431-441.

Russel, H.C. (1896) *On Periodicity of Good and Bad Seasons*, Roy. Soc. N.S.W.

Shafer, B.A., Dezman, L.E. (1982) Development of a Surface Water Supply Index (SWSI) to Assess the Severity of Drought Conditions in Snowpack Runoff Areas, *Proceedings of the Western Snow Conference*, USA, 164-175.

Selyaninov, G.T. (1930) Methods of Agricultural Climatology, in *Agricultural Meteorology,* No. 22 L.

Sly, W.K. (1970) A climatic moisture index for land and soil classification in Canada, *Can. J. Soil Sci.* 50: 291-301.

Tannehill, I.R. (1947) *Drought his causes and effects*, Princeton Univers. Press, 274 p.

Thornthwaite, C.W. (1931) The Climates of North America According to a New Classification, *Geogr. Rev.* 21: 633-655.

Thornthwaite, C.W. (1948) An approach toward a rational classification of climate, *The Geogr.Rev.* 38: 55-94.

Thornthwaite, C.W. and Mather, J.R. (1955) *The Water Balance*, Pub. In Clim. Drexel Inst. Tech Vol.8, No.1.

Trumble, H.C. (1937) *The Climatic Control of Agriculture in South Australia*, Trans. Royal Soc.S Aust. 61: 41-62.

Turc, L. (1954) Le bilan d'eau des sols. Rélations entre les precipitations, l'evaporation et l'ecoulement, *Ann. Agron Paris* 5: 491-596

Turc, L. (1955) Le bilan d'eau des sols. Rélations entre les precipitations, l'evaporation et l'ecoulement, *Ann. Agron. Paris* 6: 5-131.

Van Bavel, C.H.M. (1953) *A drought criterion and its application in evaluating drought incidence and hazard*, Agron. J., 45:167-171.

Van Bavel, C.H.M. and Verlinden, F.J. (1956) *Agricultural Drought in North Carolina, N.C.* Agric. Experimental Station Tech. Bull. No. 122.

Vjsotskii, G.N. (1905) *Steps of the European USSR*, Encyclopaedia of Russian Agriculture, Vol.9, Ed. A.Devriena.

White, R.C.L. (1955) *Drought and effective rainfall frequency in pastoral New South Wales*, Met. Study No.5, Bureau of Met., Melbourne, Australia.

WMO, World Meteorological Organization (1975) *Drought and Agriculture*.Tech.. Note No.138, WMO No. 392, Ginevra.

WMO, World Meteorological Organization (1998) *WMO Statement on the Status of the Global Climate in 1997*, WMO No. 877, Ginevra.

Part 3
Drought Risk

DROUGHT RISK IN A CHANGING ENVIRONMENT

T.E. DOWNING and K. BAKKER
Environmental Change Institute
University of Oxford
Oxford OX1 3TB, UK

Abstract. Both drought hazard and vulnerability are changing, and will change due to future global environmental change. With global warming, drought hazard, the balance of moisture supply and demand, is likely to be exacerbated in many regions. Exposure to economic losses continues, despite considerable promise to warn of forthcoming drought episodes and ameliorate impacts, especially with preparedness and insurance. Yet, the conjuncture of hazard and vulnerability only partly defines risk – the discursive construction of risk through social, economic and political institutions suggests that over the course of decades, adaptive approaches that reflect changing expectations, rather than prescriptive approaches embedded in fixed operational definitions and rules, are desirable.

1. PURPOSE

This paper reviews drought risk, as it might be affected by global environmental change. The structure of the paper follows the common definitions of natural hazard:

Hazard	Potential threat to humans and their welfare
+	
Vulnerability	Exposure and susceptibility to losses
=	
Risk	Probability of hazard occurrence
Disaster	Realisation of a risk

However, each of these compartments is problematical to some extent, particularly in the context of long-term global change. Further details of some of the arguments presented here can be found in Downing and Bakker (1999), while Downing *et al.* (1999b) present case studies of climatic hazards related to climate change.

2. CHANGING HAZARD?

Certainly, climate change will affect the frequency, duration and magnitude of drought hazard around the world. A starting point is to suggest that changes in precipitation are a reasonable measure of drought hazard. For example, table 1 shows the shift in drought hazard in the southern UK based on the differences between the distribution of

J.V. Vogt and F. Somma (eds.), Drought and Drought Mitigation in Europe, 79–90.

present precipitation (albeit based on the control run rather than actual time series) and a scenario of future climate based on the HADCM2 model (Hadley Centre Coupled GCM Model, version 2, http://www.cru.uea.ac.uk/link/experiments/2_experi_intro.html). The probability of precipitation less than 50% of normal increases from 1% at present to 12% in the 2050s. The situation changes in northern England, where wetter conditions are projected to occur, at least in the HADCM2 model.

Complementing changes in precipitation, are increases in temperature and the distribution of hot spells and warm years. For some uses, such as bathing and the quality of horticultural crop yields, this is more significant than changes in precipitation itself.

The first, reasonable extension of such estimates is to analyse the atmospheric water balance (precipitation net of potential evapotranspiration, P-PET), including a soil water component. Here a critical factor is changes in water use efficiency as increasing CO_2 changes plant respiration. Experimental and model values range from a small effect up to 50% for some crops in some conditions.

Table 1: Sensitivity of drought events to climate change in southern UK.*

	1961-90	2020s	2050s	2080s
Global dT	-	1.24	2.11	3.11
Hot August (+3.4 °)	2%	15%	32%	40%
Warm year (+1.1 °)	6%	59%	85%	99%
Summer P < 50% of average	1%	7%	12%	10%
2-year P < 90% of average	12%	11%	14%	6%

* Based on the HADCM2 Medium-high (IS92a) GGa results for southern grid boxes in the UK. Source: Hulme and Jenkins (1998).

The outcome of climate change on the average water balance depends on the interplay of changes in up to a dozen variables. What seems like a fairly straightforward calculation is quickly marked by model assumptions and the site dependence of climatic and edaphic conditions.

Figure 1 shows the effect of changes in water use efficiency (WUE) on a monthly average atmospheric water balance (P-PET). The top figure indicates that summer drought would increase with climate change, with potential water deficits of 20-40 mm per month. The lower diagramme shows that the higher end of potential values for WUE could cancel out the effect of climate change on precipitation and temperature.

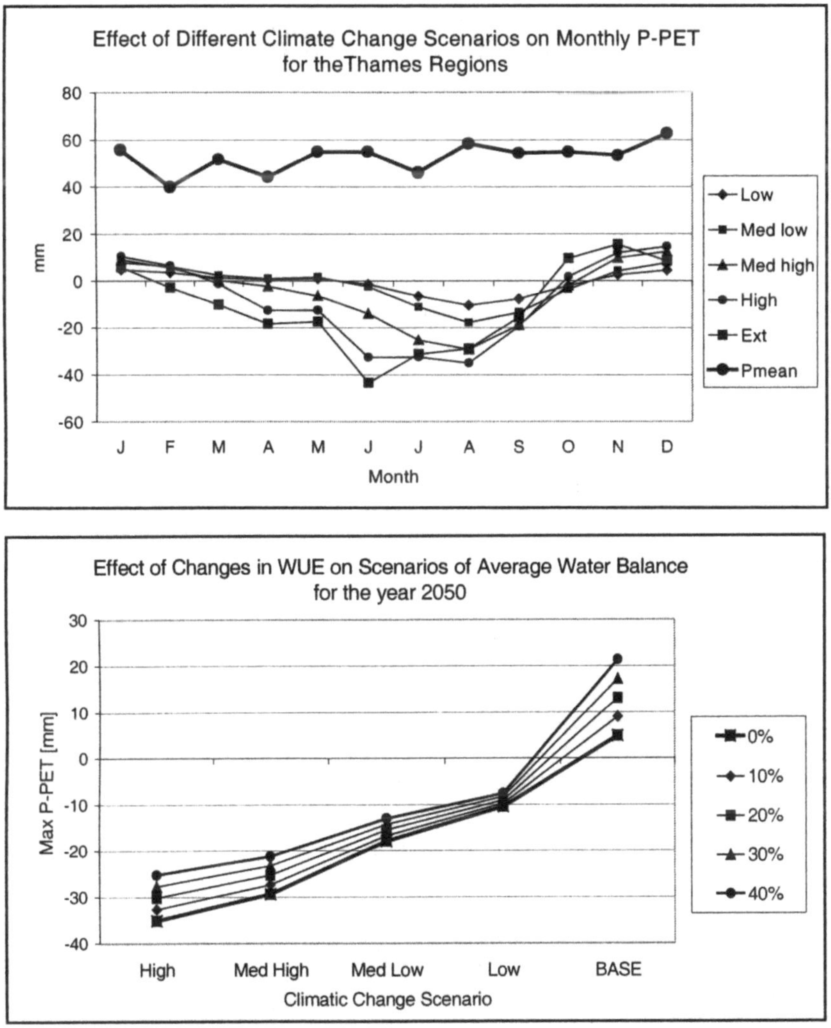

Figure 1: Effect of different climate change scenarios on monthly changes in precipitation minus potential evapo-transpiration for the Thames region, England (top) and the effect of changes in water use efficiency (WUE) on the atmospheric water balance (P-PET, bottom) for the same scenarios. Source: based on data in Hulme and Jenkins (1998).

The next logical extension is to note that the distribution of drought episodes (both within a season and over years) may be more important than changes in mean conditions. For example:

- Frequency is cited as a key determinant of risk in agricultural production: if the probability of crop failure exceeds 30%, farmers are likely to shift to other crops.

- Duration is important in water resources: a winter to summer drought is more significant than a summer-only drought.
- Magnitude influences responses in a non-linear way: a mild drought in England is often beneficial to winter wheat.

Each of these characteristics is likely to be driven by different, complex climatic changes.

A further complication is the spatial distribution of drought episodes. There is a strong dependence between the above drought measures and its spatial distribution. Estimating the effects of climate change requires explicitly spatial approaches, which are more difficult to extract from Global Circulation Models (GCMs) due to the contrasting spatial resolutions. The EU 'Climate Change, Climate Variability and Agriculture in Europe' (CLIVARA) project developed and tested methods of scaling up crop-climate models from the site to continental scale (see Downing *et al.*, 1999a). Scaling issues are likely to be more difficult for discrete components of the water balance (e.g., runoff rather than evaporation) and for patchy anomalies rather than continent-wide episodes.

Confidence in climate change forecasts of drought risk varies across these domains (Table 2). Changes in temperature and CO_2 concentrations are reasonably robust; projections tend to be of the same sign and within a narrow range (say ± 0.5 ° by 2050. Mean precipitation changes are less certain in sign (positive or negative), but tend to be within a tolerable range (say ± 10% in mean seasonal precipitation) that may or may not be critical for specific impact models. For example, the combination of temperature and CO_2 dominate effects on winter wheat, while uncertainty in precipitation scenarios leads to indeterminate forecasts for impacts on water resources in Europe (Hulme *et al.* 1999).

Table 2: Cascade of uncertainty in climate change related to drought hazard

Uncertainty	Climate Elements	Drought Impacts
Projection	Temperature	Heat stress, increased evapotranspiration
	Carbon dioxide	Increased water use efficiency
	Sea level rise	Coastal flooding, salt water intrusion
Divergent risk	Precipitation	Water balance
	Wind speed	Evapotranspiration
	Solar radiation	Evapotranspiration; photosynthesis
Variability	All elements (?)	Risk
	Many climatic hazards	
Surprise	Extreme events	Probable maximum loss

The higher order projections are more problematic - changes in the number of rain days, rainfall intensity, days between rainfall events, variability in most of the water balance variables, persistence of specific and composite variables, large scale disturbances such as ENSO. Few of these variables have been systematically analysed across a number of GCMs or related to confidence in GCM simulations of the control run.

3. CHANGING VULNERABILITY?

What determines the relationship between a hazard and its effects? Who are vulnerable? Why? These questions require a broader analysis of vulnerability than realised in an individualistic natural hazards perception framework (see for example, Sen 1981, Watts 1983, Blaikie *et al.* 1994, Bohle *et al.* 1994, Chambers 1989, Downing *et al.* 1996).

Particular vulnerabilities are the conjuncture of social, economic and political structures. Bohle *et al.* (1994) suggest a tri-partite causal structure of vulnerability (Figure 2), based on the human ecology of production, expanded entitlements in market exchanges, and the political economy of accumulation and class processes.

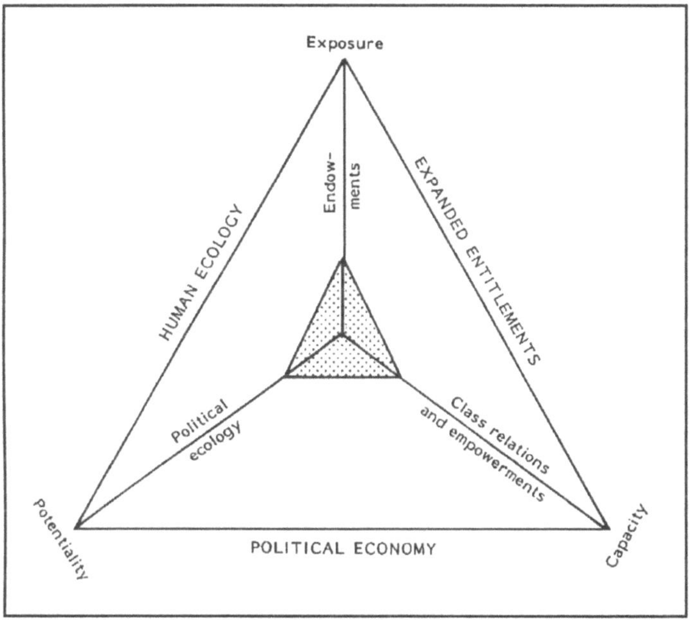

Figure 2: Three dimensions of vulnerability (after Bohle et al. ,1994).

Vulnerability *per se* is best viewed as "an aggregate measure of human welfare that integrates environmental, social, economic and political exposure to a range of harmful perturbations" (Bohle *et al.* 1994: 37-38). This conception of vulnerability shifts the focus of vulnerability away from a single hazard to the characteristics of the social system.

In a complementary view, Blaikie *et al.* (1994) regard vulnerability as a product of such characteristics as ethnicity, religion, caste membership, gender and age that influence access to power and resources (Figure 3). Dynamic pressures are processes that translate social, political and economic structures in relation to specific types of hazards into particular forms of insecurity. Regional and global pressures of rapid population growth, urbanisation, war, foreign debts, epidemic disease, export promotion, etc. affect local conditions. Some of these pressures have a universal character, others are specific to a certain region or society.

Vulnerability is not static. Rather, it changes over time incorporating social responses as well as new rounds of hazardous events. Models of vulnerability suggest specific groups of people who are vulnerable, whether based on demographic characteristics (e.g. women, children, elderly, disabled), socioeconomic class and livelihood (e.g. marginal farmers, underemployed), or resource management (e.g. junior water rights, common property users).

Regardless of the nuance of vulnerability frameworks, key concepts are:

- Everyone is vulnerable, although their vulnerability differs in its causal structure, its evolution, and the severity of the likely consequences. Not all people or regions are vulnerable to the same extent.
- Vulnerability is a relative measure – critical levels of vulnerability must be defined by the analyst, whether the vulnerable themselves, external aid worker, or various societies that include the vulnerable and policy makers.
- Vulnerability relates to the consequences of a perturbation, rather than its agent. Thus, people are vulnerable to loss of life, livelihood, assets and income, rather than to specific agents of disaster, such as floods, windstorms or technological hazards. This focuses vulnerability on the social systems rather than the nature of the hazard itself.
- The locus of vulnerability is the individual related to social structures of household, community, society and world-system. Places can only be ascribed a vulnerability ranking in the context of the people who occupy them.
- Vulnerability is spatially variable. Vulnerable groups are dispersed over space. Patterns of vulnerability depend on geographical linkages (such as competition between markets) and are often contingent on past conditions (such as the harvest in the previous season). Drought vulnerability is qualitatively different for different nations (and individuals). In the poorest households and economies, drought can still threaten lives. However, in most of the developing world, drought vulnerability constitutes a threat to livelihoods, the ability to maintain productive systems and healthy economies.
- Vulnerability is dynamic. Considerable improvements have occurred over the past several decades, some regression as well has been noted in Eastern Europe and the Former Soviet Union (FSU). Drought is a relatively short-term event (spanning several years); vulnerability changes at a variety of time scales, from within a season to decade-long trends in development.

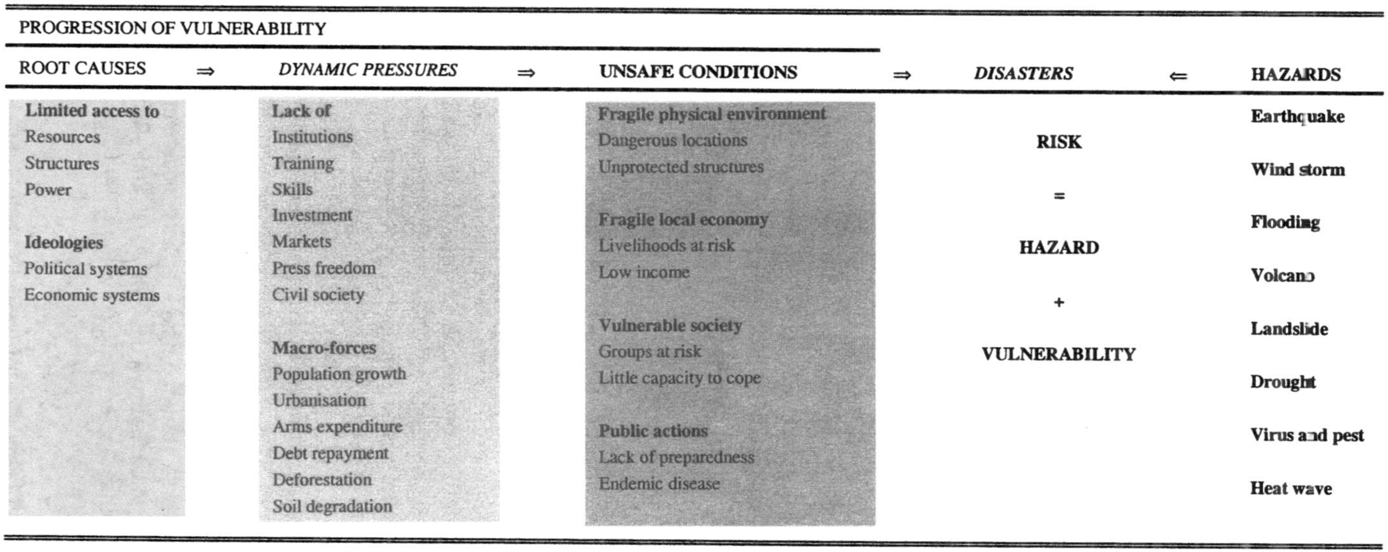

Figure 3: Structure of vulnerability and disasters. Source: after Blaikie et al. (1994).

In stark contrast to modest shifts in drought hazard (the probability of specific climatic episodes), vulnerability is dynamic and can vary between extreme crises and complete safety (in either direction) in a few months or even days. Trends in such a dynamic state are inevitably specific to individuals, communities and places.

Trends in the major determinants of vulnerability include:

- Population growth: increasing densities in hazardous areas increases the population at-risk. The effect of population growth on resource use to create marginal populations in hazardous areas is less clear.
- Settlement patterns and migration: no clear trend away from hazardous locations.
- Economic development: increasing wealth increases the amount and value of property at-risk, although it also enables greater investment in protection and mitigation.
- Health infrastructure: most countries have decreasing rates of malnourishment, although the situation among the most vulnerable (including the effect of AIDS on labour) is not clear.
- Mitigation and preparedness: some progress for specific hazards.
- Early warning, emergency assistance and recovery: increasing local and global warning and response capabilities reduce the loss of life at least for hazards where warning is effective.

The outcome of such trends can be illustrated by contrasting two situations. California, no stranger to drought, is served by the largest and likely the most complex human-made water distribution system in the world (Nevarez, 1996). In Santa Barbara, after six years of drought, residents voted in 1991 to extend the State Water Project to their county, one more link in the huge chain of hydraulic projects stretching from northern to southern California. 'Sustainable' options were abandoned by the public unwilling to change to 'drier' lifestyles in the face of severe water shortages. According to one adage, everyone comes to California, but no one brings water with them. While population growth places increased demands on water resources in the west, economic wealth and a highly organised water resource planning system reduce vulnerability to drought.

In contrast, one of the causes of desertification is population expansion in semi-arid regions, exploiting fragile resources. The result is soil erosion, reduced organic matter content and water holding capacity of soils, and highly variable crop yields due to moisture stress. Without the economic means and labour resources for protecting the soils or maintaining long fallows, resource-poor farmers accelerate the familiar processes of degradation and vulnerability (see, Blaikie and Brookfield, 1987; Kasperson *et al.,* 1995; Downing, 1992; and Downing, 1995 for examples and discussion of these processes).

In summary, vulnerability to loss of life, at least related to natural hazards, is probably reducing for most places in the world. Conversely, economic vulnerability is likely to increase, quite rapidly in many developing countries as per capita incomes rise. The situation with regard to morbidity is less clear. Where AIDS is prevalent in the work force, vulnerability to morbidity is increasing. In developed countries, vulnerability to morbidity may well depend on the orientation of health services – whether they pursue preventive health (well being) or treatment (disease diagnosis).

4. CHANGING RISK?

The conjuncture of hazard and vulnerability can be portrayed as risk. In a static environment, where vulnerability does not change and hazardous events are random in a geostationary time series, this overlay conception can be appropriate. Where past experience influences vulnerability, and present drought events are likely to modify future behaviour, the compartmentalised approach becomes problematical. Conventional dichotomies between 'natural' and 'social' break down when applied to 'hybrid' phenomena such as drought. Latour (1993, see also Haraway, 1991), for example, notes:

> *"The challenge of analysis of environmental phenomena such as drought arises from their simultaneously multidimensional character - at once "real, like nature, narrated, like discourse, and collective, like society".*

Thus, social scientists tend to focus on drought as a human-made phenomenon rather than a natural event. For example, political economists examine regulatory frameworks, the behaviour of farmers, water companies and governments, spatial distribution of industry and population, and the economics of water supply in order to explain why the amount of water available in certain places at certain times is deemed insufficient (e.g. Walker and Williams, 1982).

Emel *et al.* (1992: 38) note that "the neglect of socially constructed knowledge, ideology and institutions other than the market as mediators between humans and nature and among humans themselves, represents a major deficiency in the resource management literature." They examine the relationship between ideologies of property and the temporal and spatial variation of groundwater resources and their management in Texas and New Mexico. Discourse analysis explored the social and cognitive basis of perceptions of drought at the level of the individual and institutions. Drought can also be viewed as an emblem or sign of antagonistic social forces (Eagleton, 1991) or a "a site where the established institutions of society are put to the test" (Hajer, 1995: 39). Here, discourse analysis is used as a means of uncovering the creative potential inherent in current debates over drought and its management. In Ulrich Beck's analysis of 'risk society' (1992), the project of modernisation has shifted from the goal of the mastery of external threats to the containment of dangers emerging from the practices of industrial society. The emergence of these threats, of an 'ecological deficit', undermines institutions predicated on industrial modernisation. In Beck's theory of 'reflexive modernisation', the further modernisation proceeds, the more these institutions are undermined, forcing a radical alteration in existing institutional and political systems. Modern, industrial societies' 'unintentional self-endangerment' is the source of great conflict, but also of great potential for change, both of which are expressed in discursive practices.

The EU project on Social and Institutional Responses to Climatic Hazards and Climate Change (SIRCH) seeks to analyse the discursive construction of risk. A framework of hazards as the conjuncture of global change, institutions of risk management and stakeholder decision making (Figure 4) has been further developed and will be applied in case studies in The Netherlands, southern England and southern Spain (see Bakker, 1999; Calatrave *et al.*, 1999; Riesco Chueca, 1998).

Clearly, the social construction of risk is likely to change, quite possibly in ways that are tangential to the prospective changes in the idealised compartments of hazard, vulnerability and risk.

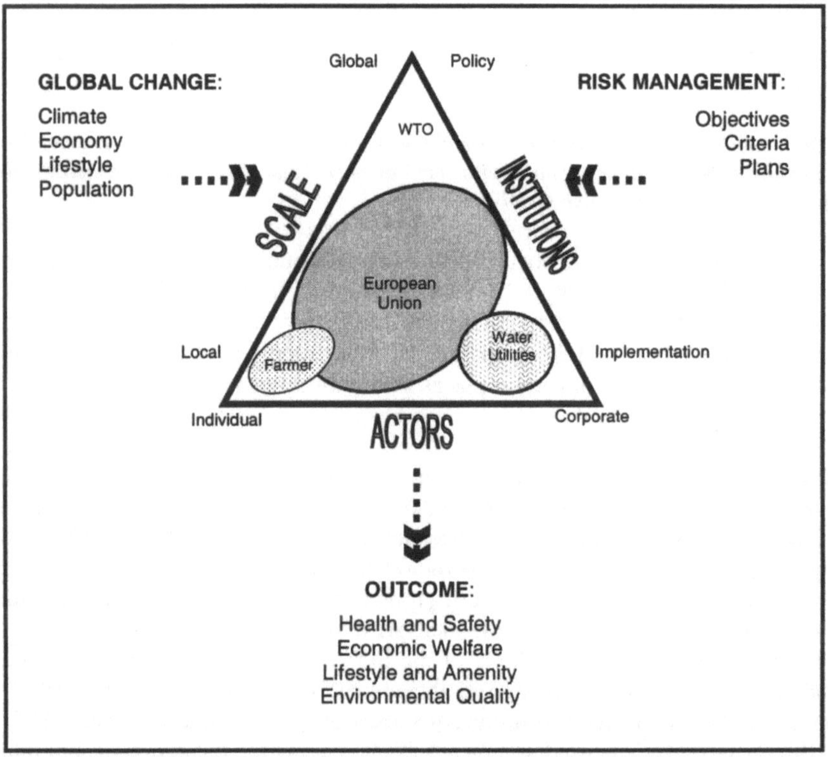

Figure 4: Stakeholders and institutions related to global change.

5. CONCLUSION

The social and cognitive bases of the construction of perceptions of drought demonstrates that the manifestation of drought and people's experience of it "are mediated at all levels by non-meteorological factors" (Solway, 1994: 474). An understanding of the discursive production of drought, in addition to the norms and vagaries of weather and climate, is thus critical to societies' preparedness, and the management of water scarcity.

Shifts in drought discourse are occurring in many regions. For example, in Europe increasing population density in some regions, increasing per capita demand for water, and an exceptionally dry period up to 1998 have forced a re-examination of water resources, water use and water management. Managed drought in Africa has to a large extent replaced the drought = food crisis sequence prevalent until the 1970s. Treating drought as a 'norm' rather than an exception requires a focus on scarcity rather than abundance, and on conservation rather than growth. In examining unquestioned assumptions and definitions, discourse analysis can stimulate creative re-thinking and requires the identification of new 'problems' before new 'solutions' are designed.

It is likely that the drought hazard will increase in many places, but the actual changes are a balance of mean moisture availability and the distribution of drought events (frequency, persistence and magnitude). The outcome of these hazards will depend on vulnerability, capability and management, related to changing exposure (lives, property, livelihood), precautionary resource management amidst conflicts over resource uses, and preparedness.

Regarding the tri-partite structure of this paper, one might conclude:

- It is hazardous to define the hazard, given the range of definitions appropriate to different geographic regions, time scales, sectors, stakeholders, and decisions. Several indicators are likely to be required to capture a broad range of circumstances.
- The nexus of vulnerability is sensitive to perturbations, which occur on similar time scales as weather and climate, from daily to monthly shocks to multi-year fluctuations and decadal trends.
- The discursive construction of risk belies most attempts to adequately conceptualise and measure drought risk, challenging facile notions of mitigation.

Is there no hope? What do we need to learn to reduce the adverse effects of drought? How should we prepare for the likely changes in drought? Understanding the hazard, taking advantage of drought forecasting (Washington and Downing, 1999), is a readily agreed strategy. Equally, understanding the evolution of societal institutions in response to changing hazard is essential (see Bakker 1999). Locating this understanding in the discursive context of drought definitions, impacts and management may well lead to more adaptive strategies that respond to changing social concerns.

REFERENCES

Bakker, K. (ed.) (1999) *Societal and Institutional Responses to Climate Change and Climatic Hazards (SIRCH): A Framework for Institutional Analysis*, SIRCH Working Paper No. 3, Environmental Change Unit, University of Oxford, Oxford (http://www.eci.ox.ac.uk).

Beck, U. (1992) *Risk Society: Towards a New Modernity*, Sage, London.

Blaikie, P., Cannon, T., Davis, I., and Wisner, B. (1994) *At Risk - Natural Hazards, People's Vulnerability, and Disasters*, Routledge, London.

Blaikie, P.M. and Brookfield, H.C. (1987) *Land Degradation and Society*, Methuen, London.

Bohle, H.G. and Watts, M.J. (1993) The Space of Vulnerability: The Causal Structure of Hunger and Famine. *Progress in Human Geography* 13(1): 43-67.

Bohle, H.G., Downing, T.E., and Watts, M.J. (1994) Climate Change and Social Vulnerability: Towards a Sociology and Geography of Food Insecurity. *Global Environmental Change* 4(1), 37-48.

Calatrave, J., Garrido, A., and Iglesias, E. (1999) *Societal and Institutional Responses to Climate Change and Climatic Hazards (SIRCH): Economics Applied to Drought Planning and Management for Damage Mitigation*, SIRCH Working Paper No. 2, Universidad Politéchnica de Madrid, Madrid.

Chambers, R. (1989) Vulnerability, Coping and Policy, *IDS Bulletin* 20(2): 1-7.

Downing, T.E. (1992) *Climate Change and Vulnerable Places: Global Food Security and Country Studies in Zimbabwe, Kenya, Senegal and Chile*, Environmental Change Unit, University of Oxford, Oxford.

Downing, T.E. (ed.) (1995) *Climate Change and World Food Security*, Springer, Heidelberg.

Downing, T.E., Watts, M.J. and Bohle, H.G. (1996) Climate Change and Food Insecurity: Toward a Sociology and Geography of Vulnerability, in T.E. Downing (Ed.), *Climate Change and World Food Security*, Heidelberg: Springer, 183-206.

Downing. T.E. and Bakker, K. (1999) Drought Vulnerability: Concepts and Theory, in D.A. Wilhite (ed.), *Drought, Volume II*. Routledge, New York, pp: 213-230.

Downing, T.E., Harrison, P.A., Butterfield, R.E., and Lonsdale, K., (eds.) (1999a) *Climate Change and Climatic Variability in Europe and Agriculture*, Final Report for the CLIVARA Project, Environmental Change Unit, University of Oxford, Oxford.

Downing, T.E., Olsthoorn, A.A., and Tol, R.S.J., (eds.) (1999b) *Climate, Change and Risk*, Routledge, London.

Eagleton, T. (1991) *Ideology*, Verso, London.

Emel, J., Roberts, R. and Sauri, D. (1992) Ideology, Property, and Groundwater Resources, *Political Geography*, 11(1): 37-54.

Hajer, M.A. (1995) *The Politics of Environmental Discourse: Ecological Modernization and the Policy Process*, Clarendon, Oxford.

Haraway, D. (1991) *Simians, Cyborgs, and Women: The Reinvention of Nature*, Free Association Books, London.

Hulme, M. and Jenkins, G. (1998) *Climate Change Scenarios for the United Kingdom*, Climatic Research Unit, Norwich.

Hulme, M., Barrow, E.M., Arnell, N.W., Harrison, P.A., Johns, T.C., and Downing, T.E. (1999) Relative Impacts of Human-induced Climate Change and Natural Climate Variability, *Nature* 397: 688-691.

Kasperson, J.X., Kasperson, R.E., and Turner, B.L. (eds.) (1995) *Regions at Risk: Comparison of Threatened Environments*, United Nations University Press, Tokyo.

Latour, B. (1993) *We Have Never Been Modern*, Harvester Wheatsheaf, London.

Nevarez, L. (1996) Just Wait until there's a Drought: Mediating Environmental Crises for Urban Growth. *Antipode* 28(3): 246-272.

Riesco Chueca, P. (1998) *Societal and Institutional Responses to Climate Change and Climatic Hazards (SIRCH): The Challenge of Climate Change for Water Technologies: An Institutional Perspective*, SIRCH Working Paper No. 1, JRC - Institute for Prospective Technological Studies – Sevilla, Brussels and Luxembourg.

Sen, A.K. (1981) *Poverty and Famines: An Essay on Entitlement and Deprivation*, Clarendon Press, Oxford.

Solway, J.S. (1994) Drought as a 'Revelatory Crisis': An Exploration of Shifting Entitlements and Hierarchies in the Kalahari, Botswana, *Development and Change* 25: 471-495.

Walker, R. and Williams, M. (1982) Water from Power: Water Supply and Regional Growth in the Santa Clara Valley, *Economic Geography* 58(2): 95-119.

Washington, R. and Downing, T.E. (1999) Seasonal Forecasting of African Rainfall Prediction: Prediction, Response and Household Food Security, *Geographical Journal* 165(3): 255-274.

Watts, M.J. (1983) *Silent Violence: Food, Famine and Peasantry in Northern Nigeria*, University of California Press, Berkeley.

SYNOPTIC SITUATIONS CAUSING DROUGHT IN THE MEDITERRANEAN BASIN

P. MAHERAS
Department of Meteorology and Climatology
University of Thessaloniki,
54006 Thessaloniki, Greece

Abstract. Anomalously dry months in the Mediterranean basin during the period 1900-1990 were identified from meteorological observations at seven stations located along the West-East axis of the Mediterranean basin (Lisbon, Madrid, Barcelona, Florence, Malta, Athens and Jerusalem). Different patterns of dry months are qualitatively identified on the basis of the spatial distribution of a Drought Index: 1) An almost uniform distribution of negative precipitation anomalies over the whole Mediterranean; 2) Strong negative anomalies to the West, but weaker or normal ones to the East, 3) Negatives anomalies to the West and East, but weaker or normal ones, or even positive anomalies in the central Mediterranean. The standardised sea level pressure at 56 grid points in the geographical domain from 35°N to 65°N and 30°W to 40°E are subjected to a T-mode Principal Component Analysis for each of the anomalously dry months. Monthly charts of mean standardised anomaly and mean sea level pressure are constructed for each principal component and each season in order to interpret the spatial distribution of the negative precipitation anomalies in terms of the mean circulation over the geographical domain. Finally, the frequencies of anticyclonic and cyclonic circulation types over Greece are compared with the mean precipitation and the mean number of days of precipitation in Greece (period 1950-1994) in order to interpret the severe drought in Greece during the period 1988-1990.

1. INTRODUCTION

While many definitions of drought exist, from the meteorological point of view a drought occurs when precipitation during a certain period is significantly lower than the average or is lower than a critical value that defines the initiation of a drought. It is very important to differentiate between aridity and drought. Both are characterised by the lack of water, however aridity and drought are two distinct concepts. Aridity is a permanent climatic feature. Drought is an extreme of a time/area process. Meteorological drought and aridity are frequently associated, because the driest regions are usually those where the variability of precipitation is the highest.

Droughts are not local phenomena. They are often associated with persistent, predominantly anticyclonic weather situations. The meteorological interpretation of drought is based on the knowledge of the physical process, which leads to very low precipitation or even to its inhibition and the concomitant synoptic patterns associated with anomalies of the general circulation. This gives the explanation of the immediate meteorological causes of droughts.

J.V. Vogt and F. Somma (eds.), Drought and Drought Mitigation in Europe, 91–102.

The Mediterranean, being a transitional zone between the continental climate areas of Asia and the Oceanic climate areas of the Atlantic, is often affected by phenomena that originate in these two areas. As the Mediterranean extends from 5°W to 35°E, very frequently climatic parameters of the east and the west Mediterranean appear fluctuating in different ways. This is very obvious in surface meteorological conditions, which are controlled by the subtropical Atlantic anticyclone, the SE Asian summer low, the Siberian winter anticyclone and the travelling depressions and anticyclones.

Climatically, the Mediterranean is characterised by generally warm temperatures, dominant winter rainfall, dry summers and a variety of microclimates.

Almost the whole Mediterranean basin suffers from two kinds of droughts: The seasonal drought or aridity, which occurs during the summer and the accidental drought with different duration and intensity, which can occur every season of the year. The aridity is more pronounced in the southern Mediterranean and generally its intensity increases from the North to the South.

The objective of this study is to establish a relationship between mean monthly negative precipitation anomalies in the Mediterranean area and the large scale circulation over Europe, the Mediterranean and the near Atlantic. This relationship will be used to identify the main synoptic situations causing accidental drought in the Mediterranean.

2. DATA AND METHODOLGY

Monthly precipitation data for the period 1900-1990 have been analysed from five stations aligned West-East along the main axis of the Mediterranean basin (Barcelona, Florence, Malta, Athens, Jerusalem), as well from Lisbon and Madrid (Maheras *et al.*, 1999). Gridded monthly sea-level pressure data at 5° latitude and 10° longitude intervals have also been used in the geographical domain from 35° to 65°N and from 30°W to 40°E. The data were taken from the CD-ROM World Climate Disk (1992). For each station, monthly standardised precipitation anomalies (SPA) have been calculated according to equation 1:

$$SPA = (x-\bar{x})/\sigma \quad (1)$$

where $\bar{x}$ is the long term mean monthly precipitation, x is the precipitation of a particular month and σ is the standard deviation of the monthly precipitation. From SPA a Drought Index (DI) has been derived for each month in the 1900-1990 period according to the scheme presented in table 1.

Table 1: Classification of SPA into a Drought Index.

	SPA < -2.1	-2.0 < SPA < -1.3	-1.3 < SPA < -0.6	-0.6 < SPA < 0.7
DI	-3	-2	-1	0

This scheme has been used to produce an index, which is similar to the Hydric Indices used by Pfister *et al.* (1994). For each station, dry months were defined as months with DI between -1 and -2. Due to the particularities of the Mediterranean climate, there has been no month with a DI of –3 for the stations analysed. At least five of the seven stations had to conform to this definition before a particular month was characterised as being anomalously dry. The months were assigned to the relevant seasons: winter (December, January, February), spring (March, April, May), autumn (September, October, November). This technique is similar to the technique used by Kutiel and Paz (1998) to differentiate between dry and wet conditions in Israel.

For each of the anomalously dry months, the monthly standardised value of sea-level pressure (SLP) was calculated for each of the 56 grid points. The resulting pressure patterns were then subjected to a T-mode principal component analysis (PCA) with Varimax rotation (Richman, 1986) in order to establish those SLP patterns associated with the monthly negative precipitation anomalies. Charts of the mean absolute SLP and of the standardised SLP anomalies were constructed for the months (in each of the three seasons) which had the greatest (positive or negative) loading on the particular significant PC (see figures 1 to 5. For a better representation, the latitude is extended to 30°N on these charts).

To enable a better discussion of the results, the mean DI for each station was calculated from all DI values for the months associated with a particular set of charts. Only the most frequent of spatial DI patterns during the anomalously dry months are analysed here. The remaining cases were relatively infrequent and, moreover, were quite similar to the analysed cases.

3. RESULTS

In table 2, the variance (per cent) explained by the significant principal components (PCs) are presented for each season. Table 3 gives the seasonal Mean Drought Indices per station for the significant PCs with positive and negative loading values. Each significant PC represents a specific pressure distribution over the study domain, which is associated with a (qualitatively identified) monthly precipitation anomaly pattern and therefore is a considerable aid in interpreting the occurrence and character of the dry and wet months.

Tables 2 and 3 show that for the winter season 16 months were selected with the first four principal components cumulatively explaining 97% of the total variance. For the spring season 16 months were selected with four principal components explaining 88.3% of the total variance. For the summer season no months have been selected because, as a rule, in the whole Mediterranean seasonal drought dominates. Finally, in autumn 19 months were selected. Again, four significant principal components explain 92% of the total variance.

Table 2: PCA-Varimax Standardised Explained Variance of the Principal Component for a Negative Drought Index.

Factor	1	2	3	4
		Winter		
Exp. Var.	53.2%	16.9%	12.7%	14.2%
Cum. Exp. Var.	53.2%	70.1%	82.8%	97.0%
		Spring		
Exp. Var.	38.2%	17.6%	16.2%	16.3%
Cum. Exp. Var.	38.2%	55.8%	72.0%	88.3%
		Autumn		
Exp. Var.	31.7%	24.6%	16.5%	19.3%
Cum. Exp. Var.	37.1%	56.2%	72.8%	92.0%

Table 3: Mean Drought Indices for the PCs with positive and negative loading-values.

PC Loadings	Lisbon	Madrid	Barcelona	Florence	Malta	Athens	Jerusalem
				Winter			
Factor 1+ (11)	-0.6	-0.9	-0.9	-1.2	-0.7	-1.3	0.0
Factor 3+ (2)	-1.0	-1.0	-1.0	0.0	-1.0	-0.5	-0.5
Factor 4+ (1)	-0.6	-1.0	-1.0	-1.0	-1.0	-1.2	-1.0
Factor 2- (2)	-1.0	-0.8	-1.0	-0.3	0.0	-1.5	-1.5
				Spring			
Factor 1+ (8)	-0.1	-0.6	-0.3	-0.8	-0.4	-0.4	0.3
Factor 2+ (2)	-1.3	-1.3	-0.7	2.0	0.7	-1.0	-0.7
Factor 3+ (2)	-0.5	0.0	-0.5	0.0	-0.5	-0.5	-0.5
Factor 4+ (2)	-0.5	0.0	-0.5	-0.5	-0.5	-0.5	0.0
Factor 2- (2)	-1.5	0.0	-1.0	-0.5	-0.5	-0.5	-1.0
				Autumn			
Factor 1+ (6)	-0.8	-1.0	-0.8	-0.7	-0.5	-1.0	-0.3
Factor 2+ (5)	-1.4	-1.2	-1.2	-1.4	-0.8	-0.2	0.0
Factor 3+ (3)	-0.7	-1.0	-1.0	-0.7	-1.0	-1.0	0.7
Factor 4+ (4)	-0.8	-0.8	-1.0	-1.0	-0.8	0.0	-0.1
Factor 2- (1)	-1.0	-1.0	-1.0	-2.0	-1.0	-1.0	0.0

+: Positive loading, -: Negative loading and (*n*): number of cases - months

The DI values associated with PCs in table 3 may be used to qualitatively group the negative precipitation anomalies into a number of general patterns across the Mediterranean. Three general patterns could be identified and are discussed below.

3.1. Negative Precipitation Anomalies Over the Whole Mediterranean Region

This distribution is most frequent in winter (Table 3; PC1 and PC4, positive loading), but also in spring (PC1, positive loading) and in autumn (PC1, positive loading). The following two examples for winter, one for spring and one for autumn are used to illustrate the links with the general circulation.

Figure 1a shows a strong positive SLP anomaly over the whole Mediterranean and Southern Europe, extending up to about 50°N, associated with the winter PC1 (positive loading). A ridge from the Açores High is the dominant feature over the Mediterranean, reflecting vigorous zonal flow over western, central and northern Europe (Figure 1b). The westerly flow, combined with anticyclonic conditions over the Mediterranean leads to negative DI values everywhere except Jerusalem (DI value of 0.0). The normal amount of rain in Jerusalem may be due to cyclogenesis, which is created over the Cyprus area (Lee depressions) because of the NE circulation over the eastern Mediterranean.

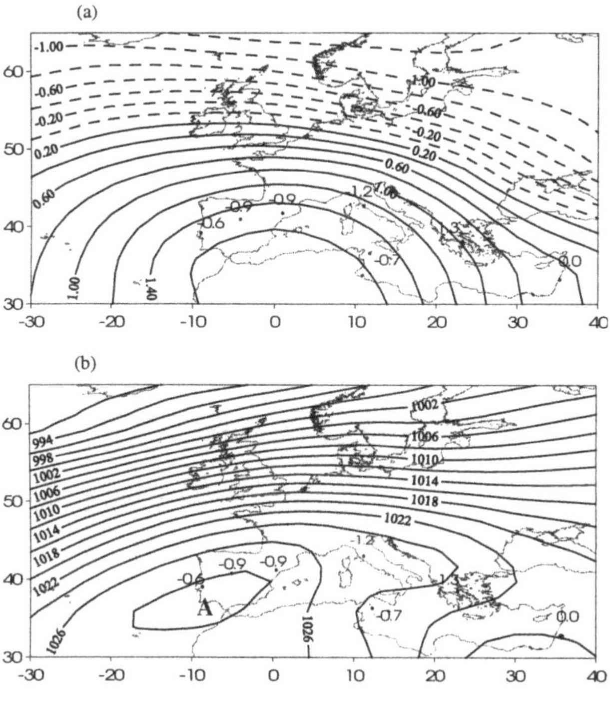

Figure 1: (a) Mean standardised pressure values for negative Drought Indices in winter for PC1 positive loading values. (b) Mean sea surface pressure values for negative Drought Indices in winter for PC1 positive loading values. Drought Indices are shown from West to East for Lisbon, Madrid, Barcelona, Florence, Malta, Athens and Jerusalem, respectively.

Figure 2 shows the SLP anomaly and SLP charts associated with the winter PC4 positive loading. There is a relatively strong zonal flow especially over the Iberian and the Western Mediterranean with an extension of the Açores High over North Africa and the Mediterranean (Figure 2b). Everywhere the DI values are equal to -1.0 (except Lisbon, DI = -0.6 and Athens, DI = -1.2), illustrating the importance of the westerly flow, combined with anticyclonic conditions over the Mediterranean.

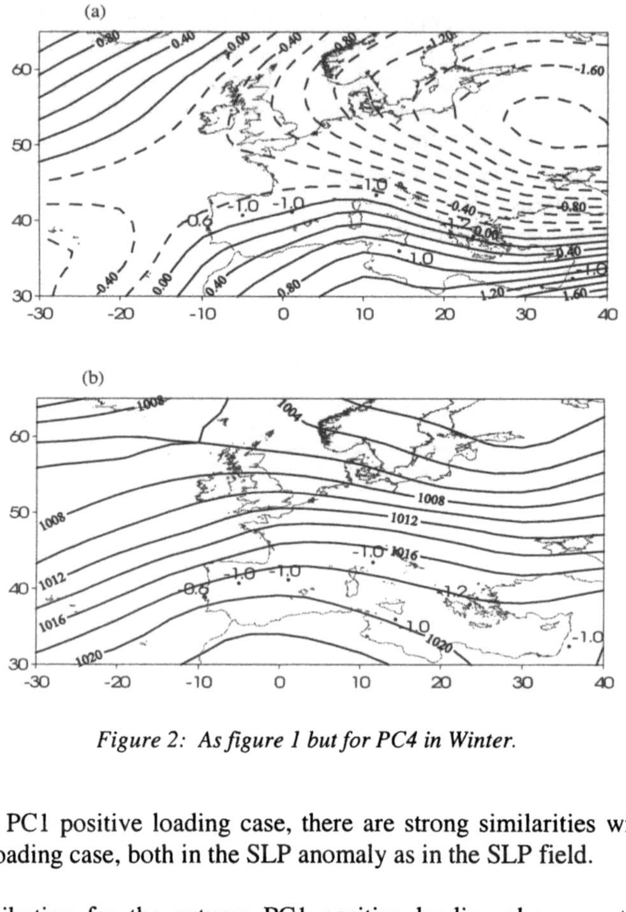

Figure 2: As figure 1 but for PC4 in Winter.

For the spring PC1 positive loading case, there are strong similarities with the winter PC1 positive loading case, both in the SLP anomaly as in the SLP field.

The SLP distribution for the autumn PC1 positive loading shows a strong positive pressure anomaly centered over North Africa (Libya and Tunisia), which extends over the Mediterranean and most of Europe (Figure 3a). This is associated with a SLP field representing an extension of the Açores High towards eastern Europe and the eastern Mediterranean. This field causes westerly / southwesterly flow extending over most the geographical domain (DI values from -0.5 to -1.0 everywhere, except Jerusalem where DI = -0.3 (Figure 3b).

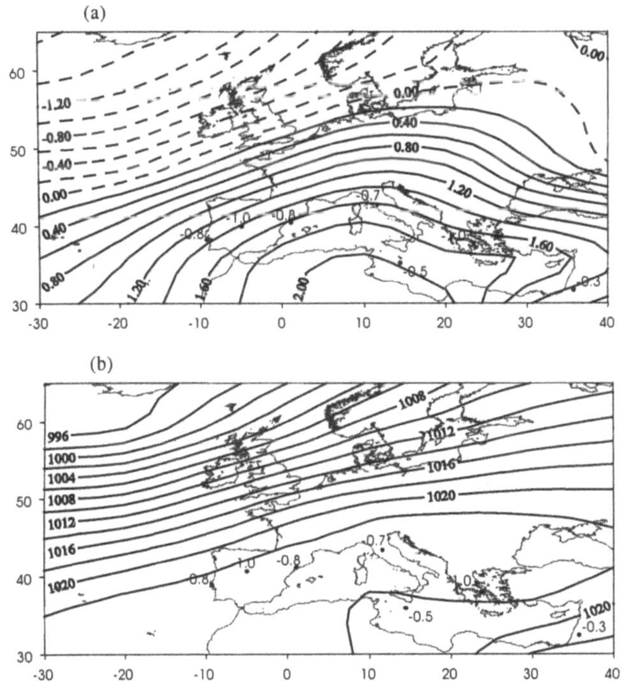

Figure 3: As figure 1 but for PC1 in Autumn.

3.2. Strong Negative Anomalies in the West Mediterranean

Here the mean synoptic condition causing strong negative anomalies of precipitation in the west Mediterranean but weaker or normal precipitation to the east Mediterranean is analysed. This is a very frequent distribution in autumn as is shown, for example, by PC2 and PC4 (both positive loading) in autumn (Table 3).

For autumn PC2 (positive loading) there is a positive SLP anomaly over the Iberian peninsula and the western Mediterranean, balanced by a weaker negative anomaly over eastern Europe and the eastern Mediterranean (Figure 4a). The SLP fields reveal a closed anticyclone over the Iberian peninsula extending eastwards to the Black Sea and the eastern Mediterranean (Figure 4b). This transports continental dry air from the continent into the western and central Mediterranean region (DI values between -0.8 and -1.4), whereas precipitation in the East are closer to normal (DI values of -0.2 and 0.0 at Athens and Jerusalem, respectively). Synoptic experience suggests that in the eastern Mediterranean this corresponds to the incursion of unstable air behind cold fronts as depressions are steered around the northern flank of the ridge (Maheras, 1983).

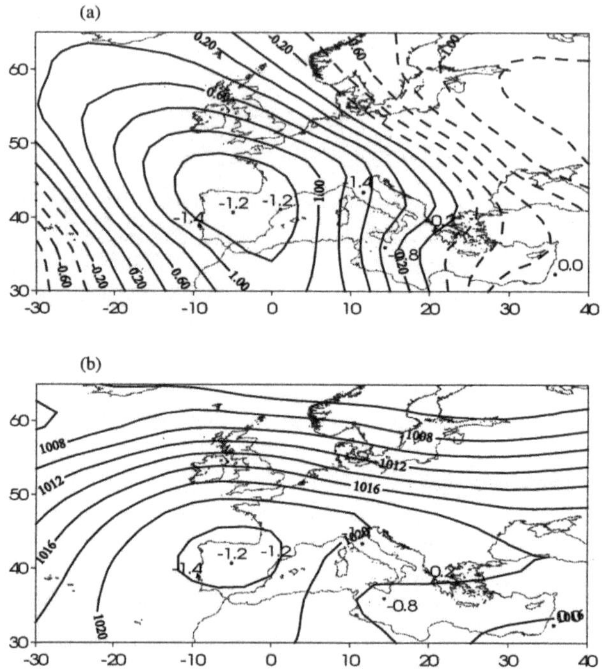

Figure 4: As figure 1 but for PC2 in Autumn.

3.3. Negative Anomalies in the West and East Mediterranean

Here the mean synoptic condition causing strong negative anomalies of precipitation in the west and east Mediterranean but weaker, normal or positive anomalies of precipitation to the central Mediterranean is analysed. Winter and spring are the seasons during which this is a relatively frequent distribution: PC2 (negative loading) in winter and PC2 (positive loading) in spring (Table 3).

In spring, for PC2 (positive loading), the SLP anomaly chart exhibits a negative zone to the west of France, affecting western Europe and a part of the western Mediterranean (Figure 5a). The northerly position of the slight flow to the West of the British Isles (Figure 5b) leads to a weak westerly flow over the Iberian peninsula, originating from colder oceanic latitudes an the southeasterly flank of the low. This is reflected in DI values equal to -1.3 at Lisbon and Madrid, –0.7 at Barcelona and Jerusalem, and –1.0 at Athens. The positives values in Florence and in Malta may be due to the secondary trough over the central Mediterranean (Figure 5b).

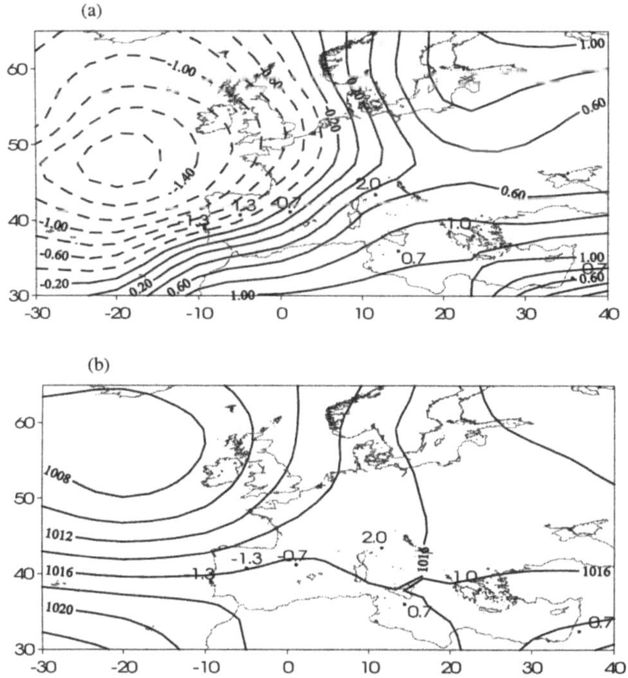

Figure 5: As figure 1 but for PC2 in Spring.

4. DROUGHT AND CIRCULATION TYPES IN GREECE

An analysis of the circulation types for every region of the Mediterranean would lead to a better explanation of the synoptic conditions that cause droughts in this region. We discuss the situation of Greece as an example. Here an objective classification of the circulation types has been based upon daily mean SLP and 500 hPa data information at 2.5°x2.5° grid points (Maheras *et al.*, 1998). It includes six anticyclonic types of circulation, 8 cyclonic types, 2 mixed types and 3 characteristic types (see table 4 and figures 6 and 7).

Table 4: Atmospheric Circulation Types Over Greece.

AnticyclonicTypes	A1, A2, A3, A4, A5, A6
Cyclonic Types	C, Cs, Csw, Cnw, Cne, Cse Cn, Cw
Mixed Types	Mt1, Mt2
Characteristic Types	MB, Dsec, Dor

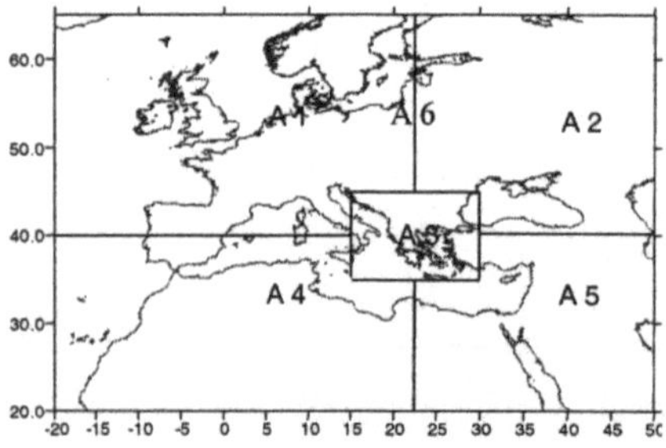

Figure 6: Location of the centre of Anticyclonic types.

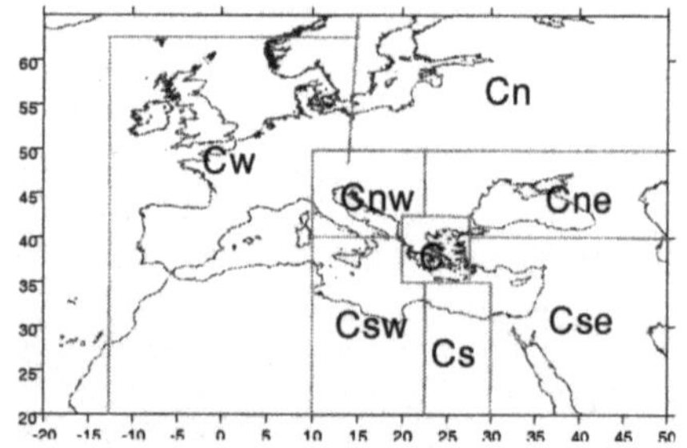

Figure 7: Location of the centre of Cyclonic types.

In figure 8 the annual and seasonal frequencies of the anticyclonic and cyclonic types are compared with the mean annual rainfall and the average number of rainfall days in Greece for the period from 1950 to 1994.

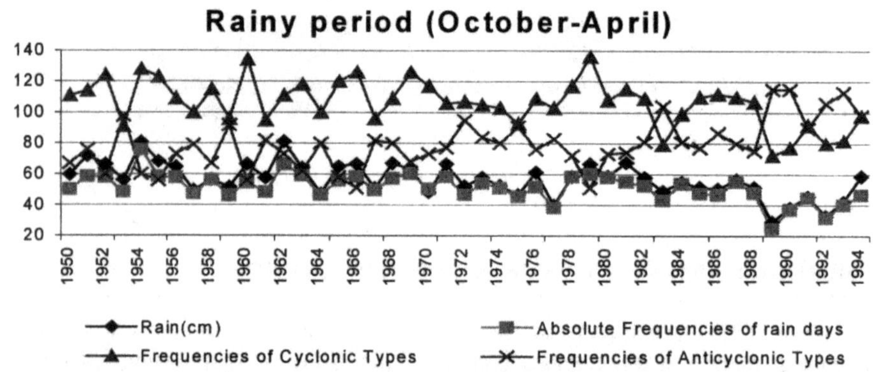

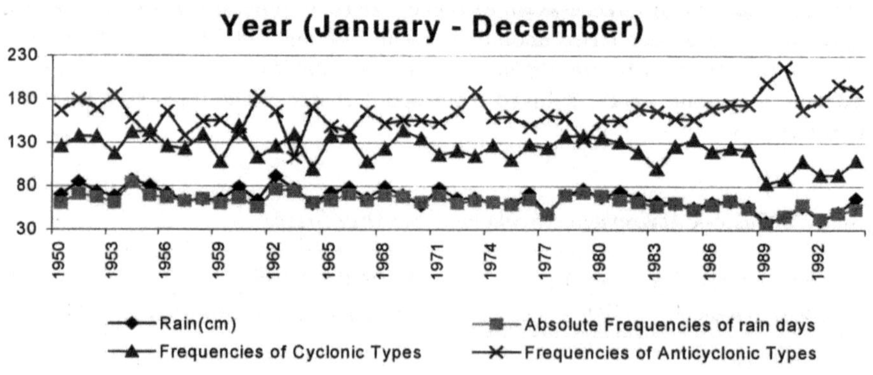

Figure 8: Number of anticyclonic and cyclonic circulation types, mean annual precipitation and number of rainy days in Greece for the period 1950-1994.

The figure shows that the significant decrease of rainfall just after 1980 comes along with an also significant fall in the frequency of occurrence of the cyclonic circulation types. On the other hand, the frequency distribution of the anticyclonic types depicts a significant increase during the same period. Particularly for the periods 1988 - 1989 and 1989 - 1990, where the maximum of the anticyclonic types and the minimum of the cyclonic types appear both for the rainy period (October - April) and for the annual situation, the rainfall amount reaches a minimum value.

5. CONCLUSIONS

The main factor for drought occurrence is the atmospheric circulation. Under conditions of an atmospheric drought, the process of relative drying takes place under the absence of precipitation and with the presence of a large radiative heat flux. The nature of drought is the same both in Europe and in the Mediterranean. It presents the process of a continuous anticyclogenesis, because drought occurs in long-term stable anticyclonic fields.

Drought does not have a local but a broader characteristic. It is connected by teleconnection with different regions of the earth and the atmosphere. The synoptic scale anomalies of the atmospheric pressure, like the North Atlantic Oscillation (NAO) (Moses *et al.*, 1987; Wanner *et al.*, 1997), are connected with the large anomalies of the regional atmospheric circulation. It also corresponds to the zonal circulation persistence, and the troughs and the ridges of large-scale planetary waves.

Among the intense drought conditions encountered in the Mediterranean, the one of 1988-89 and 1989-90 is of particular importance. It was unique both in its duration and intensity. This situation was noticed all over the Mediterranean and especially over Greece. In general, the synoptic and dynamic conditions of the aforementioned drought are similar to the ones presented already in the first analysed category (see figure 1). In such a situation large positive anomalies of the atmospheric pressure are observed over southern Europe and the Mediterranean, where the zonal circulation characteristics are pronounced.

Finally, it should be mentioned that drought is an accidental phenomenon and no connection is seen with a permanent climatic trend.

ACKNOWLEDGEMENTS

This research was funded by EC projects ADVICE under contract ENV4-CT95-129 (DG 12-ESCY) and ACCORD under contract ENV4-CT-97-0530 (DG12-EHKN).

REFERENCES

Kutiel, H. and Paz, S. (1998) Sea Level Pressure Departures in the Mediterranean and Their Relationship With Rainfall Conditions in Israel, *Theoretical and Applied Climatology* 60: 93-109.

Maheras, P. (1983) *Climatologie de la mer Egée et de ses marges continentales*, Etude de climatologie descriptive et de climatologie dynamique, Thèse d' Etat, Atelier de Reproduction de Thèses de Lille III, 787 p.

Maheras, P., Karacostas, Th., Kolyva-Machera, F., Vafiadis, M., and Patricas, I. (1998) *Annual Progress Report: Atmospheric Circulation Classification and Regional Downscaling* (ACCORD), ENV-4-CT97-0530.

Maheras, P., Xoplaki, E., and Kutiel, H. (1999) Wet and Dry Monthly Anomalies across the Mediterranean Basin and their Relationship with Circulation (1860-1990), *Theoretical and Applied Climatology* 64: 189-199.

Moses, T., Kiladis, G.N., Diaz, H.F., and Barry, R.G. (1987) Characteristics and Frequency of Reversals in Mean Sea Level Pressure in the North Atlantic Sector and Their Relationship to Long-Term Temperature Trends, *Journal of Climatology* 7: 13-30.

Pfister, C., Kington, J., Kleinlogel, G., Sehule, H., and Siffert, E. (1994) High Resolution Spatio-Temporal Reconstruction of Past Climate from Direct Meteorological Observations and Proxy Data, *Climatic Trend and Anomalies in Europe 1675-1715*, Fischer Verlag, Stuttgart, 329-375.

Richman, M.(1986) Rotation of Principal Components, *Journal of Climatology* 6: 293-335.

Wanner, H., Rickli, R., Salvisberg, E., Schmutz, C., and Schuepp, M. (1997) Global Climate Change and Variability and Its Influence on Alpine Climate - Concepts and Observations, *Theoretical and Applied Climatology* 58: 221-243.

Wold Climate Disk (1992) *Global Climatic Change Data on CD-ROM*, Chadwyck – Healey Ltd.

STREAMFLOW DROUGHT FREQUENCY ANALYSIS

L.M. TALLAKSEN
Department of Geophysics
University of Oslo
Blindern, 0315 Oslo, Norway

Abstract. An overview of methods for streamflow drought frequency analysis for a single site is presented. A streamflow drought is defined as a period when the discharge is below a given threshold level. A distinction is made between the minimum low flow discharge of a drought event, and the corresponding deficit volume and duration of the event. Different ways of defining low flow and drought events and different approaches to estimate drought characteristics and extreme events from time series of flow are discussed. The paper focuses on frequency analysis: in this case sample properties of the selected events are used as a basis for estimating design events. Finally, some alternatives to traditional at-site frequency analysis are outlined.

1. INTRODUCTION

Low flow and droughts affect many aspects of environment and society, and future increase in the demand for water will be most critical in periods of severe low flow and extensive droughts. Information on low flow and drought characteristics provides critical values for different water based activities, and strategic short and long-term plans are needed for adequate water resource management. These plans require improved technologies, as well as better monitoring and prediction capabilities for their implementation (McKay *et al.*, 1989). In terms of hydrological research there is a need to improve our ability to predict the onset, duration and severity of a drought or low flow period, thus providing a better basis for the design of drought facilities. Drought studies and frequency analysis of historical data provide in this respect an important contribution to improve our knowledge of the status and dynamics of water resources.

The present paper will limit its discussion to hydrological droughts. A hydrological drought may be defined as a deficiency in water supply over time, and this study therefore focuses on low flow events below a given threshold. A distinction is made between the minimum low flow discharge of a drought event, and the corresponding deficit volume and duration of the event. Different ways of defining low flow and drought events and different approaches to estimate drought characteristics and extreme events from time series of flow are discussed. The paper focuses on frequency analysis: in this case sample properties of the selected events are used as a basis for estimating design events. Finally, some alternatives to traditional at-site frequency analysis are outlined.

J.V. Vogt and F. Somma (eds.), Drought and Drought Mitigation in Europe, 103–117.

2. ESTIMATION OF DROUGHT CHARACTERISTICS

2.1. Selection of Drought Events

A sequence of drought events is obtained from the streamflow hydrograph by considering flow situations where the discharge is below a certain threshold level q_0. Each drought event, also referred to as a low flow spell in the literature, is characterised by its duration d_i, deficit volume (severity) s_i, minimum flow q_{min}, and time of occurrence (Figure 1). The time of drought occurrence has been given different definitions, as for instance the starting date of the drought, the mean of the onset and the termination date of the drought, or the date of the minimum flow. The threshold level is commonly defined as a percentage of the mean flow or as a percentile of the flow duration curve.

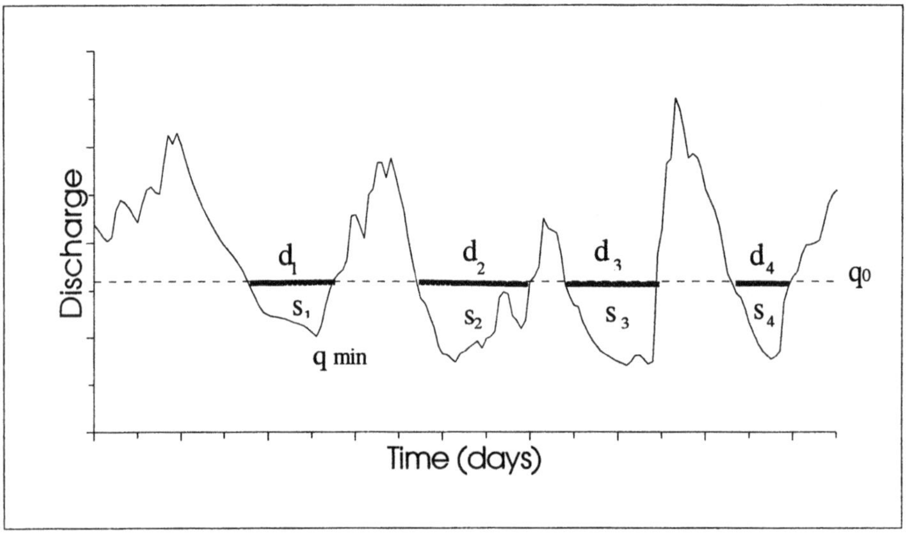

Figure 1: Definition of low flow and drought characteristics (modified from Tallaksen et al., 1997).

The choice of threshold level defines the low flow part of the hydrograph. This choice is not an arbitrary choice but a function of the type of water deficit to be studied (Dracup *et al.*, 1980; Bonacci, 1993). In a regional study, the lack of long time series over larger areas may impose restrictions to the use of very low threshold levels, since the choice of a low threshold might lead to a high number of no-drought years (the flow never falls below the chosen threshold in a year). The corresponding reduction in sample size might be significant thus implying limitations in the use of frequency analysis. On the other hand, the number of multi-year droughts (droughts lasting longer than a year) increases with increasing threshold level.

The threshold level method generally studies runs above or below a given threshold level and was originally referred to as crossing theory. The theory was first developed by Rice (1945) and later extended and summarised by Crámer and Leadbetter (1967). Early applications of crossing theory in hydrology include Yevjevich (1967) for an objective definition of drought, Nordin (1968) for describing the properties of sand waves, Rodrigues-Iturbe (1968) for predicting properties of stage and Nordin and Rosbjerg (1970) describing the applicability of the method for streamflow.

The threshold level method as described by Yevjevich (1967) was based on the statistical theory of runs for analysing a sequential time series. Statistical properties of the distributions of water deficits, run-length (drought duration) and run-sum (deficit volume) were recommended as parameters for at-site drought definition. The theory of runs was applied to discrete time series with time resolution of one month or longer, and drought duration can only take on integer values. If the purpose is to study drought events spanning less than a year, a daily or monthly time step is preferable. The use of daily time series introduces two special problems, the one of mutually dependent droughts and the one of minor droughts.

During a prolonged dry period it is often observed that the flow exceeds the threshold level in a short period of time and thereby dividing a large drought into a number of droughts that are mutually dependent (Figure 2). Thus, a consistent definition of drought events should include some kind of criterion for pooling successive events in order to define an independent sequence of droughts. Dependence does not in this context mean that the values are correlated. As pointed out by Cunnane (1979) in a flood study of peak over threshold values, dependency exists as a form of persistence in the point stochastic process producing the peaks.

On the other hand, minor droughts are a result of the fluctuation of the flow around the threshold level, and these droughts might distort the extreme value modelling. Tallaksen *et al.* (1997) address these problems and recommend two different methods for pooling dependent droughts into a single event. The use of a moving average procedure (MA) and a method based on the sequent peak algorithm (SPA) both proved satisfactory with respect to pooling dependent droughts and reducing the number of minor droughts. The MA procedure is applied to the time series prior to selecting the droughts. In this case the time series is smoothed and minor peaks above the threshold level removed. The SPA procedure is closely related to the analysis of water resources systems in which case the threshold level may be interpreted as the desired yield from a reservoir, and drought deficit volume defines the required storage in a given period.

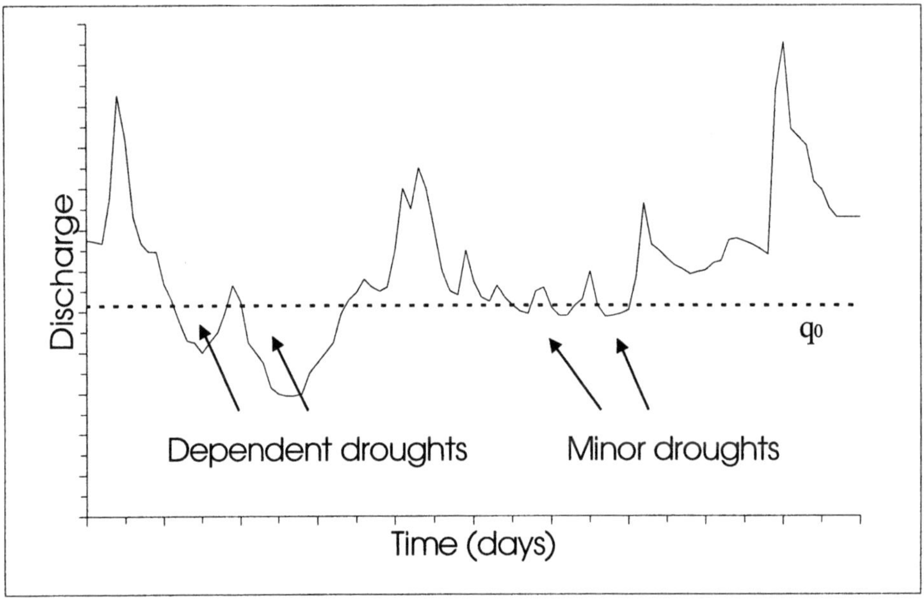

Figure 2: Daily time series of flow illustrating the problems of mutual dependence and minor droughts.

Drought characteristics and risk estimates can be derived using historical data in combination with a stochastic model, traditional frequency analysis, or a data generation scheme. A stochastic model is a model having parameters that are obtained from properties of the original time series of flow, and estimates of extreme events are based on the statistical properties of the original time series. Alternatively, drought characteristics can be derived from the sample properties of the selected events, and the design events estimated using frequency analysis. Although this procedure is straightforward, the reliability of the sample parameters depends on the sample size. The presence of zero-drought years and multi-year droughts will reduce the sample size, and in this case estimation based on the original time series of flow might be more efficient. However, if, the focus is on within-year droughts (droughts related to the low flow season), the sample estimation procedure is often preferable. Data generation analysis where drought characteristics are estimated from generated samples of time series of drought characteristics will not be discussed in this paper. Synthetic streamflow generation models often fails to accurately reproduce extreme historical events (Askew *et al.*, 1971). All models depend on the assumptions of stationarity of series and representativeness of sample.

2.2. Estimation Based on Time Series Properties

Classical stochastic models used within drought studies are the Bernoulli and the Poisson process for independent processes and different types of Markov models for dependent processes. Theoretical probability distributions and estimates of extreme events are derived based on the statistical properties of the original time series of flow, *X(t)*. Estimation of design events based on the time series of *X(t)* requires that simplifying assumptions are made to the distribution of the flow. Annual or monthly time steps are used and *X(t)* is most commonly assumed identically or normally distributed. By introducing a threshold level the series is divided into two distinct events, surpluses and deficits. The time series is classified as a series of surpluses (positive runs) and deficits (negative runs). A run theory model based on independent Bernoulli trials must have a sufficient small correlation between values to allow the independence assumption as a reasonable approximation (Moyé and Kapandia, 1995). This assumption is often satisfied in annual time series. Persistence, i.e. observations are not independent, however, is likely to be present in hydrological time series, particularly using a monthly time step. In order to describe the dependence structure Sen (1976) introduced a two-state lag-one Markov model, where the state at time *i* is dependent of the state at time *i-1*.

Analytical expressions have been derived for drought characteristics assuming both independent and dependent processes, such as drought duration (Rosbjerg, 1977; Saldarriaga and Yevjevich, 1970; Sen, 1976), drought deficit (Sen, 1977; Phien, 1982), maximum drought duration (Millan and Yevjevich, 1971; Sen, 1980a; Todorovich, 1970; Moyé and Kapadia, 1995), and maximum drought deficit (Güven, 1983; Troutman, 1978). Extreme values, rather than averages, of hydrological phenomena are the values most critical for design purposes and risk estimation. Drought estimation based on the time series *X(t)* is restricted, as there is a limited number of distributions for which analytical expressions can be derived.

2.3. Estimation Based on Sample Properties

Estimation of design events can alternatively be based on the sample properties of the drought events using frequency analysis. The drought characteristics are selected from the time series *X(t)*, and the data fitted to a distribution function. Estimates of extreme events are often based on extrapolation in the tail of the distribution. Although it is possible to perform a frequency analysis by merely plotting the data and without making any distributional assumptions, hydrological design often requires extrapolation beyond the range of observations. Therefore, one of the main purposes of the frequency analysis is to find the most suitable probability distribution to describe the data.

3. DROUGHT FREQUENCY ANALYSIS

Frequency analysis is one of the most frequent and earliest applications of statistics within hydrology. When a disastrous event occurs, the question of assessing its frequency or probability is raised. In terms of water management the probability of occurrence refers to the question of responsibility and may lead to re-evaluation of design guidelines, policy standards and risk acceptance (Kundzewich *et al.*, 1993). Frequency analysis involves the definition and selection of the type of hydrological event to be studied, selection of an appropriate distribution to describe the data, estimation of parameters of the distribution and calculation of return periods or risk estimates for the given problem.

Frequency analysis of hydrological data requires that the data are homogeneous (from the same population) and independent (no serial correlation in the data). Low flow data are commonly influenced by human impacts, e.g. withdrawals for irrigation and water supply. Prior to performing a low flow frequency analysis, an effort should therefore be made to correct the series and obtain 'natural' streamflow series. If sufficiently long records are available and no change over time can be assumed, the frequency distribution for a given data set could be precisely defined. In most cases, however, the record length is insufficient to precisely define the probability or frequency of an extreme event. Given a data sample from an unknown distribution, the problem is to estimate accurately the tail of the distribution. Three important issues in this context are: i) there are few observations in the tail of the distribution, ii) estimates are often required above the largest observed value and, iii) standard probability distributions can be severely biased in estimating tail probabilities (Coles, 1999).

3.1. Extreme Value Model

The two most applied approaches for extreme value analysis are the annual maximum/minimum series (AMS) model, and the partial duration series (PDS) model. In the AMS approach the largest/smallest event within a year is selected, whereas the PDS considers all events below/above a given threshold. The PDS model ensures that the most severe events are selected for the extreme value analysis, independent of the time of occurrence, and hence intuitively provides a more consistent definition of the extreme value region than the AMS approach. Minimum flows in very wet years might not belong to the low extreme population. A drawback of the PDS model is that independent events have to be identified, thus making the selection of events more complicated than in the AMS. Alternatively, one might select the n largest events in each period (Coles, 1999).

The drought events selected may not belong to the same population, allowing for the fact that different processes may give rise to drought. Snow and ice-affected regions frequently experience their lowest flow during the winter months, due to precipitation being stored as snow and river discharge reduced by freezing. Summer droughts are primarily caused by the lack of precipitation. Other regions may experience different types of weather conditions dominating in different seasons. If the historical streamflow data can be divided into separate subsets, either by season or meteorological information, several AMS or PDS can be obtained, one for each generating process. If *a priori* subdivision of the data is not feasible, a two-component extreme value distribution might be applicable as suggested for flood frequency analysis by Rossi *et al.* (1984).

3.2. Exceedance Probability

Let X denote a random variable, and x a real number. The cumulative distribution function (cdf), $F_X = Prob(X \leq x)$ is the non-exceedance probability for the value x. The probability density function (pdf) is defined as the derivative of the cdf, and describes the relative likelihood that the continuos random variable X takes on different values. The percentiles or quantiles of a distribution is often used as a design quantity. The $100p$ percentile, x_p, is the value with cumulative probability p, i.e. $F_X(x_p) = p$, which will have an exceedance probability of $(1-p)$. The exceedance probability is frequently expressed in terms of the return period, T. When the right hand tail is of interest, $T = 1/1-p$, which is the average number of time intervals between occurrence of an event $X>x_p$. For minimum flow values, $T=1/p$, is the average number of time intervals between occurrence of an event $X \leq x_p$. For a time step of one year, annual exceedance probabilities are defined and the corresponding T-year event is given by the value for x_p. Estimates of extreme quantiles of AMS is obtained by inverting the expression for $F_X(x_p)$ for a given value of p. It is common to extrapolate this relationship beyond the range of the data to which the distribution has been fitted. Confidence intervals (C.I.) for a quantile are often calculated using the standard error of the quantile. A 90 or 95 percent C.I. is commonly used, which in repeated sampling will contain the parameter 90 respectively, 95 percent of the time. Expressions for the C.I. of different distributions are given by e.g. Stedinger *et al.* (1993).

There are several expressions that account for the general relationship between the probability distribution for the AMS and the frequency of events in a PDS. Annual exceedance probabilities can be estimated from the PDS provided the average number of events per year is known (Stedinger *et al.*, 1993).

3.3. Choice of Distribution Function

The choice of distribution function can be made on a theoretical base, as demonstrated by Zelenhasic and Salvai (1987) and Mathier *et al.* (1992) for example. Following the theory of a random number of random variables (Todorovic, 1970), Zelenhasic and Salvai (1987) presented a stochastic model that considers a number of drought characteristics, including the largest streamflow drought deficit and duration in a time interval using daily time series of low flow. The method is based on the assumption that the droughts are independent, identically distributed random variables and that their occurrence is subject to the Poisson probability law. A good agreement was found between the theoretical and empirical distribution functions for all analysed drought components. Mathier *et al.* (1992) analysed monthly flow data and derived theoretically a discrete distribution function for drought duration by accounting for the fact that the variable can only take on positive integer values.

In most cases, however, it is not possible to analytically determine which probability distribution should be used, and it is necessary to apply prior knowledge and statistical judgement to select the distribution that is best suited for the data being studied. Relatively short return periods that do not greatly exceed the length of hydrological records have often been sufficient in low flow design, and this has led to a rather unrestricted use of various distributions. Such an approach may give acceptable results if a prediction of low flow events with a low return period is needed. However, the estimates of events with higher return periods will always depend on the behaviour of the tail of the fitted distribution.

Fisher and Tippet (1928) distinguished between three different types of extreme value (EV) distributions. The EV I distribution, commonly referred to as Gumbel's extreme value distribution, is unbounded in the direction of the desired extreme and frequently applied for analysis of maximum values (Gumbel, 1958). The Weibull distribution is an EV type III distribution for minima which is bounded at the direction of the extreme (Weibull, 1961). The three parameter generalised extreme value distribution (GEV) independently derived by von Mises (1936) and Jenkinson (1955), is a general mathematical form which encompasses the three types of limiting distributions (Coles, 1999). Whereas the GEV distribution describes the limit distribution of normalised maxima, the generalized Pareto distribution (GPD), which goes back to Pickands (1975), appears as the limit distribution of scaled excesses over high thresholds (Embrechts *et al.*, 1999). In hydrology, the three-parameter GPD distribution has been used in association with the peak over threshold approach (Ashkar *et al.*, 1996) and for describing events exceeding a lower bound, such as flows above zero. Stedinger *et al.* (1993) present an overview of several families of distribution functions commonly used in hydrology.

Probability plots are an effective way to judge whether the fitted distributions appear consistent with the data. The data are ranked, assigned a probability using a plotting position formula, and plotted on probability paper. Analytical goodness of fit criteria can be used to estimate whether a departure from a theoretical distribution function is statistical significant, e.g. the Kolmogorov-Smirnov test or the probability plot correlation test (Filliben, 1975). In many cases several distributions will provide statistically acceptable fits to the data, and goodness of fit tests can not be used to discriminate between the different distributions (e.g. Cunnane, 1985). The likelihood ratio test is an example of a test that can be used to compare distributions with different number of parameters. L-moment skewness-kurtosis and coefficient of variation-skewness diagrams are valuable for investigating what families of distributions are consistent with available data sets for a region (Vogel and Fennessey, 1993).

3.4. Parameter Estimation

Having chosen a probability distribution, the estimation of a design event with a given probability is reduced to the problem of fitting the distribution to the sample. A standard procedure to estimate the parameters of a distribution function is to apply statistical inference using sample moments. The method of moments relates the theoretical moments (e.g. mean value, variance, coefficient of skewness) of the distribution exactly to the sample estimates. Ordinary product moments, probability weighted moments (PWM) and L-moments can be used. L-moments are weighted linear sums of the expected order statistics and do not involve squaring or cubing the observations, as product moment estimators do. L-moments have been demonstrated to suffer less than conventional moments from the effects of sample variability (Hosking, 1990). Values of different moments for several distributions can be found in e.g. Stedinger *et al.* (1993) and Embrechts *et al.* (1997). There are also other approaches available for estimating the parameters of a distribution, the most common ones being graphical techniques and the method of maximum likelihood.

3.5. Risk Estimation

The concept of risk has many interpretations in hydrological literature. Here risk is understood to be equal to the probability of exceedance (non-exceedance) of a certain maximum (minimum) level within a given design period. This is often referred to as a risk of failure. If a construction is designed to withstand a T-year event, the probability of failure during n years equals the probability that the event is exceeded (for maximum values) at least once in the n year period, and given by: $1-(1-1/T)^n$. The use of historical data provides no guarantee that the design will perform satisfactorily in the future because the calculation of risk assumes that the underlying process can be considered stationary, and that the sample is a representative one.

4. MINIMUM LOW FLOW DISCHARGE OF A DROUGHT EVENT

In the traditional approach of drought hazard assessment, low flows are defined in terms of the minimum annual n-day average discharge, and the selection of events is based on the lowest value each year (AMS). Some studies have reported a break in the probability plots of AMS, which has been interpreted as representing a point where the higher frequency flows are not considered belonging to the extreme low flow region. The Institute of Hydrology (1980) suggested a break in the curve for an exceedance probability of 65 %, whereas Nathan and McMahon (1990) adopted the 80 percentile. A systematic lower skewness and coefficient of variation was found for AMS of minimum discharge using the 70 percentile as a threshold (Gottschalk *et al.*, 1997). Imposing a threshold level implies that the AMS may contain years without droughts, and the information content of the series may accordingly be reduced. A conditional probability procedure is recommended for frequency analysis at sites whose record of AMS is truncated by omission of data above a threshold or contains years with zero flow (Stedinger *et al.*, 1993). The procedure is not recommended when the record is reduced by more than 25%. The problem of zero flow is discussed amongst others by Nathan and McMahon (1990), Bulu (1997) and Gordon *et al.* (1992).

4.1. Low Flow Distribution Functions

It is generally recommended that the theoretical probability distributions for low flow have lower limits equal to or greater than zero and no more than three parameters (Matalas, 1963). The more parameters a distribution has, the better it will adapt to the data sample, but the lower the reliability in the estimate of the parameters will be. Low flow values generally show a range of variations of the skewness (Gottschalk *et al.*, 1997), which should be accounted for in the choice of distribution function. The log-Pearson type 3 is a popular choice for frequency analysis of low flow. Its application in hydrology is, however, strictly limited (Kite, 1977). The Weibull distribution is a limiting distribution of extremes with the parent distribution bounded at the direction of the extremes. The flexibility of the distribution, as well as its theoretical basis, makes it a favorable choice for low flow frequency applications (Gottschalk *et al.*, 1997). It has been applied in a large number of low flow prediction studies around the world: Australia (Nathan and McMahon, 1990), Asia (Ewart and Brutsaert, 1972; Al-Mashidani *et al.*, 1980; Institute of Hydrology, 1988), Africa (Drayton *et al.*, 1980; Zucchini and Adamson, 1984), North America (Gumbel, 1958; Matalas, 1963; Joseph, 1970; Pilon, 1990), Europe (Institute of Hydrology, 1980; Gustard *et al.*, 1989; Polarski, 1989).

5. DURATION AND DEFICIT VOLUME OF A DROUGHT EVENT

The minimum low flow discharge involves only one drought characteristics, the drought magnitude, which may be insufficient in some applications. The threshold level method allows the characterisation of the drought both in terms of duration and severity; in terms of water management these constitute important parameters for the design and operation of water supply systems.

Zelenhasic and Salvai (1987) investigated the stochastic behaviour of droughts using daily time series. Duration and deficit volume were found to follow an exponential distribution, whereas corresponding maximum values for a given time interval (usually chosen as the year) followed a double exponential distribution. Clausen and Pearson (1995) applied L-moments diagrams for 44 catchments in New Zealand, and the three-parameter log-Normal distribution was selected as the best distribution for both severity and duration. The L-moment diagram was also applied by Tallaksen and Hisdal (1997) in a study of 52 daily discharge series from the Nordic region. It was concluded that the generalized three-parameter Pareto distribution gave the overall best fit to the annual and summer drought samples of duration and deficit (Figure 3). Madsen and Rosbjerg (1995) found that the two-parameter General Pareto distribution was suitable to describe drought duration but not deficit.

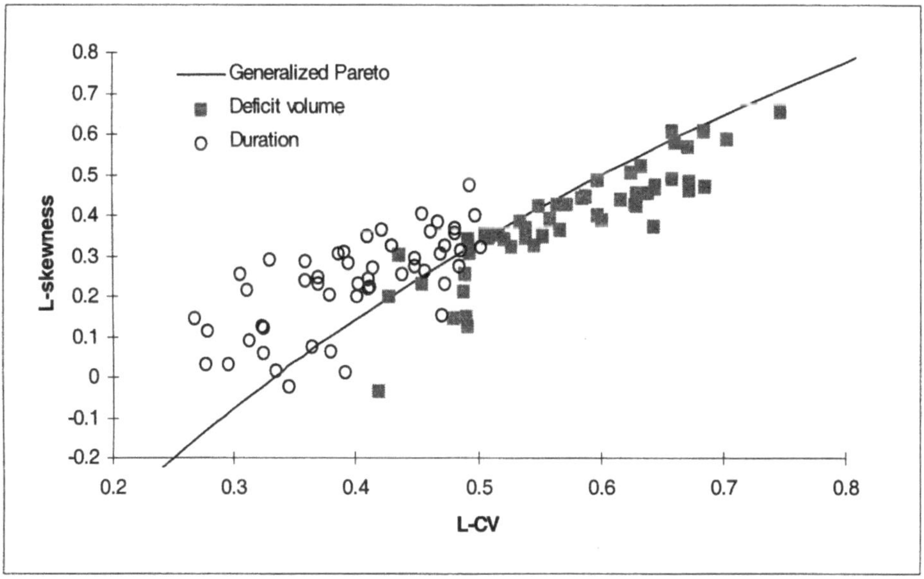

Figure 3: L-moment ratio diagram for annual values of drought duration and deficit volumes using the 90 percentile of the flow duration curve as a threshold level (from Tallaksen and Hisdal, 1997).

Kjeldsen and Lundorf (1997) tested seven distribution functions and found that the two-component distribution (Rossi, 1984) gave the best description of the sample of drought duration and deficit volume. The analysis was done on drought events extracted using a monthly varying threshold level on daily time series of streamflow in Zimbabwe. They experienced a problem in fitting a distribution due to the presence of many small events, and suggested removing 40% of the smallest events to obtain a satisfactory fit. Woo and Tarhule (1994) identified a similar problem with small droughts. They distinguished between three types of streamflow droughts extracted from daily time series of flow in Nigeria; short droughts that cluster in the rainy season, long droughts that span the dry season and multi-year droughts. They suggested separating the events into sub populations prior to performing a frequency analysis.

In a study by Mathier *et al.* (1992) daily values were averaged to give monthly time series, and a monthly varying threshold level was applied. The threshold level was defined as a percentile of the mean discharges of a given month. A theoretical base is given for the choice of the geometric distribution for drought duration, and the exponential distribution was found to give reasonable good results for the distribution of drought deficit volumes. In both cases there is only one parameter to be determined. Approximate expressions of the standard error of quantiles of duration and deficit were also derived.

Many hydrological phenomena are characterised by the joint occurrence of two or more random variables, implying that the frequency analysis of these variables must be based on their joint probability distribution (bivariate or multivariate). High correlations are generally found between drought duration and deficit volume (Chang and Stenson, 1990; Clausen and Pearson, 1995; Woo and Tarhule, 1994), and Sen (1977) deduced an analytical expression for the correlation between drought duration and deficit. Ashkar *et al.* (1998) demonstrated the use of two types of bivariate distributions for drought duration and deficit volume. They concluded, however, that the models presented certain difficulties when a very strong correlation between the random variables existed, as is the case of duration and deficit. Drought frequency analysis based on a bivariate or multivariate distribution requires further empirical information regarding the drought measures (Matalas, 1991).

6. CONCLUDING COMMENTS

The methods and studies presented have been limited to at-site streamflow frequency analysis. Data from a single site is a random sample from an unknown distribution, and not only are the parameters of this population unknown, but the frequency distributions that generated the sample are also unknown (Haan, 1977). The high variability encountered in hydrological variables and the short records often available have led the hydrologists to develop alternatives to the traditional method of frequency analysis, in an effort to provide more robust estimates of design quantities.

In a regional analysis data sets from several sites are included and homogenous regions in terms of the variable being studied are identified. A distribution function for a region can then be derived assuming that the data of single sites follow the same distribution except for a scale parameter (usually the mean flow). In addition to provide more robust estimates, the method can be used for estimation at ungauged sites. For the latter purpose regression analysis are also commonly used and the drought variable is related to physical and climate characteristics of the catchment.

Klemes (1993) criticised the standard approach to frequency analysis, which he argues to contribute merely to the art of curve fitting and to the theory of small samples from known distributions, rather than provide better information about the probability of extreme events. Historical evidence suggests that extreme events occur more due to unusual combinations of causative factors than to unusual magnitudes of the factors themselves. He therefore suggested to treat the variable of interest as a compound event, based on an explicit consideration of the physical processes involved. This is also the philosophy behind the derived distribution approach, which was first introduced by Bernier (1964) for low flow frequency analysis. The approach allows knowledge about the hydrological processes generating streamflow to be considered, and was applied by Gottschalk *et al.* (1997) to derive theoretical expressions for low flow distribution functions. The parameters of the distributions are given a physical interpretation that offers an alternative for more robust parameter estimation. The usefulness of the physical parameter interpretation for regionalisation purposes has not yet been tested and is a topic for further research.

As low flow and droughts are regional in nature and commonly cover large regions and extend for long time periods, a regional approach should be adopted for drought frequency studies. More information is required on the regional behaviour of severe droughts, and also regional variables like the area covered by drought and total deficit over the drought area ought to be considered (Tase, 1976; Sen, 1980b; Santos, 1983; Rossi *et al.*, 1992). To improve our capability of predicting future droughts, the relationship between atmospheric circulation patterns and both meteorological and streamflow droughts should be further investigated (e.g. Stahl and Demuth, 1999).

ACKNOWLEDGEMENTS

This review was carried out within the framework of the ARIDE Project (Assessment of the Regional Impact of Droughts in Europe) supported by the European Commission under Contract no. ENV4-CT97-0553.

REFERENCES

Al-Mashidani, G., Lal, B.B., and Quadri, I. (1980) Drought Flow Analysis of River Tigris in Baghdad, *Hydrological Science Journal* 25(4): 453-459.

Ashkar, F., Jabi, N.El., and Issa, M. (1998) A Bivariate Analysis of the Volume and Duration of Low-Flow Events, *Stochastic Hydrol. Hydraul.* 12: 97-116.

Ashkar, F. and Ouarda, Taha B.M.J. (1996) On some Methods of Fitting the Generalized Pareto Distribution, *Journal Hydrol.* 177: 117-141.

Askew, A.J., Yeh, W.W-G., and Hall, W.A. (1971) A Comparative Study of Critical Drought simulation, *Water Resources Research* 7(1): 52-62.

Bernier, J. (1964) La prévision statistique des bas débits, in *Symposium on Surface Waters* (General Assembly of Berkley), IAHS Publ.63, 340-351.

Bonacci, O. (1993) Hydrological Identification of Drought, *Hydrol. Proc.* 7: 249-262.

Bulu, A. (1997) Statistical Analysis of Low Flow with Zero Discharges, in *FRIEND: Flow Regimes from international experimental and network data*, Third report, Cemagref, 167-170.

Chang, T.J. and Stenson, J.R. (1990) Is it realistic to define a 100-year drought for water management? *Water Resources Bulletin* 26(5): 823-829.

Clausen, B. and Pearson, C.P. (1995) Regional Frequency Analysis of Annual Maximum Streamflow Drought, *Journal of Hydrology*, 173: 111-130.

Coles, S. (1999) Extreme Value Theory and Applications, Course notes, http://www.maths.lancs.ac.uk/~coless

Cramér, H. and Leadbetter, M.R. (1967) *Stationary and Related Stochastic Processes*, New York, John Wiley and Sons.

Cunnane, C. (1979) A Note on the Poisson Assumption in Partial Duration Series Models, *Water Resources Research* 15(2): 489-494.

Cunnane, C. (1985) Factors Affecting Choice of Distribution for Flood Series, *Hydrological Science Journal* 30(1): 25-36.

Dracup, J.A., Lee, K.S., and Paulson, E.G. (1980) On the Definition of Droughts, *Water Resources Research* 16(2): 297-302.

Drayton, R.S., Kidd, C.H.R., Mandeville, A.N., and Miller, J.B. (1980) *A Regional Analysis of River Floods and Low Flows in Malawi*, Report No 72, Institute of Hydrology, Wallingford, UK.

Embrechts, P., Klüppelberg, C., and Mikosch, T. (1997) *Modelling Extremal Events for Insurance and Finance*, Springer.

Ewart, Ch.J. and Brutsaert, W. (1972) Some Generalized Characteristics of the Floods and Droughts of the Lower Mekong, *Hydrological Science Journal* 17(3): 323-338.

Filliben, J.J. (1975) The Probability Plot correlation Test for Normality, *Technometrics* 17(1): 111-117.

Fisher, R.A. and Tippett, L.H.C. (1928) Limiting Forms of the Frequency Distribution of the Largest or Smallest Member of a Sample, *Proc. Cambridge Philosophical Society* 24(2): 180-190.

Gordon, N.D., McMahon, T.A. and Finlayson, B.L. (1992) *Stream Hydrology (An Introduction for Ecologists)*, J. Wiley and Sons.

Gottschalk, L., Tallaksen, L.M., and Perzyna, G. (1997) Derivation of Low Flow Distribution Functions Using Recession Curves, *Journal Hydrol.* 194: 239-262.

Gumbel, E.J. (1958) *Statistics of Extremes*, Columbia Univ. Press. New York.

Gustard, A., Roald, L.A., Demuth, S., Lumadjeng, H.S., and Gross, R. (1989) *Flow Regimes from Experimental and Network Data (FRIEND)*, vol.1 Hydrological Studies, Institute of Hydrology, Wallingford, UK.

Güven, O. (1983) A Simplified Semi-empirical Approach to Probabilities of Extreme Hydrologic Drought, *Water Resources Research* 19(2): 441-453.

Haan, Ch.T. (1977) *Statistical Methods in Hydrology*, The Iowa State University Press, Ames.

Hosking, J.R.M. (1990) L-moments: Analysis and Estimation of Distributions Using Linear Combinations of Order Statistics, *Journal Royal Statistical Society* 52(1): 105-124.

Institute of Hydrology (1980) *Low Flow Studies Report*, Wallingford, UK.

Institute of Hydrology (1988) Water Balance Study, Phase 3 Report: Investigation of Dry Season Flows (Lower Mekong Basin) Wallingford, UK.

Jenkingson, A.F. (1955) The Frequency Distribution of the Annual Maximum (or Minimum) Values of Meteorological Elements, *Quat. Journal Royal Met. Soc.* 87(158).

Joseph, E.S. (1970) Probability Distribution of Annual Droughts, *Journal Irrig. Drain. Div.*, ASCE 96(IR4), 461-474.

Kite, G.W. (1977) Frequency and Risk Analyses in Hydrology, *Water Res. Publ.*, Fort Collins, Colorado.

Kjeldsen, T.R. and Lundorf, A. (1997) Drought Management and Modelling – Zimbabwe Case, MSc Thesis, DTU, Technical University of Denmark.

Klemes, V. (1993) Probability of Extreme Hydrometeorological Events - a Different Approach, in *Extreme Hydrological Events: Precipitation, Floods and Droughts (Proc. Yokohama Symp.*, July, 1993), IAHS Publ. 213.

Kundzewicz, Z.W., Rosbjerg, D., Simonovic, S.P., and Takeuchi, K. (1993) Extreme Hydrological Events in Perspective, in *Extreme Hydrological Events: Precipitation, Floods and Droughts (Proc. Yokohama Symp.*, July, 1993), IAHS Publ. 213.

Madsen, H. and Rosbjerg, D. (1995) A Regional Bayesian Method for Estimation of Extreme Streamflow Droughts, Paper presented at the Conference on Statistical and Bayesian Methods in Hydrology, Paris, UNESCO, September 11-13, 1995.

Matalas, N.C. (1963) Probability Distribution of Low Flows, *U.S.G.S. Prof. Pap.* 434-A, Washington, D.C.

Matalas, N.C. (1991) Drought Description, *Stochastic Hydrol. Hydraul.* 5: 255-260.

Mathier, L., Perreault, L., and Bobeé B. (1992) The Use of Geometric and Gamma Related Distributions for Frequency Analysis of Water Deficit, *Stochastic Hydrol. Hydraul.* 6: 239-254.

McKay, G.A., Godwin, R.B., and Maybank, J. (1989) Drought and Hydrological Drought Research in Canada, An Evaluation of the State of the Art, *Canadian Water Res. Journal* 14(3): 71-84.

Millan, J. and Yevjevich, V., (1971) Probabilities of Observed Droughts, *Colorado State University*, Hydrology Paper 50, Fort Collins, Colorado, USA.

Mises, R. von (1936) La distribution de la plus grande de *n* valeurs, in Ph. Frank, S. Goldstein, M. Kac, W. Prager, G. Szegö, and G Birkhoff (eds.), *Selected papers of Richard von Mises, vol II: Probability and Statistics, General*, Amer. Math. Soc., Providence, R.I. (1964), 271-294.

Moyé, L.A. and Kapadia, A.S. (1995) Predictions of Drought Length Extreme Order Statistics Using Run Theory, *Journal Hydrol.* 169: 95-110.

Nathan, R.J. and McMahon, T.A. (1990) Practical Aspect of Low-Flow Frequency Analysis, *Water Resources Research* 26(9): 2135-2141.

Nordin, C.F. (1968) The Statistical Properties of Dune Profiles, *U.S.Geol. Survey Open File Report.*

Nordin, C.F. and Rosbjerg (1970) Application of Crossing Theory in Hydrology, *Bull. Int. Ass. Sci. Hydrol.*, XV 1(3): 27-43.

Phien, H.N. (1982) Moments of the Run-sum for Independent Sequences, *Journal Hydrol.* 54: 389-400.

Pickands, J. III (1975) Statistical Inference Using Extreme Order Statistics, *Ann. Statist.* 3: 119-131.

Pilon, P.J. (1990) The Weibull Distribution Applied to Regional Low Flow Frequency Analysis, in

Regionalization in Hydrology (Proc. Ljubljana Symp., April 1990), *IAHS Publ.* 191: 227-237.
Polarski, M. (1989) Fitting Distributions to Annual Minimum Flows of Different Duration, in *FRIENDS in Hydrology,* IAHS Publ. 187: 97-104.
Rice, S.O. (1945) Mathematical Analysis of Random Noise, *Bell System Tech. Journal* 24: 46-156.
Rodriguez Iturbe, I. (1968) A Modern Statistical Study of Monthly Levels of the Orinoco River, *Bull. Int. Assoc. Scientific Hydrology* 13(4): 25-40.
Rosbjerg, D., (1977) Crossing and Extremes in Dependent Annual Series, *Nordic Hydrol.* 8: 257-266.
Rossi, F., Fiorentino, M. and Versace, P. (1984) Two-component Extreme Value Distribution for Flood Frequency Analysis, *Water Resources Research,* 20(7): 847-856.
Rossi, G., Benedini, M., Tsakiris, G., and Giakoumakis, S. (1992) On Regional Drought Estimation and Analysis, *Wat. Res. Man.* 6: 249-277.
Saldarriaga, J. and Yevjevich, V. (1970) Application of Run-lengths to Hydrologic Series, *Colorado State University, Hydrology Paper* 40, Fort Collins, Colorado, USA.
Santos, M.A. (1983) Regional Droughts: a Stochastic Characterisation, *Journal Hydrol.* 66: 183-211.
Sen, Z. (1976) Wet and Dry Periods of Annual Flow Series, *Journal Hydraul. Div., ASCE* 102(HY10): 1503-1514.
Sen, Z. (1977) Run-sums of Annual Flow Series, *Journal Hydrol.* 35: 311-324.
Sen, Z. (1980a) Statistical Analysis of Hydrological Critical Droughts, *Journal Hydraul. Div., ASCE* 106(HY1): 99-114.
Sen, Z. (1980b) Regional Drought and Flood Frequency Analysis: Theoretical Consideration, *Journal Hydrol.* 46: 265-279.
Stahl, K. and Demuth, S. (1999) Investigating the Influence of Atmospheric Circulation Patterns on Regional Streamflow Drought in Southern Germany, Proceedings of XXII General Assembly of the International Union of Geodesy and Geophysics, Birmingham, UK, 18-30 July 1999, IAHS Publ. 255: 19-27.
Stedinger, J.R., Vogel, R.M., and Foufoula-Georgiou, E. (1993) Frequency Analysis of Extreme Events, in D.R. Maidment (ed.), *Handbook of Hydrology*, McGraw-Hill.
Tallaksen, L.M., Madsen, H., and Clausen, B. (1997) On the Definition and Modelling of Streamflow Drought Duration and Deficit Volume, *Hydrological Science Journal* 42 (1).
Tallaksen, L.M. and Hisdal, H. (1997) Regional Analysis of Extreme Streamflow Drought Duration and Deficit Volume, *Proceedings of the 3rd International conference on FRIEND*, 1-4 Oct. 1997, Postojna, Slovenia, IAHS Publ. 246: 141-150.
Tase, N. (1976) Area-deficit-intensity Characteristics of Droughts, *Hydrology Paper* 87, Colorado State University, Fort Collins, Co., USA.
Todorovic, P., (1970) On Some Problems Involving Random Number of Random variables, *Ann. Math. Statist.* 41(3): 1059-1063.
Troutman, B. M. (1978) Reservoir Storage with Dependent, Periodic Net Inputs, *Water Resources Research* 14(3): 395-401.
Vogel, R.M. and Fennessey, N.M. (1993) L-moment Diagrams Should Replace Product Moment Diagrams, *Water Resour. Res.* 29(6): 1745-1752.
Weibull, W. (1961) Fatigue Testing and Analysis of Results, Pergamon Press.
Woo, M-K. and Tarhule, A. (1994) Streamflow Droughts of Northern Nigerian Rivers, *Hydrological Science Journal* 39(1): 19-34.
Yevjevich, V. (1967) An Objective Approach to Definition and Investigations of Continental Hydrologic Droughts, *Hydrology paper* 23, Colorado State University, Fort Collins, Co., USA.
Zelenhasic, E. and Salvai, A. (1987) A method of Streamflow Drought Analysis, *Water Resources Research* 23(1): 156-168.
Zucchini, W. and Adamson, P.T. (1984) Assessing Risk of Deficiencies in Streamflow, *Water Res. Comm. Rep.* 91(2), Univ. of Stellenbosch, Stellenbosch, S.A.

DROUGHT PREPAREDNESS IN THE UNITED STATES: RECENT PROGRESS

D.A. WILHITE
National Drought Mitigation Center
International Drought Information Center
Lincoln, Nebraska 68583-0749, USA

Abstract. Since 1986, the United States has experienced numerous droughts causing billions of dollars in losses in many economic, social, and environmental sectors. To lessen the risks associated with drought, governments at all levels have taken greater interest in drought planning, with the greatest progress coming at the state level. However, state plans, and recent actions taken by states in response to drought, have been largely reactive, emphasizing short-term actions over long-term planning. The National Drought Mitigation Center's program is directed toward helping governments and others lessen societal vulnerability to drought. The NDMC provides a national drought information clearinghouse; assists state, federal, and regional entities in drought planning; advises on policy issues; and interacts with scientists on an international level. The establishment of the National Drought Policy Commission by the U.S. Congress in 1997 indicates a new commitment to a national drought policy that emphasizes risk management, but the accomplishment of this commission is uncertain at this time.

1. INTRODUCTION

Drought continues to be a common visitor to the American landscape despite a trend of increasing precipitation for most portions of the country (Karl *et al.*, 1995). The period from 1986 to the present witnessed numerous severe and extended drought periods that resulted in significant impacts in nearly all portions of the nation. Drought conditions in 1986 affected most of the Southeast, and these conditions continued into 1987 for a portion of the region. The West experienced widespread drought in 1987, with conditions worsening in 1988. Drought persisted through 1992 for a large portion of the West. Some portions of this region, particularly California and Nevada, experienced six consecutive drought years from 1987 to 1992.

The most widespread drought year during the period from 1986 to 1992 was 1988, when nearly 40% of the nation experienced severe to extreme drought conditions. Drought extended across most of the West through the northern Great Plains and into the Midwest and parts of the Southeast. Costs and losses associated with this drought have been estimated at nearly $40 billion (Riebsame *et al.*, 1991), making it the costliest disaster in American history. Impacts affected many sectors, including agriculture, transportation, energy, and recreation and tourism. Environmental impacts were also significant as forest and rangeland fires were rampant throughout the West and water shortages put wildlife at risk. Federal government responded with more than $5 billion in drought relief programs

J.V. Vogt and F. Somma (eds.), Drought and Drought Mitigation in Europe, 119–131.

(Riebsame *et al.*, 1991). Drought continued in 1989 but was confined mostly to the western states; it resulted in serious reductions in agricultural yields in the Great Plains.

This series of drought years ended abruptly in 1993 with one of the wettest years on record. Flooding occurred over much of the Missouri and Mississippi river basins. However, drought conditions emerged again in 1994 over large portions of the West. In 1995, a short-term drought affected the northeastern states, followed by severe drought in the Southwest and southern Great Plains states in 1996. This drought resulted in impacts of nearly $6 billion in Texas alone (Boyd, 1996). Drought revisited this region in 1998. Drought conditions also existed in the Southeast in 1998; in particular, Louisiana, Georgia, Florida, and South Carolina experienced serious impacts on agriculture, tourism, and municipal water supply. Wildfires were especially serious in portions of Florida. Drought impacts were estimated to be $2 billion and $5 billion in Oklahoma and Texas, respectively (Thurman, 1998; Chenault and Parsons, 1998). Drought conditions developed in the northeastern and mid-Atlantic states during the 1998-99 winter season, resulting in drought warnings in portions of New York, Pennsylvania, and the Delaware River Basin. These conditions eased in early February. Other portions of the United States, in particular the Southwest and southern Great Plains states, are preparing for the possibility of severe drought during the spring and summer months of 1999 as a result of La Niña.

During the period since 1986, interest in drought planning increased dramatically as a strategy to reduce the risk associated with extended periods of water shortage, and political will to make substantive changes in the way governments at all levels in the United States deal with drought is being sustained. Historically, government actions have focused on response, with little attention to mitigation. This crisis management approach to drought management has been largely ineffective, poorly coordinated, and untimely (Wilhite *et al.*, 1986), as recent drought assessments (e.g., Western Governors' Association, 1996) have noted.

The purpose of this paper is to review recent progress in drought planning and mitigation in the United States. Discussion will focus on progress made by state government in drought planning, examples of recent mitigation actions, the role of the National Drought Mitigation Center (NDMC), and the mission of the newly formed National Drought Policy Commission (NDPC). The paper will conclude by speculating on some recommended next steps. Some of the lessons learned in the United States may be of value to Europe as it tries to initiate a more coordinated approach to drought mitigation.

2. PROGRESS IN STATE DROUGHT PLANNING

The most notable progress in drought planning in the United States since 1982 has occurred at the state level. Before the widespread and severe drought of 1976-77, no state had prepared a formal plan. In 1982, only three states had developed plans: South Dakota (1981), Colorado (1981), and New York (1982). During the past decade, however, an additional twenty-five states have developed and implemented formal drought plans.

Presently, 28 states have drought plans (Wilhite, 1997a) (Figure 1). Three of these states (New Mexico, Nebraska, and Utah) are in the process of adopting a more proactive approach to drought planning, with an emphasis on mitigation. In the case of New Mexico and Nebraska, the methodology (i.e., *How to Reduce Drought Risk*) developed by the NDMC and the Western Drought Coordinating Council (WDCC) is being employed. In addition, Texas is placing added emphasis on drought planning at both state and municipal levels and Arizona is also developing a drought plan. The NDMC has discussed drought plan development recently with Kansas, Louisiana, and Vermont, and each is considering further action. Clearly, states took the initiative early and have been referred to as policy innovators in drought planning (Wilhite, 1997a). The NDMC is trying to encourage all states to place more emphasis on mitigation.

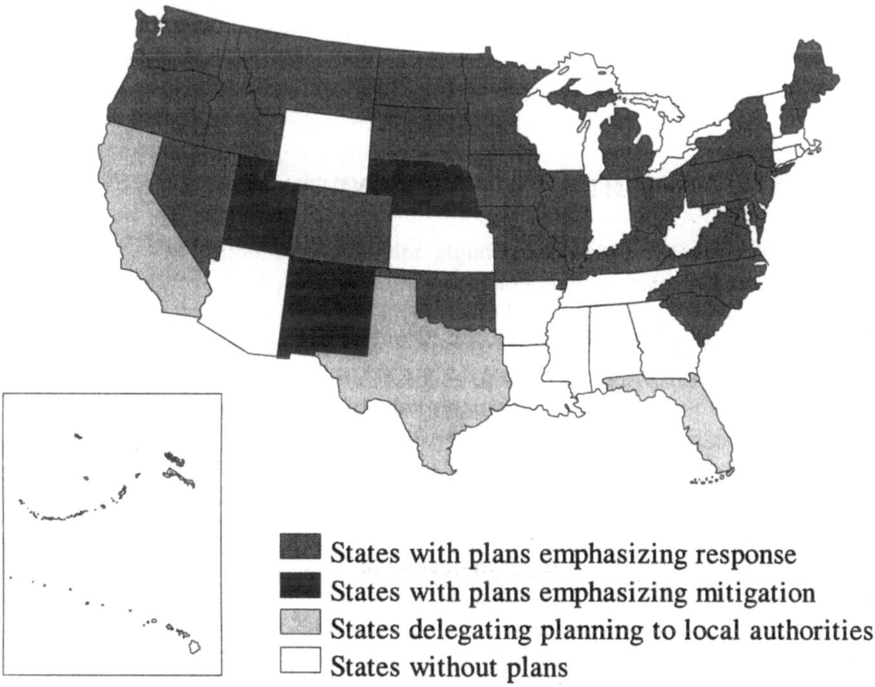

Figure 1: Status of state drought plans in the United States, February 1999.

Although the increased number of states with drought plans is encouraging, these plans are still largely reactive, emphasizing improved coordination in an emergency response mode.

The basic goal of state drought plans is to improve the effectiveness of state response efforts by enhancing monitoring and early warning, vulnerability and impact assessment, and response and mitigation programs. These plans are also directed at improving coordination within agencies of state government and between state and federal government. The growth in the number of states with drought plans suggests an increased concern about the potential impact of extended water shortages and an attempt to address those concerns through planning. Drought plans are the foundation for improved drought management in the United States.

The pattern of state-level drought planning is quite complex and cannot be explained adequately on the basis of drought climatology alone. A state's decision to develop (or not develop) a drought plan is based on specific climatological, political, economic, environmental, and demographic factors. Wilhite and Rhodes (1994) constructed a typology of state behavior in an attempt to explain the pattern of drought plans that existed in the early 1990s and found that social, political, and institutional influences may be as important as or more important than recent drought experiences. This conclusion has been altered somewhat in the years since this study was completed, since the occurrence of drought in the Southwest and southern Great Plains in 1996 prompted planning actions by Oklahoma, Texas, New Mexico, Utah, and Arizona.

3. MITIGATING DROUGHT IMPACTS: RECENT STATE ACTIONS

Ongoing federal, interstate, and state drought mitigation technologies, programs, and policies in the United States were assessed by Wilhite (1993; 1997a). These data were collected through a survey of states and key federal agencies with responsibility for the management of water and other natural resources. The survey was directed at specific drought mitigation actions taken or programs adopted during the period from 1986 to 1992. This study was based on the assumption that the roles of federal and state governments in drought mitigation needed to be reexamined, given the severity of drought experienced in the United States between 1986 and 1992; the economic, social, and environmental costs associated with these droughts; and the mitigation actions and policy efforts underway at all levels of government. One of the goals of the study was to identify opportunities to improve the effectiveness of drought mitigation efforts by the Natural Resources Conservation Service of the U.S. Department of Agriculture and other federal agencies. A premise of this study was that the nation's ability to cope with and manage water shortages resulting from drought would only be improved if an integrated approach within and between levels of government, involving regional organizations and the private sector where appropriate, were adopted.

State mitigation actions used to address issues during recent droughts are clustered into nine primary areas in Table 1. These actions represent a full range of possible mitigative actions, from monitoring and assessment programs to the development of drought contingency plans. Some of the actions included were adopted by many states, while others may have been adopted only in a single case.

Table 1: Drought-related mitigative actions taken by states during recent droughts (Wilhite, 1997a).

Category	Specific Action
Assessment Programs	• Developed criteria or triggers for drought-related actions • Developed early warning system, monitoring program • Conducted inventories of data availability • Established new data collection networks • Monitored vulnerable public water suppliers
Legislation/Public Policy	• Prepared position papers for legislature on public policy issues • Examined water rights statutes for possible modification during water shortages • Passed legislation to protect instream flows • Established a water banking program
Water Supply Augmentation/ Development of New Supplies	• Issued emergency permits for water use • Provided pumps and pipes for distribution • Proposed and implemented program to rehabilitate reservoirs to operate at design capacity • Undertook water supply vulnerability assessments • Inventoried self-supplied industrial water users for possible use of their supplies for emergency public water supplies • Inventoried and reviewed reservoir operation plans
Public Awareness/ Education Programs	• Organized drought information meetings for the public and the media • Implemented water conservation awareness programs • Published and distributed pamphlets to individuals, businesses, and municipalities on water conservation techniques and agricultural drought management strategies • Organized workshops on special drought-related topics • Prepared sample ordinances on water conservation for municipalities and domestic rural supplies
Technical Assistance on Water Conservation	• Provided advice on potential new sources of water • Evaluated water quantity and quality from new sources • Advised water suppliers on assessing vulnerability of existing supply system • Recommended the adoption of water conservation measures to suppliers

Table 1: continued

Category	Specific Action
Demand Reduction/ Water Conservation Programs	• Established stronger economic incentives for private investment in water conservation • Encouraged voluntary water conservation • Improved water use and conveyance efficiencies • Implemented water metering and leak detection programs
Emergency Response Programs	• Established alert procedures for water quality problems • Stockpiled supplies of pumps, pipes, water filters, and other equipment • Established water hauling programs for livestock from reservoirs and other sources • Compiled list of locations for livestock watering • Established hay hotline • Provided funds for improvement of water systems, developing new systems, and digging wells • Provided funds for recovery programs for drought and other natural disasters • Lowered well intakes on reservoirs for rural water supplies • Extended boat ramps and docks in recreational areas
Water Use Conflict Resolution	• Acted to resolve emerging water use conflicts • Negotiated with irrigators to gain voluntary restrictions on irrigation in areas where domestic wells were likely to be affected • Clarified state law regarding sale of water • Clarified state law on changes in water rights • Suspended water use permits in watersheds with low water levels • Investigated complaints of irrigation wells interfering with domestic wells
Drought Contingency Plans	• Recommended to water suppliers the development of drought plans • Established statewide contingency plan • Evaluated worst-case drought scenarios for possible further actions

Assessment programs adopted by states range from the development of improved criteria or triggers for the initiation of specific actions in response to drought to the establishment of new data collection networks. Automated weather data networks such as those that exist in Nebraska, California, and Oklahoma have significantly improved state monitoring and early warning capabilities (one of the three critical components of a drought plan is a comprehensive early warning system). Parameters that must be monitored to detect the early onset of drought include temperature and precipitation, stream flow, reservoir and ground water levels, snow pack, and soil moisture. Each of these parameters represents different components of the hydrologic system and, therefore, different impact sectors (e.g., agriculture, energy, transportation, recreation and tourism).

To assess emerging drought conditions, these data must be integrated to provide a comprehensive snapshot of water availability and outlook. Many recommendations for the development of a national drought watch (Riebsame *et al.*, 1991) or integrated climate monitoring system (U.S. Congress, OTA 1993; Wilhite and Wood, 1994; FEMA, 1996) have been offered, but none have been implemented. Some states have also undertaken vulnerability assessments of public water supplies in conjunction with drought planning efforts. This is an especially critical issue in states with many small water supply systems that may be quite sensitive to extended periods of water shortage. It is important to identify vulnerable systems in advance so that adequate mitigation measures can be adopted.

Legislative actions taken by states included the passage of measures to protect instream flows and guarantee low-interest loans to farmers. Low-interest loans, a common federal response to drought, are not generally available from states. Many states have been reexamining aspects of water rights doctrine in response to growing water use and associated conflicts. Water banks have been used in some states (e.g., California) as a means of temporarily modifying water allocation procedures during water shortages. The California Drought Water Bank program is an example of an innovative and successful mitigation action (California Department of Water Resources, 1992). MacDonnell *et al.* (1994) have reviewed water banks in the West.

Augmentation of water supplies during recent droughts included rehabilitating reservoirs to operate at design capacity and reviewing reservoir operation plans. Cities also worked with self-supplied industrial users on programs to reallocate some water for emergency public water supplies. One of the key responsibilities of state government during periods of drought is to keep the public aware of the severity of the situation through timely reports. These reports must provide a clear rationale for mitigative actions that are being imposed on either a voluntary or mandatory basis. During recent droughts, states organized informational meetings for the media and the public, implemented water conservation awareness programs, prepared and distributed informational materials, and organized workshops on drought-related topics. Sample ordinances on water conservation were also prepared and distributed to municipalities and rural suppliers.

Most states lack the financial resources necessary to provide drought relief to individual citizens during times of emergency. However, it is often within the mission and capacity of state agencies to provide technical assistance to municipalities and others. During recent droughts, states assisted by providing advice on potential new sources of water and evaluating the quality and quantity of those supplies. Agencies also assisted municipalities in assessing the vulnerability of water supply systems. States encouraged the adoption of voluntary water conservation measures and established stronger economic incentives for water conservation within the private sector. Water metering and leak detection programs were implemented.

Emergency response programs would not be considered by some to be mitigative. However, if these measures are implemented to reduce immediate impacts or the risk of future impacts as part of a long-term mitigation program, they represent a proactive approach to drought management. State responses included a wide range of measures such as lowering of well intakes on reservoirs for rural water supplies, establishing water hauling programs for livestock, extending boat ramps in recreational areas, and creating a tuition assistance program to enable farmers to participate in farm management classes.

Conflicts between water users increase during water-short periods. Timely intervention to resolve these conflicts will become increasingly necessary as demands on limited water supplies continue to expand in number and complexity. The best approach is to anticipate these conflicts well in advance of drought and initiate appropriate actions to avoid conflict. Many of the actions the states took focused on the growing conflicts between municipal and agricultural water use.

As mentioned previously, the growing number of states with drought plans is an indication of greater concern about the impacts of drought and the acceptance by states of the role that planning can play in reducing some of drought's most adverse effects. The optimal time to plan for drought is during nondrought periods; however, considerable progress in establishing a basic response framework is often accomplished during the period of peak severity, as occurred in several drought-stricken states in the Southwest in 1996. The challenge is to transform this framework into a response/mitigation plan during the post-drought period. A brief window of opportunity usually exists to initiate a longer-term mitigation program immediately following a drought.

Many of the mitigative programs implemented by states during recent droughts can be characterized as emergency or short-term actions taken to alleviate the crisis at hand, although these actions can be successful, especially if they are part of a preparedness or mitigation plan. Other activities, such as legislative actions, drought plan development, and the development of water conservation and other public awareness programs, are considered actions with a longer-term vision. As states gain more experience assessing and responding to drought, future actions will undoubtedly become more timely and effective and less reactive. Viewed collectively, the mitigative actions of states in response to recent drought conditions are numerous, but most individual state actions were quite narrow. Table 1 is illustrative of the arsenal of mitigation programs and actions currently available to states. In the future, state drought plans need to address a broader range of mitigative actions, including provisions for expanding the level of intergovernmental coordination. One of the goals of the NDMC is to facilitate this process by identifying and disseminating potential mitigative programs to governments at all levels in the United States and elsewhere. Improved coordination will require a greater commitment by federal, state, local, and tribal governments to work together to promote an integrated approach to drought planning.

4. NATIONAL DROUGHT MITIGATION CENTER

The NDMC was established in 1995 with funding from the U.S. Congress through the U.S. Department of Agriculture. The NDMC is located at the University of Nebraska-Lincoln in the School of Natural Resource Sciences. The NDMC's program is aimed at lessening societal vulnerability to drought by promoting and conducting research on drought mitigation and preparedness technologies, improving coordination of drought-related activities and actions within and between levels of government, and assisting in the development, dissemination, and implementation of appropriate mitigation and preparedness technologies in the public and private sectors. Emphasis is directed toward research and outreach projects and mitigation/management strategies and programs that stress risk management measures rather than reactive, crisis management actions.

The objectives of the NDMC are:

(1) To conduct research to maintain and enhance a national information clearinghouse.

(2) To conduct and foster collaborative research.

(3) To assist state and federal agencies and regional organizations in developing integrated assessments of drought severity and impacts, including current climate/drought and water supply assessments.

(4) To foster coordination and cooperation within and between levels of government and with the private sector on drought-related activities.

(5) To serve in an advisory capacity to policy makers and others.

(6) To interact and conduct collaborative research with international scientists and facilitate the timely exchange of information with foreign governments, international and nongovernmental organizations, and regional organizations.

Since 1995, a major portion of the NDMC's activities have been focused in the following areas:

- **Information Clearinghouse**. The NDMC's web site serves as an electronic textbook of information on drought and is linked to more than 200 web sites with additional information on a wide range of topics. More than 900 web sites are hotlinked to the NDMC's web site. The content of the web site is expanded continuously and frequently updated. Content needs are determined through continuous interaction with users through various channels. User sessions and user hits have increased dramatically in the past two years, averaging more than 15,000 user sessions, and 60,000 user hits per month. Approximately 15% of user sessions are from international sources.

- **Research projects**. The NDMC has developed a range of research activities aimed at developing new tools and methodologies that governments and others can use to reduce vulnerability to drought. The projects in progress include: (1) an analysis of state-level drought plans; (2) modification of the 10-step planning methodology; (3) an evaluation of the effectiveness of the Standardized Precipitation Index (SPI) in detecting and tracking drought conditions and other moisture anomalies; (4) the development of the Standardized Vegetation Index to track vegetation anomalies; and (5) the development of a risk assessment methodology to identify areas of vulnerability and potential mitigation actions to reduce risk.

- **Integrated Climate Monitoring/Drought Watch**. The NDMC has been networking with state and federal agencies and universities to develop an integrated climate/drought monitoring system. The emergence of the Internet and the availability of data and information products on the WWW provided the NDMC with the opportunity to create an electronically based drought watch system by linking the informational products developed by the NDMC with products available from other agencies and organizations. The NDMC has also been collaborating on an integrated drought watch system for the western United States under the framework of the Western Drought Coordinating Council (WDCC).

- **Advisor on Policy Issues**. The NDMC regularly serves in an advisory capacity to governments at all levels in the United States and internationally. The NDMC has served on various national and regional task forces and currently is the technical headquarters of the WDCC, formed in 1997 under the auspices of the Western Governors' Association. The NDMC will also serve in an advisory capacity to the National Drought Policy Commission (NDPC).

- **Training Workshops**. The NDMC organizes and conducts training workshops on drought contingency planning. Since 1997, the Center has conducted 4 regional workshops in the U.S. and co-conducted one workshop with Mexico for the U.S./Mexico border states region. The NDMC also conducted workshops in northeast Brazil in 1998 to assist them with regional and state planning efforts. During 1999, the NDMC will conduct workshops in Hawaii and southern Africa and assist with workshops in the western United States.

- **International Drought Mitigation Program**. The NDMC continuously interacts with many scientists in foreign governments and with United Nations agencies on technical issues associated with drought monitoring and planning. In addition, the NDMC is currently participating in projects in Hungary, Mexico, Brazil, and sub-Saharan Africa.

5. NATIONAL DROUGHT POLICY COMMISSION

As a result of the 1996 drought and its effects in the Southwest and southern Great Plains, policy initiatives were developed to improve federal and state drought management efforts. One of the most significant of these policy initiatives was the National Drought Policy Act, introduced in the U.S. Senate in January 1997. Both the Federal Emergency Management Agency (FEMA) and the Western Governors' Association drought task forces recommended the development of a comprehensive, integrated national drought policy to reduce the risks associated with future drought events and improve response to drought emergencies when they occur. The National Drought Policy Act was passed by the U.S. Congress in July 1998 and signed into law.

The major tenet of this bill is the establishment of an advisory commission (National Drought Policy Commission/NDPC) "to provide advice and recommendations on the creation of an integrated, coordinated Federal policy designed to prepare for, mitigate the impacts of, respond to, and recover from serious drought emergencies." The NDPC will function under the leadership of the U.S. Department of Agriculture and will be composed of 16 members, including federal agencies; representatives of states, counties and cities; and six at-large members from groups affected by drought. The NDPC will begin meeting in spring 1999 and conclude its study within 18 months. The commission will submit a thorough study of drought policy needs to the U.S. Congress and the president at that time.

During its existence, the NDPC is required to: (1) determine the needs that exist at all levels to prepare for and respond to drought emergencies; (2) review existing federal, local, state, and tribal laws and programs relating to drought; (3) determine what differences exist between the needs of those affected by drought and the federal laws and programs designed to mitigate the impacts of and respond to drought; (4) collaborate with appropriate entities to consider appropriate regional drought initiatives and the application of such initiatives at the national level; (5) make recommendations on how federal drought laws and programs can be integrated with local, state, and tribal laws and programs into a comprehensive national policy; (6) make recommendations on improving public awareness of the need for drought mitigation and prevention; and (7) recommend whether all federal drought preparation and response programs should be consolidated under one existing federal agency.

6. FUTURE DIRECTIONS AND CHALLENGES

A review of drought management and policy needs for the western United States was recently completed by Wilhite (1997b; 1997c) at the request of the Western Water Policy Review Advisory Commission. In this report, Wilhite reviewed major studies that had evaluated federal and state government's role in drought management and had offered recommendations to improve future management efforts. These studies were reviewed to identify common themes that might help to highlight future needs or actions that the federal government could take to improve drought management in the western United States. The common themes identified included:

- Create a national drought policy and plan
- Develop a comprehensive, integrated national climate monitoring system
- Incorporate drought in the National Mitigation Strategy
- Conduct post-drought audits of federal/state response efforts
- Establish regional drought forums
- Encourage development of state drought mitigation plans

The development of a national drought policy is progressing under the leadership of the NDPC. I am cautiously optimistic that the commission will produce recommendations that will move the nation to a greater emphasis on risk management. There is considerably less optimism that these recommendations will be acted on in a prompt manner by the U.S. Congress.

A regional integrated climate monitoring system is currently operational under the leadership of the WDCC and the NDMC. The goal should be to expand the geographical coverage of this system to include all portions of the United States. This system should also be integrated with systems established by the states and regional climate centers to deliver better and more timely information to decision makers at local, state, and national levels.

The National Mitigation Strategy (FEMA 1995) includes all major natural hazards, except drought. The FEMA-led drought task force recommended that steps be taken to incorporate drought in this strategy. No action has taken place on this recommendation.

Post-drought audits of federal and state drought response and mitigation efforts are not routinely conducted to determine successes and failures. The NDPC should make specific recommendations to address this issue. Results from these studies could significantly improve national and state-level policies and plans.

The WDCC is the only existing regional forum, and it has experienced only moderate success, largely because of budgetary constraints. Regardless of the limited success, it is imperative that approaches to drought management be region specific, and regional forums could facilitate that process. Regional forums could also be useful in tailoring federal response and mitigation programs to better address regional needs.

All states should be encouraged to develop comprehensive drought mitigation plans through the provision of incentives. Methodologies are now available for states to follow in placing emphasis on mitigation actions and programs. Examples from Utah, New Mexico, and Nebraska should further facilitate this process. The lack of methodologies and models has been one of the constraints to the adoption of an approach to drought management that emphasizes mitigation.

7. CONCLUSIONS

Although considerable progress has been made in several key areas in drought management in the United States in the past 15 years, there has been little progress in the development of an integrated national approach that incorporates the needs of local, state, federal, and tribal governments. Although the impacts of drought occur mainly at the local, state, and regional level, it is imperative for the federal government to provide the leadership necessary to improve the way the United States prepares for and responds to drought. The federal role should be one of facilitating the development of a national policy and plan through a participatory process involving all levels of government, regional organizations, the private sector, and other interests. The establishment of the National Drought Policy Commission by the U.S. Congress is an indication that a more coordinated approach within and between levels of government is needed to address this issue. Whether the NDPC will be able to identify needs and assemble the resources necessary to accomplish this objective is open to speculation at this time.

REFERENCES

Boyd, J. (1996) *Southwest Farmers Battle Record Drought*, United Press International.

California Department of Water Resources (1992) *The 1991 Drought Water Bank*, Department of Water Resources, Sacramento, California.

Chenault, E.A. and Parsons, G. (1998) *Drought Worse than 96: Cotton Crop's One of Worst Ever*, http://agnews.tamu.edu/stories/AGEC/Aug1998a.htm.

FEMA (1995) *National Mitigation Strategy, Partnerships for Building Safer Communities*, Washington, DC.

FEMA (1996) *Drought of 1996: Multi-State Drought Task Force Findings*, Federal Emergency Management Agency, Region VI, Denton, Texas.

Karl, T.R., Knight, R.W., Easterling, D.R., and Quayle, R.G. (1995) Trends in U.S. Climate During the Twentieth Century, *Consequences* 1(1): 2-12.

MacDonnell, L.J., Howe, C.W., Miller, K.A., Rice, T.A., and Bates, S.F. (1994) *Water Banks in the West*, Natural Resources Law Center, University of Colorado, Boulder, Colorado.

Riebsame, W.E., Changnon, S.A. Jr, and Karl, T.R. (1991) *Drought and Natural Resources Management in the United States: Impacts and Implications of the 1987-89 Drought*, Westview Press, Boulder, CO. 174 p.

Thurman, J.N. (1998) Oklahoma in Grip of New Dust Bowl, *Christian Science Monitor*, 24 August, http://www.csmonitor.com/durable/1998/08/24/pls3.htm.

U.S. Congress, Office of Technology Assessment (1993) *Preparing for an Uncertain Climate*, Washington, DC.

Western Governors' Association (WGA) (1996) *Drought Response Action Plan*, Denver, Colorado.

Wilhite, D.A., Rosenberg, N.J. and Glantz, M.H. (1986) Improving Federal Response to Drought, *J. of Clim. and Appl. Meteorol.* 25: 332-342.

Wilhite, D.A. (1993) *Drought Assessment, Management, and Planning: Theory and Case Studies*, Kluwer Academic Publishers, Boston, MA.

Wilhite, D.A. and Rhodes, S.R. (1994) State-level Drought Planning in the United States: Factors Influencing Plan Development, *Water International* 19: 15-24.

Wilhite, D.A. and Wood, D.A. (1994) Drought Management in a Changing West: New Directions for Water Policy, Proceedings of a Conference held in May in Portland, Oregon, *IDIC Technical Report Series 94-1*. University of Nebraska, Lincoln.

Wilhite, D.A. (1997a) State Actions to Mitigate Drought: Lessons Learned, *J. Am. Water Res. Assoc.* 33: 961-968.

Wilhite, D.A. (1997b) *Improving Drought Management in the West: The Role of Mitigation and Preparedness*, Report to the Western Water Policy Review Commission, Denver, CO.

Wilhite, D.A. (1997c) Responding to Drought: Common Threads from the Past, Visions for the Future, *J. Am. Water Res. Assoc.* 33: 951-959.

THE ESTIMATION OF DROUGHT RISK IN EUROPE FROM SOIL AND CLIMATIC DATA

R.J.A. JONES, P. ZDRULI and L. MONTANARELLA
European Soil Bureau, Space Applications Institute
EC Joint Research Centre, 21020 Ispra (Va), Italy

Abstract. This paper proposes a methodology for using soil and climatic data to assess the risk of drought impacts in Europe. In the case of agricultural drought, it is the soil water available to plants (SWAP) that is the most important factor in assessing this risk and a simple model for estimating this is described. This model can be linked to spatial and point data from the European Soil Database and a preliminary map of SWAP in Europe has been produced using a pedotransfer rule. The study concludes that more precise modelling of droughtiness, based on interactions of soil available water with the average soil moisture deficit, estimated from meteorological data, is needed, to support policy making today but that currently the necessary soil physical data are lacking to fully implement this approach.

1. INTRODUCTION

A drought is a *decrease in water availability in a particular period over a particular area* (Wilhite, 1993). Research in the early 1980s has revealed more than 150 published definitions of drought (Dracup *et al.*, 1980; Wilhite and Glantz, 1985). A combination of meteorological factors, mainly precipitation and temperature, produces drought conditions. For example, Thomasson (1979) describes drought as a lack of rain, and drought conditions result from a lack of an expected amount of precipitation.

There are different types of drought, namely meteorological, agricultural and hydrological droughts. In addition, the term soil droughtiness has also been applied in an agricultural context. An *agricultural drought*, sometimes referred to as an agro-meteorological or agroclimatic drought, occurs when there is insufficient moisture in the root zone to sustain plant growth. It can be estimated and/or quantified by considering three aspects: climate (or weather), soil water conditions and plant (or crop) requirements.

Water plays an important part in many soil-forming processes and in the subsequent modification and use of soil (Buckman and Brady, 1960: 162-190). It is an essential constituent of the soil and, as a consequence, has been the subject of intensive studies over the past fifty years, especially by soil physicists and hydrologists. The source of water in the soil is one or a combination of meteorological, fluvial, or ground-water.

This paper proposes a methodology for using soil and climatic data to assess the risk of agricultural drought in Europe. Assessing drought risk is now an important objective for

J.V. Vogt and F. Somma (eds.), Drought and Drought Mitigation in Europe, 133–146.

the support of policy making which is increasingly focussing on desertification in southern European countries (Kosmas *et al.*, 1999).

2. ASSESSMENT OF AGRICULTURAL DROUGHT RISK

Zdruli *et al.* (2000) have examined the potential of soil maps for approximately delineating areas at risk from drought. However, in the case of agricultural drought, it is the water available to plants that is the most important soil factor that needs to be quantified, if the areas at risk are to be identified precisely enough for policy making. Soil physicists have combined with agronomists to develop a number of models for estimating the amount of soil available water. Linking the output from these models with spatial soil information (in the form of maps) and climatic (or weather) data is a development (Proctor *et al.* 1989) that has been made possible by the advance in Geographical Information Systems (GIS) and database management (Heineke *et al.*, 1998). The following sections describe a methodology for calculating the soil water available to plants and how this can be used to estimate drought risk.

Soil available water

Thomasson (1995) has comprehensively reviewed a range of models for estimating the soil water available to plants, often called *available water capacity* (AWC) in the literature. He concludes that a simple 'capacity' model is the most appropriate approach given the current availability of data in Europe. Thomasson also noted that the term AWC has been used mostly for the water held between two suctions, in a fixed depth of soil, for example 1m (Salter and Williams 1965), without consideration of the processes or possibilities for extracting this water. Recognising the need for more flexibility, Thomasson (1979) defined a new term the *soil water available to plants* (SWAP) for the total amount of water that can be considered to be extractable by the roots of different crops. SWAP is therefore crop dependent and AWC, the available water in a 1m depth of soil, is just one measure of SWAP.

The calculation of SWAP employs the basic soil water balance concept shown in figure 1, with rainfall (R) as input, potential evapo-transpiration (PE), drainage (D) and run-off (F) as outputs, and locally perhaps lateral or vertical fluxes of groundwater (G). A basic component of the model is the soil profile and this acts as a storage reservoir, supplying plant roots with moisture during periods of meteorological water deficit.

Water is held or adsorbed by soil particles at varying suctions (negative pressures). Several forms of soil water are recognised. The water *unavailable* to plants is held in the mineral and organic structures of the soil at suctions of 1500 kPa or greater. This suction (1500kPa) is regarded as the wilting point for most plants. The processes involved here include absorption and adsorption. The *SWAP* is held at a range of suctions from wilting point (1500kPa) to field capacity.

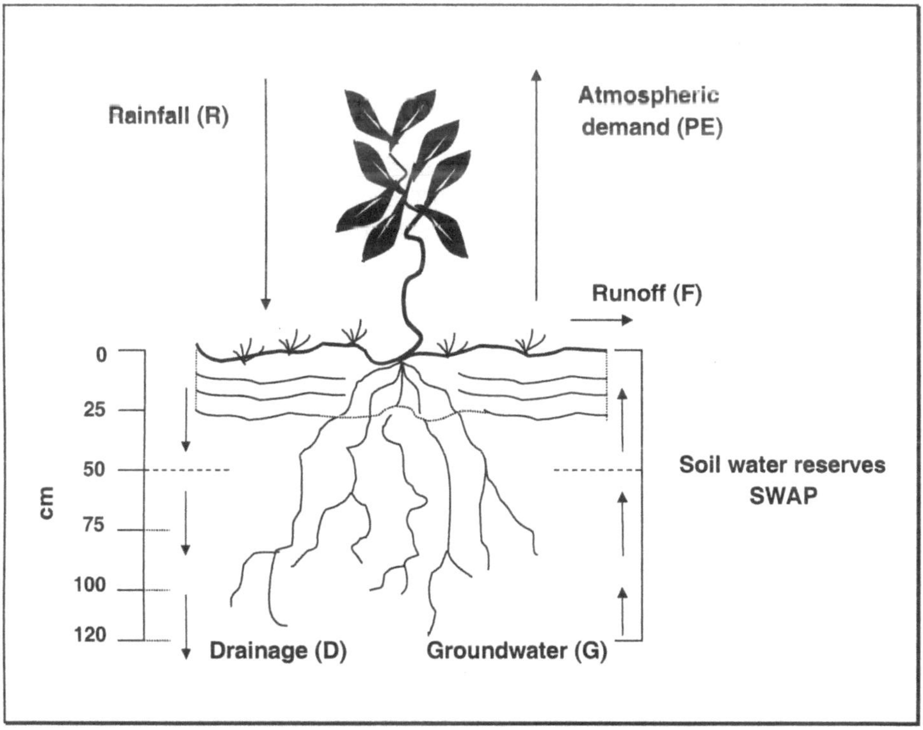

Figure 1: Simple soil water balance model.

Field capacity has been defined as the soil moisture state when, 48 hours after saturation or heavy rain, all downward movement of water has ceased (Veihmeyer and Hendrikson, 1931). Therefore field capacity is obtainable only by direct measurement and generally this is not feasible in a soil survey context. Field capacity must be estimated from measurements of the water retained by soil at low suctions. A variety of suctions (5, 6, 10, or 33 kPa) have been proposed as the lower limit for available water (Hall *et al.*, 1977). From field investigations under British conditions, Webster and Beckett (1972) demonstrated that field capacity can occur between 0.03kPa and 33kPa depending on the texture and structure of the soil.

After analysing soil water retention data from more than 2000 undisturbed soil cores, Hall *et al.* (1977) proposed that 5kPa be used as the most appropriate suction approximating to field capacity in British soils. This is justified from results showing that the difference between the water content at field capacity, measured as proposed by Veihmeyer and Hendrikson (1931), and the water retained at 5kPa is very small except in coarse textured soils with a shallow ground water table and in peaty soils. Salter and Haworth (1961) have established that for British conditions the water content at 1500kPa is a good approximation for wilting point.

We can, therefore, consider the difference between the two limits of 5 and 1500kPa as the *total available water* (TAW). Thomasson (1979) proposed separating easily available water (EAW), water held at suctions less than 200 kPa, from the *restricted available water* (RAW), which is more tightly held at the higher suctions (200–1500 kPa) and therefore difficult to abstract, especially from deeper subsoils. This procedure for partitioning soil available water has been adopted by Hough and Jones (1997) for computing actual moisture deficits in the UK using the Meteorological Office rainfall and evaporation calculation system (MORECS). The approach is shown graphically in figure 2 and expressed in equation (1).

$$\text{TAW} = \text{EAW} + \text{RAW} \qquad (1)$$

The model facilitates the calculation of SWAP in the soil profile, because the available water held in each of the different soil layers or horizons that comprise the rooting zone can be integrated accordingly.

The soil information needed to calculate SWAP can be summarised as:

1. Horizon thickness;
2. Water content at field capacity (5kPa), 200kPa and wilting point (1500kPa);
3. In the absence of water retention measurements, pedotransfer functions, as regression equations or tabulated classes, to calculate water content based on:
   i) particle size composition (PSC);
   ii) PSC and organic matter;
   iii) PSC, bulk density and horizon class (e.g. A, B or C).

In general:

$$\text{SWAP}_{\text{crop}} = \{\sum_{d1}^{0} (\text{TAW}) + \sum_{d2}^{0} (\text{TAW}) + \sum_{d3}^{0} (\text{EAW})\}/100 \qquad (2)$$

Where d_1 is the thickness of the topsoil, d_2 is the thickness of the upper subsoil and d_3 is the thickness of the lower subsoil. The values of d_1, d_2 and d_3 depend on the crop but can be deduced from figure 2. The resulting SWAP value is expressed in *mm* rainfall equivalent.

3. ASSESSMENT OF DROUGHTINESS IN SOILS

The first step in calculating the droughtiness of a particular soil is the estimation of the amount of water available to plants (SWAP) stored in the rooting zone and the extent to which this water is extractable by plant roots. Some water held at suctions below 1500kPa, normally considered available to plants, may not be extractable in practice because there are insufficient plant roots.

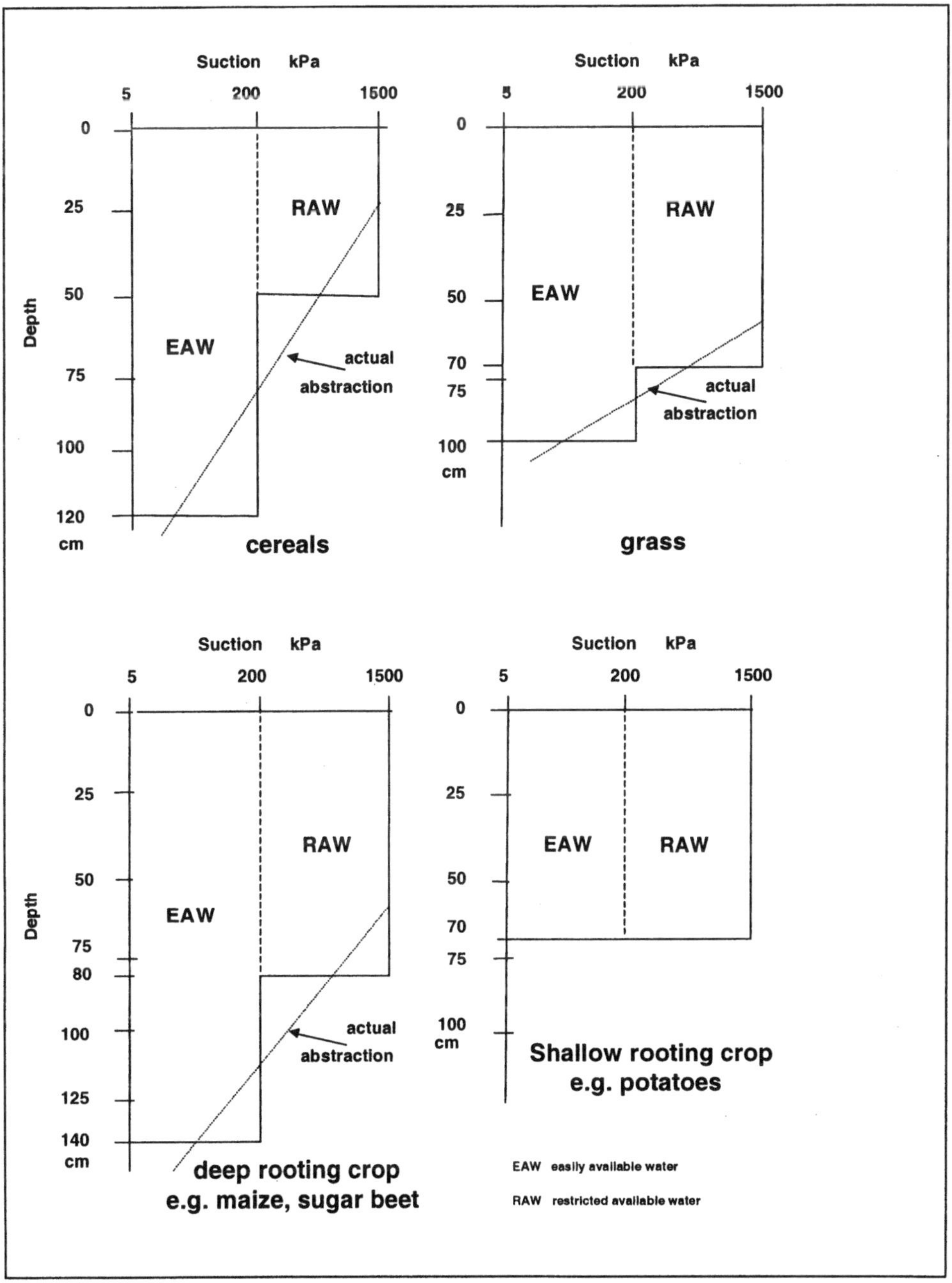

Figure 2: Simplified root abstraction models for temperate arable crops in northern Europe.

Figure 2 is essentially a graphical representation of a series of simplified crop rooting models originally proposed by Thomasson (1979). They are based on idealised rooting depth patterns for arable crops (grass, cereals, sugar beet, maize and potatoes) in northern Europe. The models take into account the fact that a high proportion of plant roots, in some cases as much as 70%, is found in the upper part of the soil profile (0–50cm). These roots are normally able to extract all the water held by soil particles in the range 5–1500kPa. However, plant root densities are often much smaller in the deeper subsoil (e.g. >50cm depth) where, in practice, only the easily available water (5-200kPa) is extractable.

Soil water retention data are often lacking, because the measurements are complicated, time consuming and expensive (Hall *et al.*, 1977), and thus pedotransfer functions (Bouma and Van Lanen, 1987) have to be used to calculate water contents. These functions normally use numerical data on soil texture, packing density (PD), calculated from bulk density and clay content or estimated from soil structure and morphology (Renger, 1971; Hall *et al.*, 1977), and organic carbon content. The option to use packing density is advantageous because, in the absence of direct measurements, it is an easier measure of density to estimate in the field than bulk density (Hodgson 1976, p.39).

At the simplest level, a standardised depth of soil (1m or to an obstruction such as rock) with suction limits between 5 and 1500 kPa can be assumed to apply universally. This has been the traditional approach for calculating the available water capacity (AWC) of soils (Salter and Williams 1965). Zdruli *et al.* (2000) show the distribution of SWAP integrated to 1m depth (equivalent to AWC) for European soils based on a pedotransfer concept that is described below.

The SWAP approach of Thomasson (1995) is more sophisticated than this simple calculation of soil water available to plants, because it incorporates soil depth and/or suction limits that can be adjusted for different patterns of water abstraction by crops (Figure 2). Thomasson (1995) further describes the computation of SWAP in a European context.

Although this approach does involve simplistic assumptions about plant root behaviour, it does have the advantage of defining water reserves unambiguously as a soil property comparable to cation exchange capacity or organic matter. It also provides a common basis for comparing the available water contents of different soils.

However, the information to estimate SWAP at this level of sophistication does not exist on a consistent basis throughout Europe. In particular, there is a lack of adequate data on bulk density and depth of water abstraction for some regional soil and climatic landscapes. Also, there are generally insufficient soil water retention measurements to construct accurate pedotransfer functions for estimating the water content, at the two suction limits of available water (5 and 1500kPa), from particle size, organic matter and density data.

Use of pedotransfer rules

The European Soil Bureau (ESB), based at the Joint Research Centre, Ispra (Italy), has been sponsoring the collection of soil information throughout Europe for the past ten years (Montanarella and Jones, 1999). This has culminated in the compilation of a European Soil Database (Heineke *et al.*, 1998). The database contains spatial data at 1:1,000,000 scale for the whole continent and analytical data for standard soil profiles (Madsen and Jones, 1995, 1998) that can be linked to the spatial soil data. The spatial data are largely derived from the European Soil Map (CEC, 1985; FAO-UNESCO, 1974). The collection of analytical data that can be related to the soil map units promises to provide much needed data sometime in the future but, until these data become more comprehensive and readily available, simplistic methods based on pedotransfer rules - PTRs (simplified pedotransfer functions) - must be adopted.

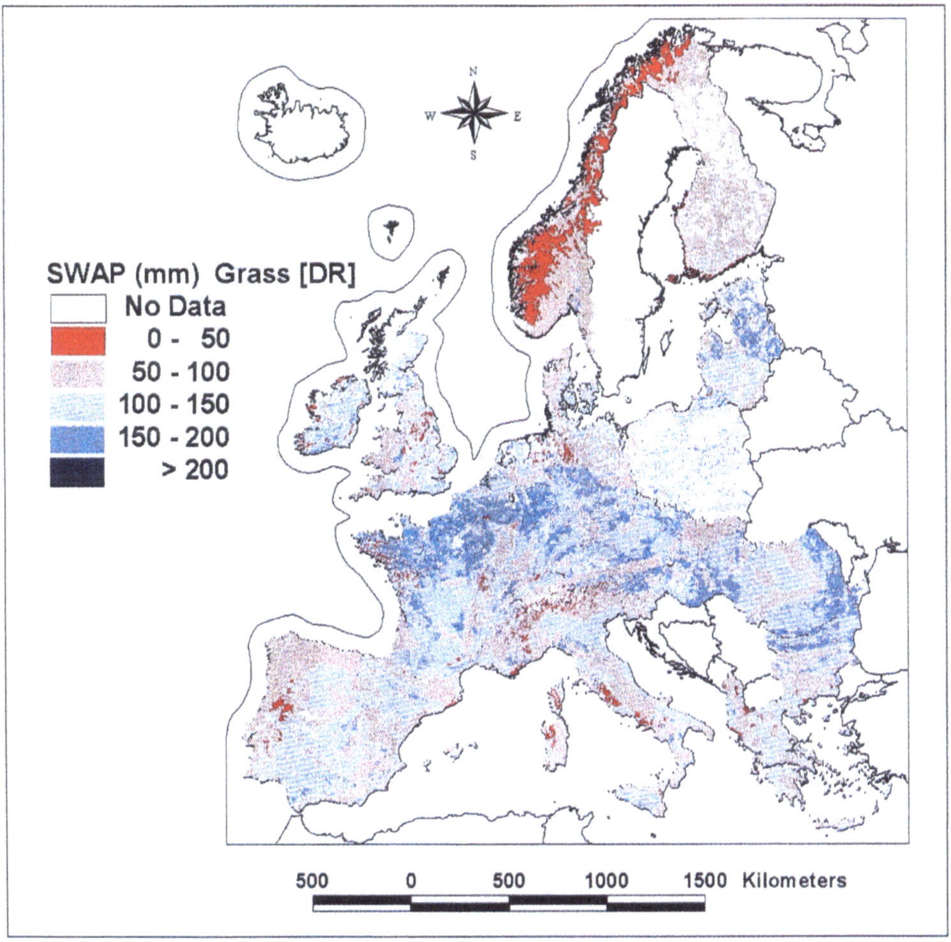

Figure 3: Soil water available to plants, SWAP, in mm, for grass.

Use of PTRs with standardized data is all that can be applied on a continent-wide basis at present (King *et al.* 1995; Van Ranst *et al.*, 1995). The ESB and the European Commission are now planning to extend the European Soil Database to encompass the whole Mediterranean Basin (including the Middle and the Near East, and North Africa) where the problems of drought are particularly acute. At the same time the existing data will be improved. The resulting geo-referenced soil database could also be used in the first instance to delineate areas most at risk for the purposes of management and mitigation of drought. However, for more precise estimation of drought risk, the European Soil Database provides a starting point for delineating SWAP on a European level.

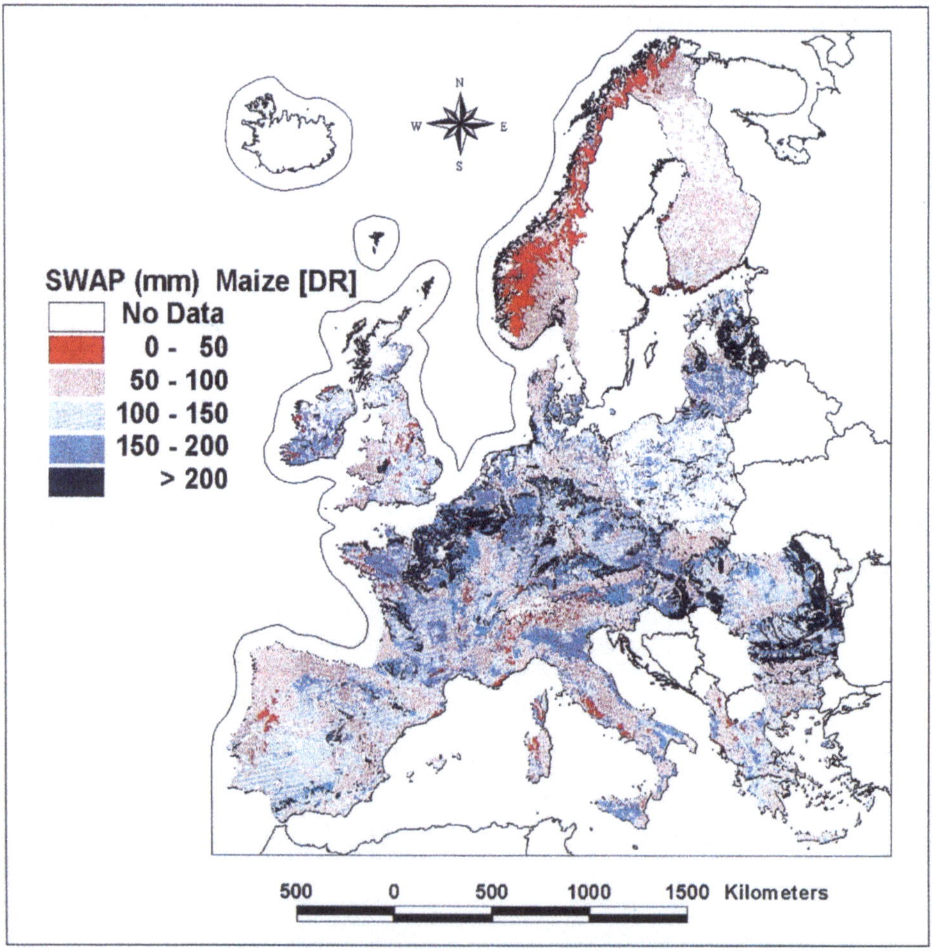

Figure 4: Soil water available to plants, SWAP, in mm, for a deep rooting crop such as maize.

Figures 3 and 4 show the distribution of SWAP classes for grass and a deep rooting crop such as maize, with limits defined in mm. The depth to a rooting restriction or to rock (DR) was used in calculating SWAP. In the absence of sufficient water retention data in Europe, the SWAP values were estimated using pedotransfer rules (Daroussin and King, 1997) that transform the Soil Typological Units (STU) of the European Soil Database into AWC or SWAP classes for topsoil and subsoil. These data were then integrated according to the crop rooting models for grass and maize in figure 2.

Using pedotransfer rules is a much more simplistic way of estimating SWAP than using water retention measurements or estimates of water content based on accurate pedotransfer functions (Tielje and Tapkenhinrichs, 1993). However, it does provide a consistent basis for estimating the soil water available to plants throughout Europe.

The general structure of a pedotransfer rule can be visualised by:

INPUT ATTRIBUTES → PEDOTRANSFER RULE → OUTPUT ATTRIBUTE + CONFIDENCE LEVEL

The values of the input attributes determine the value (class) of the output attribute to which a confidence level is linked (very low, low, medium, high). An inferred attribute, obtained through the application of a rule, can be introduced as an input attribute in another rule. The pedotransfer rule for SWAP (Jones and Hollis, 1996) consists of a table, in which each line is a possible occurrence, meaning a combination of input and output data. For each occurrence the author and the date have been stored. It is important to emphasise here that such a rule is heavily dependent on expert judgement. Part of the rule for SWAP in the topsoil is given as an example in table 1. Other rules define SWAP in the subsoil.

Table 1. Tabular representation of part of the pedotransfer rule for topsoil water available to plants (SWAP_TOP)

INPUTS	OUTPUTS				
Texture	**PD_TOP**	**SWAP_TOP**	**SWAP_TOP_CL**	**Authors**	**Update**
*	*	#	V (Very low)	DK	06/01-94
1	*	M (100-140)	M (Moderate)	DK-AT-MJ	10/01/93
2	*	H (141-190)	M (Moderate)	DK-AT-MJ	10/01/93
3	*	VH (>190)	M (Moderate)	DK-AT-MJ	10/01/93
4	*	H (141-190)	M (Moderate)	DK-AT-MJ	10/01/93
5	*	H (141-190)	M (Moderate)	DK-AT-MJ	10/01/93

Texture → see CEC (1985) PD → Packing Density CL → Confidence Limit
TOP → Topsoil SWAP_TOP → SWAP in the Topsoil, units mm/m

Wild cards (*), implying that a particular input is undifferentiated, are used systematically. In the first line, wildcards are given to all input attributes, taking care of all possible missing data and producing an output for all STUs. Moving down the table, the input attributes are defined progressively more precisely, resulting in an output with a higher confidence level. Daroussin and King (1997) describe implementation of pedotransfer rules European Soil Database.

The term *confidence level* (CL) is used qualitatively, and does not imply a calculated statistic. Definition is therefore difficult. The confidence level seeks to describe the likelihood that the inferred attribute has been correctly predicted for the STU, based on the current level of expert knowledge and information. Clearly where an attribute is part of the definition of the STU the confidence level will normally be high, e.g. depth to a gleyed horizon. However, SWAP data are known to depend on properties partially, or even dominantly, influenced by land use, e.g. organic carbon and bulk density, and the confidence level is therefore reduced.

More details of the SWAP and other pedotransfer rules, that form part of the European Soil Database, are given in Van Ranst *et al.* (1995), Jones and Hollis (1996) and Daroussin and King (1997). King *et al.* (1995) describe an earlier attempt to map the *SWAP* in European soils.

However, it must be emphasised that the maps presented in figures 3 and 4, showing the estimated SWAP for grass and maize, are based on the idealised crop rooting models in figure 2. These were developed mainly for use in temperate conditions that prevail in Europe north of the Alps. Furthermore, there is considerable scope for improving the estimation of SWAP for central and southern Europe where hydrological conditions are different from areas further north. Improving the SWAP estimates throughout Europe could easily be achieved by using better information on soil depth.

4. SOIL-CLIMATE INTERACTIONS FOR ASSESSING DROUGHT RISK

Estimating the risk of agricultural drought should take into account the frequency and amount of effective precipitation, as well as the SWAP, and how it can affect the actual water balance. Examining the interaction between SWAP and the soil moisture deficit (MD), estimated from meteorological data (Jones and Thomasson, 1985), provides a better basis for accurately predicting drought risk than using either SWAP or MD alone.

During periods when potential evapotranspiration (PE) exceeds rainfall (R), i.e. (PE>R), soil dryness can be quantified as the accumulated negative sum of (R-PE). This has been defined by Jones and Thomasson (1985) as the potential soil moisture deficit (PSMD), after the work of Smith (1967). The PSMD can be calculated from basic meteorological data, such as daily, weekly or monthly R and PE, and accumulated over the normal growing season of agricultural crops.

The average risk of drought in an area can then be estimated using a simple 'capacity approach'. The average potential soil moisture deficit (mm), PSMD, can be adjusted for different crops (MD-crop) using generalised data for crop cover and growth period (Thomasson, 1979). The resulting crop adjusted MD values, in mm, are then subtracted from the respective SWAP values (mm). If the result is negative the area is regarded as droughty for that crop, if positive then the area is as non-droughty in average years.

Thomasson (1979) describes the methodology studying soil-climate interactions in detail and provides a table of soil droughtiness classes defined in terms of (SWAP-MD). The data flows for this capacity model are shown in figure 5. The approach is operational in the UK (Hall *et al.*, 1977; Thomasson and Jones, 1989), Denmark (Jensen *et al.*, 1998), and Germany (Renger and Strebel, 1980), where there are adequate soil and climatic data sets (Jones and Thomasson, 1985; Thomasson and Jones, 1989, 1991). For example, Proctor *et al.* (1989) have computed droughtiness for spring-sown barley in England and Wales using the Thomasson methodology and both national and European soil data sets. It is also used by the MORECS system, developed by the UK Meteorological Office, to estimate soil moisture deficits for use in the agriculture and water supply sectors (Hough and Jones, 1997).

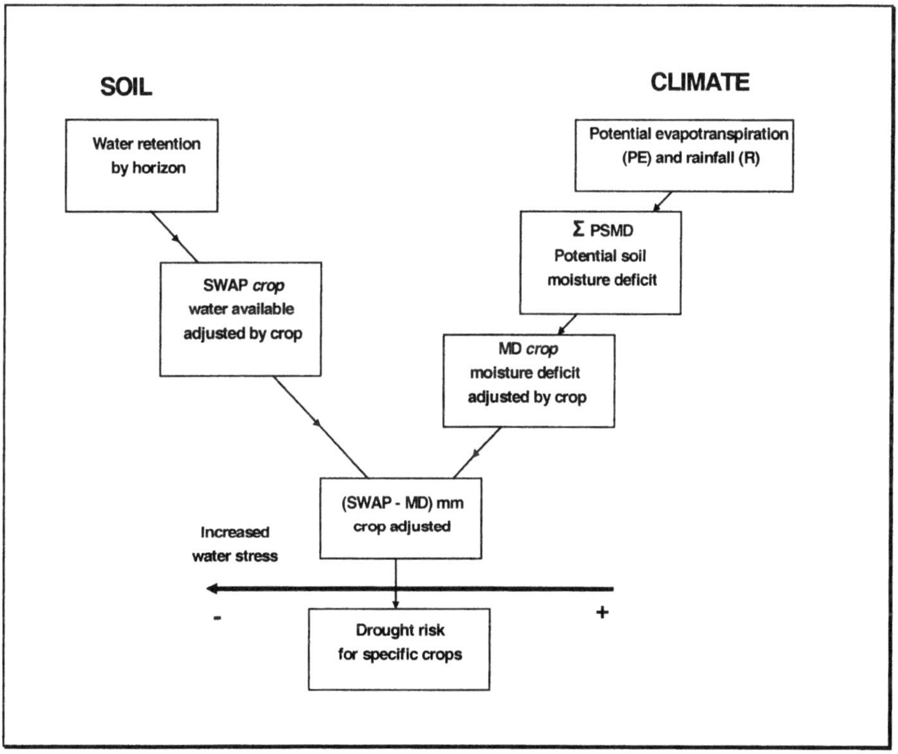

Figure 5: Information flows to assess crop water balance and drought risk.

The next stage at a European level would be to calculate climatic moisture deficits, preferably on a grid basis as has been done for the UK (Ragg *et al.*, 1988). A distribution map of average droughtiness based on SWAP, calculated from the European Soil Database, and average MD data adjusted for specific crops, calculated from the Monitoring Agriculture with Remote Sensing (MARS) agroclimatic database, could then be produced. The current MARS agroclimatic database provides information on a 50×50 km grid.

Ideally, the necessary climatic parameters are needed on a 25×25 km or 20×20 km grid for the whole of Europe.

5. CONCLUSIONS

It is likely during the next few years that techniques to measure or derive data to quantify the complex soil water relations will improve. However, currently there is a serious lack of reliable soil water retention measurements, essential for calculating SWAP for soils in Europe. In the meantime, simple pedotransfer rules (King *et al.,* 1995; Van Ranst *et al.,* 1995; Daroussin and King, 1997) and mathematically based pedotransfer functions (Bruand *et al.*, 1997; Wösten *et al.*, 1998) offer the only alternative for estimating the soil component of drought risk. Crucially there is also a lack of weather and climatic data needed to generate accurate spatial moisture deficit data.

The approach described here could be useful as 'first filter' for land evaluation to identify the ecological limits for a particular crop in a specific area or the degree of risk that could result from adopting a particular cropping system. However, more precise predictions of droughtiness, based on soil-climate interactions quantified according to the models described above, are needed to support policy making today. The relevance of this type of modelling, applied through a soil map at 1:1,000,000 scale, may be questionable. It is more appropriate at scales of 1:50,000 or larger, where real crop performance in specific fields, or where detailed management interventions, are being evaluated. It is clear that the basic data to run such models at scales larger than 1:1,000,000 will be lacking for some parts of Europe for many years to come. In the absence of these data the approach described in this paper offers the best chance of achieving results that can help broad policy making in the immediate future.

ACKNOWLEDGEMENTS

The authors are grateful to several colleagues for their help in the preparation of this paper. In particular we wish to thank Mr. Joel Daroussin (INRA-Orleans, France) for GIS elaboration of the European Soil Database; Professor Arthur Thomasson for commenting on the manuscript; and Mr. Jean Meyer-Roux, Head of Agricultural and Regional Information Systems (ARIS) Unit and Deputy Director of the Space Applications Institute, JRC, for his continued support in all the activities of the European Soil Bureau.

REFERENCES

Bouma, J. and Van Lanen, H.A.J. (1987) Transfer Functions and Threshold Values: from Soil Characteristics to Land Qualities, in K.J. Beek, P.A. Burrough, and D.E. McCormack (eds.), *Quantified Land Evaluation Procedures*, ITC Publication No. 6:106-110

Bruand, A., Duval, O., Wosten, J.H.M., and Lilly, A. (eds.) (1997) *The Use of Pedotransfer in Soil Hydrology Research in Europe,* Workshop Proceedings, INRA Orleans, France, 10-12 October 1996, European Commission: Human Capital and Mobility (DG XII) CHRX-CT94-0639, 211 p.

Buckman, H.O. and Brady, N.C. (1960), *The Nature and Properties of Soils*, Macmillian, New York.

CEC (1985) *Soil Map of the European Communities*, 1:1,000,000. CEC Luxembourg, 124 p, 7 maps.

Daroussin, J. and King, D. (1997), A Pedotransfer Rules Database to Interpret the Soil Geographical Database of Europe for Environmental Purposes, in A. Bruand, O. Duval, H. Wosten, and A. Lilly (eds.), *The Use of Pedotransfer in Soil Hydrology Research in Europe*, Proceedings of the Second Workshop of the Project 'Using Exisitng Soil Data to Derive Hydraulic Parameters for Simulation Modelling in Environmental Studies and in Land Use Planning', EUR 17307 EN, INRA, Orleans, France, 25-40.

Dracup, J.A., Kil Seong, Lee, Paulson, Jr. E.G., (1980) On the Definition of Droughts, *Water Resources Research*, 16(2): 297-302.

FAO-UNESCO (1974) *FAO-UNESCO Soil Map of the World: Vol. 1*, Legend, UNESCO, Paris.

Hall, D.G.M., Reeve, M.J., Thomasson, A.J., and Wright, V.F. (1977) *Water Retention, Porosity and Density of Field Soils*, Soil Survey Technical Monograph No. 9, Harpenden, UK, 75 p.

Heineke, H.J., Eckelmann, W., Thomasson, A.J., Jones, R.J.A., Montanarella, L., and Buckley, B. (eds.) (1998) *Land Information Systems – Developments for Planning the Sustainable Use of Land Resources*, European Soil Bureau Research Report No. 4. EUR 17729 EN, Office of the Official Publications of the European Communities, Luxembourg. 546 p.

Hodgson, J.M. (ed.) (1976) *Soil Survey Field Handbook*, Soil Survey Technical Monograph No. 5, Harpenden, UK, 99 p.

Hough, M.N. and Jones, R.J.A. (1997) The United Kingdom Meteorological Rainfall and Evaporation Calculation System: MORECS Version 2.0 - An Overview, *European Journal of Hydrology*, 1(2): 227-239.

Jensen, N.H., Balstroem, Th., and Madsen, H. Breuning (1998) Root Zone Capacity Maps for Denmark Based on the EU Soil Profile Analytical Database, in H.J. Heineke, W. Eckelmann, A.J. Thomasson, R.J.A. Jones, L. Montanarella, and B. Buckley (eds.), *Land Information Systems: Developments for Planning the Sustainable Use of Land Resources*, EUR 17729 EN, Office for Official Publications of the European Communities, Luxembourg, 421-434.

Jones, R.J.A. and Hollis, J.M. (1996) Pedotransfer Rules for Environmental Interpretations of the EU Soil Database, in C. Le Bas and M. Jamagne (eds.), *Soil Databases to Support Sustainable Development,* EUR 16371 EN, Commission of the European Communities and INRA-Orleans, 125-133.

Jones, R.J.A. and Thomasson, A.J. (1985) *An Agroclimatic Databank for England and Wales*, Soil Survey Technical Monograph No. 16, Harpenden, UK, 45 p.

King, D., Le Bas, C., Daroussin, J., Thomasson, A.J., and Jones, R.J.A. (1995) The EU Map of Soil Water Available for Plants, in D. King, R.J.A. Jones, and A.J. Thomasson (eds.), *European Land Information Systems for Agro-environmental Monitoring,* EUR 16232 EN, Office for Official Publications of the European Communities, Luxembourg, 131-142.

Kosmas, C., Kirkby, M., and Geeson, N. (1999) The Medalus Project: Mediterranean Desertification and Land Use, Project (ENV4 CT 95 0119) Report, Directorate-General Science, Research and Development, Commission of the European Communities, EUR 18882, 87 p.

Montanarella, L. and Jones, R.J.A. (1999) The European Soil Bureau, in P. Bullock, R.J.A. Jones, and L. Montanarella (eds.), *Soil Resources of Europe,* European Soil Bureau Research Report No.6, EUR 18991 EN, Office for Official Publications of the European Communities, Luxembourg, p.3-14.

Madsen, H. Breuning and Jones, R.J.A. (1995) Soil Profile Analytical Database for the European Union, *Danish Journal of Geography*, 95: 49-57.

Madsen, H. Breuning and Jones, R.J.A. (1998) Towards a European Soil Profile Analytical Database, in H.J. Heineke, W. Eckelmann, A.J. Thomasson, R.J.A. Jones, L. Montanarella, and B. Buckley (eds.), *Land Information Systems: Developments for Planning the Sustainable Use of Land Resources,* EUR 17729 EN, Office for Official Publications of the European Communities, Luxembourg, 43-50.

Proctor, M.E., Jones, R.J.A., and Thomasson, A.J. (1989) Interpretations of the Digital EC Soil Map for Agricultural and Environmental Applications in the U.K, in J. Feyen (ed.), *Agriculture: Application of Computerized EC Soil Map and Climate Data*, EUR 12039 EN, Office for Official Publications of the

European Communities, Luxembourg, 107-118.

Ragg, J.M., Jones, R.J.A., and Proctor, M.E. (1988) The Refinement and Representation of Spatial Data in an Information System Using Statistical and DBMS Procedures, and Trend Surface Analysis, *Geol. Jb.* A104: 295-308, Hannover.

Renger, M. (1971) Die Ermittlung der Porengrössenverteilung aus der Körnung, dem Gehalt an organischer Substanz und der Lagerungsdichte, *Z. Pfl Ernähr. Bodenk* 130: 53-67.

Renger, M. and Strebel, O. (1980) Beregnungsbedarf landwirtschaftlicher Kulturen, in *Abhängigkeit vom Bodenwasser und Boden* 32: 572-576.

Salter, P.J. and Haworth, F. (1961) The Available-Water Capacity of a Sandy Loam Soil. II. The Effects of Farmyard Manure and Different Primary Cultivations, *J. Soil Sci.* 12: 335-342.

Salter, P.J. and Williams, J.B. (1965) The Influence of Texture on the Moisture Characteristics of Soils. II. Available-Water Capacity and Moisture Release Characteristics, *J. Soil Sci.* 16: 310-317.

Smith, L.P. (1967) Potential Transpiration, *Tech. Bull. Minist. Agric. Fish. Fd*, Lond. No. 16.

Tietje, O. and Tapkenhinrichs, M. (1993) Evaluation of Pedo-Transfer Functions, *Soil Sci. Soc. Am. J.* 57(4): 1088-1095.

Thomasson, A.J. (1979) Assessment of Average Soil Droughtiness, in M.G. Jarvis and D. Mackney (eds.), *Soil Survey Applications, Soil Survey Technical Monograph No. 13*, 43-50, Harpenden, UK.

Thomasson A.J. (1995) Assessment of Soil Water Reserves Available for Plants (SWAP): a Review, in D. King, R.J.A. Jones, and A.J. Thomasson (eds.), *European Land Information Systems for Agro-environmental Monitoring,* EUR 16232 EN, Office for Official Publications of the European Communities, Luxembourg, 115-130.

Thomasson, A.J. and Jones, R.J.A. (1989) Land Evaluation at Regional Scale, in J. Bouma and A.K. Bregt (eds.), *Land Qualities in Space and Time*, Pudoc, Wageningen, The Netherlands, 231-240.

Thomasson, A.J. and Jones, R.J.A. (1991) An Empirical Approach to Crop Modelling and the Assessment of Land Productivity, *Agricultural Systems* 34: 351-367.

Van Ranst, E., Thomasson, A.J., Daroussin, J., Hollis, J.M., Jones, R.J.A., Jamagne, M., King, D., and Vanmechelen, L. (1995) Elaboration of an Extended Knowledge Database to Interpret the 1:1,000,000 EU Soil Map for Environmental Purposes, in D. King, R.J.A. Jones and A.J. Thomasson (eds.), *European Land Information Systems for Agro-environmental Monitoring*, EUR 16232 EN, Office for Official Publications of the European Communities, Luxembourg, 71-84.

Veihmeyer, F.J. and Hendrickson, A.H. (1931) The Moisture Equivalent as a Measure of the Field Capacity of Soils, *Soil Sci.* 32: 181-193.

Vossen, P. and Meyer-Roux, J. (1995) Crop Monitoring and Yield Forecasting Activities of the MARS Project, in D. King, R.J.A. Jones and A.J. Thomasson (eds.), *European Land Information Systems for Agro-environmental Monitoring,* EUR 16232 EN, Office for Official Publications of the European Communities, Luxembourg, 11-29.

Webster, R. and Beckett, P.H.T. (1972) Matric Suctions to Which Soils in South Central England Drain, *J. Agric. Sci. Camb.* 78, 379-387.

Wilhite, D.A. (ed.) (1993) *Drought Assessment, Management and Planning: Theory and Case Studies*, Kluwer Academic Publishers.

Wilhite, D. A. and Glantz, M.H., (1985) Understanding the Drought Phenomenon: The Role of Definitions, *Water International* 10: 111-20.

Wösten, J.H.M., Lilly, A, Nemes, A. and Le Bas, C. (1998) *Using Existing Soil Data to Derive Hydraulic Parameters for Simulation Models in Environmental Studies and Land Use Planning*, SC-DLO Report 156, Wageningen, The Netherlands, 106 p.

Zdruli, P, Jones, R.J.A. and Montanarella, L. (2000) Assessing the Risk of Agricultural Drought in Europe Using Soil and Climate Data, *Agronomie* (in press).

Part 4
Drought Monitoring

DROUGHT MONITORING AND ASSESSMENT: STATUS AND TRENDS IN THE UNITED STATES

D.A. WILHITE, M.J. HAYES and M.D. SVOBODA

National Drought Mitigation Center
International Drought Information Center
Lincoln, Nebraska 68583-0749, USA

Abstract. Drought is a normal, recurring feature of climate. The characteristics of this hazard, along with its far-reaching impacts, make its effects on society, economy, and environment difficult to identify and quantify. A national drought watch system has been recommended as a necessary component of drought assessment and management. The National Drought Mitigation Center, in collaboration with others, has undertaken a number of drought watch activities for the past few years, using a variety of resources available over the Internet. These resources include tools such as the Standardized Precipitation Index, a new drought index that is more reliable for detecting emerging drought conditions.

1. INTRODUCTION

The economic, social, and environmental costs and losses associated with drought in the United States are significant. In 1995, the U.S. Federal Emergency Management Agency (FEMA) estimated annual losses attributable to drought at US $6-8 billion. A critical component of planning for drought is the provision of timely and reliable seasonal forecasts and informational products that help decision makers at all levels reduce the impacts of drought events. A comprehensive, integrated national climate monitoring or drought watch system has been recommended for some time (Wilhite *et al.*, 1986; Riebsame *et al.*, 1991; Wilhite and Wood, 1994), but no action on these recommendations has taken place. The Internet and the wide range of data and information that is readily accessible to users have made the development of an integrated climate monitoring system a more executable task.

The purpose of this paper is to document the current status of drought monitoring and assessment in the United States. This paper begins with a brief overview of some of the concepts of drought and how it differs from other natural hazards. Understanding the unique characteristics of drought is crucial to establishing an effective and comprehensive monitoring and early warning system. The paper will then discuss more specifically the drought watch activities of the National Drought Mitigation Center (NDMC) and its collaborative activities with the Monitoring, Assessment, and Prediction (MAP) Working Group of the Western Drought Coordination Council (WDCC). The WDCC was formed in 1997 in response to the severe drought that affected many western states in 1996. The WDCC was developed as a result of a Memorandum of Understanding between the Western

J.V. Vogt and F. Somma (eds.), Drought and Drought Mitigation in Europe, 149–160.

Governors' Association and key federal agencies. The value of the newly developed Standardized Precipitation Index (SPI) in monitoring the onset, development, and severity of drought conditions will also be included in this discussion.

2. DROUGHT: THE CONCEPT

Drought is a normal, recurring feature of climate; it occurs in virtually all climatic regimes. It occurs in high as well as low rainfall areas. It is a temporary aberration, in contrast to aridity, which is a permanent feature of the climate and is restricted to low rainfall areas. Drought is the consequence of a natural reduction in the amount of precipitation received over an extended period of time, usually a season or more in length, although other climatic factors (such as high temperatures, high winds, and low relative humidity) are often associated with it in many regions of the world and can significantly aggravate the severity of the event. Drought is also related to the timing (i.e., principal season of occurrence, delays in the start of the rainy season, occurrence of rains in relation to principal crop growth stages) and the effectiveness (i.e., rainfall intensity, number of rainfall events) of the rains. Thus, each drought year is unique in its climatic characteristics and impacts.

Drought differs from other natural hazards in several ways. First, since the effects of drought often accumulate slowly over a considerable period of time and may linger for years after the termination of the event, the onset and end of drought is difficult to determine. Because of this, drought is often referred to as a creeping phenomenon (Tannehill, 1947). Although Tannehill made this statement more than fifty years ago, climatologists continue to struggle with recognizing the onset of drought and scientists and policy makers continue to debate the basis (i.e., criteria) for declaring an end to a drought.

Second, the absence of a precise and universally accepted definition of drought adds to the confusion about whether or not a drought exists and, if it does, its degree of severity. Realistically, definitions of drought must be region and application (or impact) specific. This is one explanation for the scores of definitions that have been developed. Wilhite and Glantz (1985) analyzed more than 150 definitions in their classification study, and many more exist. Although the definitions are numerous, many do not adequately define drought in meaningful terms for scientists and policy makers. The thresholds for declaring drought are arbitrary in most cases (i.e., they are not linked to specific impacts in key economic sectors). These types of problems are the result of a misunderstanding of the concept by those formulating definitions and the lack of consideration given to how other scientists or disciplines will eventually need to apply the definition in actual drought situations (e.g., assessments of impact in multiple economic sectors, drought declarations or revocations for eligibility to relief programs).

Third, drought impacts are nonstructural in contrast to floods, hurricanes, and most other natural hazards. Its impacts are spread over a larger geographical area than are damages that result from other natural hazards. For these reasons, the quantification of impacts and the provision of disaster relief are far more difficult tasks for drought than they are for other natural hazards. Emergency managers, for example, are more accustomed to dealing with impacts that are structural and localized, responding to these events by restoring communication and transportation channels, providing emergency medical supplies, ensuring safe drinking water, and so forth.

These characteristics of drought have hindered the development of accurate, reliable, and timely estimates of severity and impacts and, ultimately, the formulation of drought contingency plans by most governments. Drought severity is dependent not only on the duration, intensity, and spatial extent of a specific drought episode, but also on the demands made by human activities and vegetation on a region's water supplies. The characteristics of drought, along with its far-reaching impacts, make its effects on society, economy, and environment difficult to identify and quantify. This continues to be a formidable challenge to those scientists involved in operational climate assessments.

Many people consider drought to be largely a natural event. In reality, the risk associated with drought for any region is a product of both the region's exposure to the event (i.e., probability of occurrence at various severity levels) and the vulnerability of society to the event. The natural event (i.e., meteorological drought) is a result of the occurrence of persistent large-scale disruptions or anomalies in the global circulation pattern of the atmosphere. Exposure to drought varies spatially and there is little, if anything, that we can do to alter drought occurrence. Vulnerability, on the other hand, is determined by social factors such as population, demographic characteristics, technology, policy, and social behavior. These factors change over time, and thus vulnerability is likely to increase or decrease in response to these changes. Subsequent droughts in the same region will have different effects, even if they are identical in intensity, duration, and spatial characteristics, because societal characteristics will have changed. However, much can be done to lessen societal vulnerability to drought.

3. DROUGHT CHARACTERISTICS AND SEVERITY

Droughts differ from one another in three essential characteristics: intensity, duration, and spatial coverage. Intensity refers to the degree of the precipitation shortfall and/or the severity of impacts associated with the shortfall. It is generally measured by the departure of some climatic index from normal and is closely linked to duration in the determination of impact. Many indices of drought are in widespread use today, such as the decile approach (Gibbs and Maher, 1967; Lee 1979; Coughlan, 1987) used in Australia, the Palmer Drought Severity Index and Crop Moisture Index (Palmer, 1965 and 1968; Alley, 1984) in the United States, and the Yield Moisture Index (Jose *et al.*, 1991) in the Philippines and elsewhere. A relatively new index that is gaining popularity in the United States and worldwide is the Standardized Precipitation Index (SPI), developed by McKee *et al.* (1993 and 1995). The SPI will be discussed in greater detail in later sections of this paper.

Another distinguishing feature of drought is its duration. Droughts usually require a minimum of two to three months to become established but then can continue for months or years. The magnitude of drought impacts is closely related to the timing of the onset of the precipitation shortage, its intensity, and the duration of the event.

Droughts also differ in terms of their spatial characteristics. The areas affected by severe drought evolve gradually, and regions of maximum intensity shift from season to season. During the drought of 1934 in the United States, the area affected was approximately 65% of the country (see Figure 1). By contrast, drought affected more than 95% of the Great Plains region in 1934. In the United States, it is unusual for drought not to exist in a portion of the country each year. Figure 1 illustrates that in the United States, the percent area affected by drought is commonly greater than 10%. A recent analysis of drought occurrence by the U.S. National Drought Mitigation Center (NDMC) for the 48 contiguous states in the United States demonstrated that severe and extreme drought affected more than 25% of the country in one out of four years.

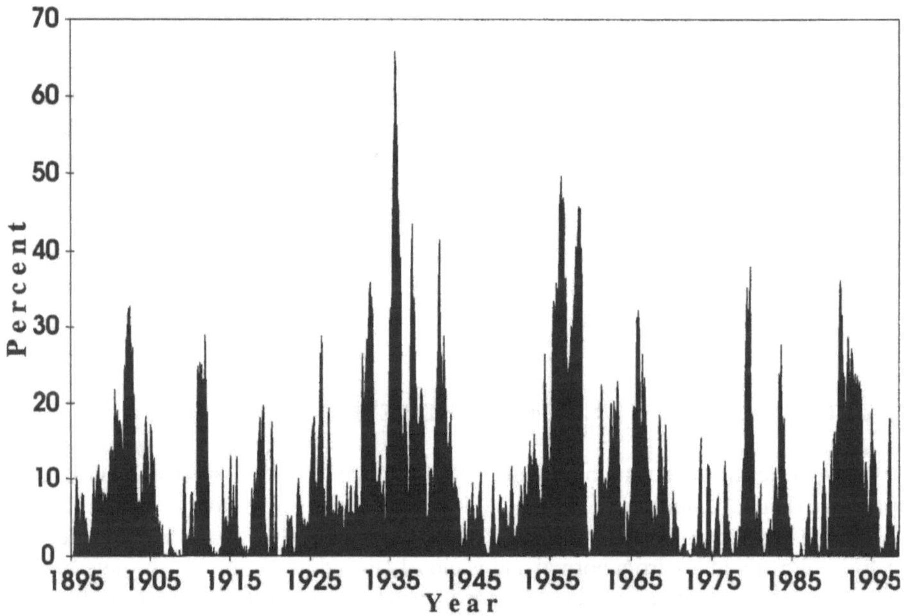

Figure 1: Percentage area of U.S. in severe and extreme drought, 1895 – 1995.

4. INTEGRATED CLIMATE MONITORING/NATIONAL DROUGHT WATCH SYSTEM

The need for a national drought watch system has been acknowledged for some time in the literature as part of a more comprehensive approach to drought assessment and management. Following the creation of the NDMC, one of our first goals was to create a "one-stop shopping" section of our web site that would provide users with access to all of the information necessary to develop a timely and reliable climatic assessment for their state or region. The development of the "drought watch" section was undertaken because no routine national or regional assessment was available from federal agencies. However, all of the components necessary to assess current climate conditions and the long-range outlook were becoming readily available on the World Wide Web (WWW). The goal of the Drought Watch Section of the NDMC's web site is to provide first-hand assessments using the SPI and also to link with other sites that provide information on snow pack, soil/crop moisture conditions, ground water and reservoir levels, stream flow, fire danger, and forecasts. Users can assemble the necessary data and informational products to assess current climate conditions and longer-range climate and water supply outlooks. This has been a very effective tool and is the most often visited section of the NDMC web site.

In 1997, the formation of the MAP working group of the WDCC led to the development of a quarterly climate assessment report for the western United States. The purpose of this report was to provide an overview of current climate conditions and identify potential areas where drought conditions were likely to emerge in the months ahead. The MAP working group membership consists of a cross-section of state, regional and national interests from many different agencies. Some of the organizations involved in the WDCC MAP group represent the NDMC; Western Regional Climate Center; Colorado State Climatologist; Arizona Department of Water Resources; U.S. Department of Agriculture's (USDA) Agricultural Research Service and Natural Resources Conservation Service; National Oceanic and Atmospheric Administration's (NOAA) National Climatic Data Center, Climate Prediction Center, and National Weather Service; U.S. Geological Survey; and Joint Agricultural Weather Facility of USDA and NOAA.

The goal of the MAP working group is to develop an integrated drought monitoring system that provides timely recognition of the occurrence of drought to local, state, tribal, and federal officials responsible for implementing drought response and mitigation measures. The system also tries to track and report physical impacts during droughts, which helps others assess socio-economic impacts on various water users and economic sectors. The main product developed by the MAP group is a quarterly report called *Western Climate and Water Status*. During times of drought the group produces reports on a more frequent basis. An explanation of the report follows a discussion of how the report is generated.

Today's technology allows us to tap data resources that once took weeks or months to obtain. The WWW is the backbone of the MAP group's information and dissemination system. Many hard-copy reports in the climate arena are out of date by the time they reach a user's desk. In reality, reports put on the WWW are also out of date by the time they are put up on the web, but there is much less of a time lag. No other delivery tool lends itself to an assessment report like the Internet does. Users can be notified as soon as this report is available and they can download the product as needed.

The MAP group has undertaken the task of identifying and organizing literally hundreds of links dealing with climate and, more specifically, drought. This is an ever-evolving list as new sites are constantly being added. These sites give us timely access to relevant data and information products. It is also imperative, however, that these products are not used before analyzing the products and the source of data on which these products are based. This can be a rather time consuming task.

This process does allow for near real-time gathering of data and products across all agencies and at all levels of detail on a nationwide basis. To provide access to more timely monitoring information, a separate set of coordinating activities has been initiated by MAP to improve the flow and timeliness of climate and water condition information and to synthesize this information into products in an automated manner on a daily or weekly basis. Most of the product links are housed in the NDMC's *Drought Watch* section (*http://enso.unl.edu/ndmc/watch*) or at the Western Regional Climate Center's web site (*http://www.wrcc.sage.dri.edu/monitor/wdccmon.html*).

Each quarter, a review of the products is done by the members of the MAP group. The most relevant products are gleaned from the various sites and are incorporated as either text or graphic information. This approach is subjective, but usually a consensus is reached that takes into account where the trouble spots are or may emerge in the future. One person takes the lead in drafting the report and all members of the MAP group have an opportunity to edit the report. The NDMC's communication specialist also edits the report for clarity before distribution. Final content edits and reviews are then incorporated into two formats: 1) a full-length, comprehensive web-embedded .html document with hyper-links to all relevant information located on the WDCC home page (*http://enso.unl.edu/wdcc/quarterly/index.html*) and (2) a 2-page color briefing version created in a .pdf format, which is available on the web and distributed to more than 400 people, including all members of Congress and the governors of the 18 western states and other relevant personnel.

The full report consists of an **Executive Summary,** a **Climate Recap** for the region (by month), **Water Supply Conditions** (snow pack, current and projected stream flow conditions, and reservoir levels), **Impacts** occurring or expected, **Special Conditions** when relevant (El Niño or La Niña events, for example), and **Future Forecasts.** The **Briefing Version** contains a condensed **Executive Summary,** areas experiencing **Dry** and/or **Wet** conditions, and an **Outlook** discussion. This product also contains two to four relevant maps/products depicting current conditions and/or expected conditions in the region.

This process has also highlighted a number of shortcomings and some unevenness in the level of accessibility to information; these need to be addressed. We see these from state to state, agency to agency, sector to sector, element to element, and so forth.

To help in policy or decision makers' resource allocation decisions, it would help to have timely and systematic reports of social, economic and environmental impacts, in addition to the available climate data. Currently, the NDMC produces the closest product there is to a detailed national impacts report. *Drought in the U.S.* is produced monthly and located on the NDMC's web site at *http://enso.unl.edu/ndmc/impacts/us/usimpact.htm.*

The MAP working group has been successful in bringing technical experts together across federal and state agencies to produce a quality text/graphic-based product that is timely and useful to a wide range of decision makers at all levels of government. The status of the WDCC (i.e., co-chaired by the Secretary of Agriculture and the Governor of New Mexico with representation from key federal agencies and functioning under the authority of the governors of western states) gave weight to the effort and encouraged cooperation from appropriate government entities in order to meet the MAP working group's tight time line.

The formation of the National Drought Policy Commission (NDPC) provides the opportunity to extend the geographical coverage of this report to include the entire United States (background on the NDPC is included in Wilhite, *Drought Preparedness in the United States: Recent Progress*, this volume). The severe drought that developed during the spring and summer of 1998 in the Southeast and the drought conditions that evolved in the Northeast during the fall and winter months of 1998-99 further emphasized the importance of routine national climate assessments. The NDMC will encourage the NDPC to implement a national integrated climate monitoring system that produces routine national climatic assessments.

5. CLIMATE/DROUGHT INDICES

Climatic indices are important elements of any monitoring and assessment system because their purpose is to simplify complex interrelationships between many climate and climate-related parameters. Indices make it easier to communicate information about climate anomalies to diverse user audiences and allow scientists to quantitatively assess climate anomalies in terms of their intensity, duration, and spatial extent. This allows for the analysis of the historical occurrence of drought and its probability of recurrence. This information is extremely useful for planning and design applications in agriculture and numerous other sectors.

The most widely used drought index in the United States was developed by Palmer (1965). Palmer recognized the need for a better monitoring tool that could identify drought in terms of its intensity, duration, and spatial extent. Palmer's index became known as the Palmer Drought Severity Index (PDSI). Although there are several variations of the index, each variation has characteristics similar to the PDSI (Palmer, 1965; Karl and Knight, 1985; Heddinghaus and Sabol, 1991). Palmer based the PDSI on anomalies in the supply and demand concept of the water balance equation. Inputs into weekly or monthly calculations include precipitation, temperature, and the local antecedent soil moisture conditions. The data are standardized to account for regional differences so that the PDSI values can be compared from one location to another. Therefore, identical PDSI values, in theory, in the Midwest and Texas indicate the same severity of drought, even though actual rainfall deficiencies would be different at the two locations.

Weekly maps of a modified PDSI (Heddinghaus and Sabol, 1991) are produced by the Climate Prediction Center (CPC)/NOAA and are frequently used in assessment of agricultural conditions around the United States. The PDSI is also used in USDA policy decisions on requests for drought relief and by states as triggers for response actions as part of state drought plans. However, the PDSI has limitations that diminish its application and bring into question the practice of basing policy decisions solely on the PDSI. These limitations have been well documented (Alley, 1984; Karl and Knight, 1985; Smith *et al.*, 1993; Willeke *et al.*, 1994; Kogan, 1995; McKee *et al.*, 1993; Guttman, 1998). The most significant limitations of the PDSI for monitoring drought include: (1) an inherent time scale that makes the PDSI better suited for monitoring agriculturally related impacts rather than longer-term hydrological impacts; (2) the tendency to treat all precipitation as rain so that snowfall, snow cover, and frozen ground are not accounted for, making real-time winter PDSI values of questionable reliability; (3) the tendency to ignore the natural lag between precipitation and runoff and the belief that no runoff occurs until the water capacities of the surface and sub-surface soil layers are full, leading to an underestimation of runoff; and (4) a wide variance in the "extreme" and "severe" classifications of the PDSI values, depending on location in the country. If a drought index is going to be spatially comparable and useful for policy decisions, extreme and severe classifications must occur consistently and with low frequency (Guttman *et al.*, 1992). An additional concern is that the PDSI does not do well in the mountainous west, especially since a majority of that region's precipitation falls during the winter as snowfall. The PDSI also can respond slowly to developing drought conditions and, once a region is in drought, the PDSI can retain values reflecting drought well after a climatological recovery from drought has occurred (McKee, 1996). All of these limitations reveal the importance of caution when using the PDSI for drought monitoring and policy decisions. The Standardized Precipitation Index (SPI) was developed to address some of the problems inherent in the PDSI.

McKee and his colleagues at Colorado State University (McKee *et al.*, 1993 and 1995) designed the SPI to be a relatively simple, year-round index applicable to the water supply conditions important to Colorado and as a supplement to information provided by the PDSI and a second drought index, the Surface Water Supply Index (Shafer and Dezman, 1982). The SPI is based on precipitation alone. Its fundamental strength is that it can be calculated for a variety of time scales. This versatility allows the use of the SPI to monitor short-term water supplies, such as soil moisture important for agricultural production, and longer-term water resources such as ground water supplies, streamflow, and lake and reservoir levels. The ability to examine different time scales also allows droughts to be readily identified and monitored for the duration of the drought. Colorado has now used the SPI information as part of the climatic assessment done routinely by their Water Availability Task Force since 1994, including 1996, when drought affected portions of the state.

Calculation of the SPI for the specified time period for any location requires a long-term monthly precipitation data base with 30 years or more of data. The probability distribution function is determined from the long-term record by fitting a function to the data. The cumulative distribution is then transformed using equal probability to a normal distribution with a mean of zero and standard deviation of one so the values of the SPI are really in standard deviations (Edwards and McKee, 1997). A particular precipitation total for a specified time period is then identified with a particular SPI value consistent with probability. Positive SPI values indicate greater than median precipitation, while negative values indicate less than median precipitation. The magnitude of departure from zero represents a probability of occurrence so that decisions can be made based on this SPI value. Because SPI values fit a typical normal distribution, one can expect these values to be within one standard deviation approximately 68% of the time, within two standard deviations 95% of the time, and within three standard deviations 99% of the time. A related interpretation would be that a SPI value of less than -1.0 occurs 16 times in one hundred years and a SPI of less than -2.0 occurs 2-3 times in one hundred years.

McKee *et al.* (1993 and 1995) originally used an incomplete gamma distribution to calculate the SPI. Efforts are now in progress to standardize the SPI computing procedure so that common temporal and spatial comparisons can be made by SPI users (Guttman, 1998).

The SPI has a number of advantages over the PDSI. First, it is a simple index and is based only on precipitation. The PDSI calculations are complex (i.e., 68 terms are actually defined as part of the calculation procedure (Soulé, 1992). In spite of the complexity of the PDSI, McKee (1996) believes that the main driving force behind the PDSI is precipitation. Second, the SPI is versatile. It can be calculated on any time scale, which gives the SPI the capability to monitor conditions important for both agricultural and hydrological applications. This versatility is also critical for monitoring the temporal dynamics of a drought, including its onset and end, which have typically been a difficult task for other indices. Third, because of the normal distribution of SPI values, the frequencies of extreme and severe drought classifications for any location and any time scale are consistent. McKee *et al.* (1993) suggest an SPI classification scale (Table 1). An extreme drought according to this scale (SPI ≤ -2.0) occurs approximately 2-3 times in 100 years, an acceptable frequency for water planning. Fourth, because it is based only on precipitation and not on estimated soil moisture conditions, as is the PDSI, the SPI is just as effective during the winter months.

Table 1: SPI Values

SPI Values	
2.0 and above	**extremely wet**
1.50 to 1.99	**very wet**
1.00 to 1.49	**moderately wet**
-0.99 to 0.99	**near normal**
-1.00 to -1.49	**moderately dry**
-1.50 to -1.99	**severely dry**
-2.00 and less	**extremely dry**

Although developed for use in Colorado, the SPI can be applied to any location with a data set of 30 years or longer. Montana, Wyoming, New Mexico, South Carolina, and Nebraska have investigated or are using the SPI as part of their statewide efforts to monitor drought. The SPI is also used routinely at the national level by the NDMC and federal agencies. Researchers in Mexico, Costa Rica, Argentina, Brazil, Chile, Turkey, Hungary, South

Africa, and Kenya have either considered or are using the SPI for projects in their respective countries.

The NDMC has created national maps of the SPI by climatic division for the continental United States since February 1996. The precipitation data set comes from CPC and the SPI values are calculated by the Western Regional Climate Center (WRCC). These are the same data used to create near real-time PDSI values used in weekly maps created by the CPC. The NDMC makes its maps available over the WWW, with links to other federal agencies. NDMC maps of the SPI are produced at 1-, 3-, 6-, and 12-month time scales, as well as for the year to date. These maps are archived on the NDMC's web site.

The NDMC's experience has been that the SPI more quickly detects emerging drought conditions than the PDSI, a characteristic that is extremely critical in the timely implementation of mitigation and response actions by individuals and government. A growing number of states are now using the SPI to trigger various drought-related actions. Additional research needs to be done to determine the relationships between SPI values and impacts in specific sectors.

6. SUMMARY

Drought is an insidious natural hazard that is a normal part of the climate of virtually all regions. It should not be viewed as merely a physical phenomenon. Rather, drought is the result of an interplay between a natural event and the demand placed on water supply by human-use systems. Drought should be considered relative to some long-term average condition of balance between precipitation and evapotranspiration.

The three characteristics that differentiate one drought from another are intensity, duration, and spatial extent. Intensity refers to the degree of precipitation shortfall and/or the severity of impacts associated with the departure. Intensity is closely linked to the duration of the event. Droughts normally take two to three months to become established but may then persist for months or years, although the intensity and spatial character of the event will change from month to month or season to season. Droughts, unlike other natural hazards, can affect large areas and complicate assessment and response efforts.

Because of the slow-onset nature of drought, monitoring and early warning is an essential part of a comprehensive drought management plan. The development of an integrated climate/drought monitoring system in the United States has been recommended for some time. Recently, the development of the Internet and the WWW has provided the technology to improve information sharing on those critical elements of the hydrologic cycle required to monitor drought and water supply. The NDMC has taken advantage of this technology by developing a drought watch section on their web site, with links to other agencies and organizations. The WDCC has provided the necessary support and motivation to develop a comprehensive regional drought watch system for the western states. This system can serve as a model for a national system in the United States that could be transferable to other regions. New indices, such as the SPI, have provided an important tool in assessing climate conditions and initiating mitigation and response actions.

REFERENCES

Alley, W.M. (1984) The Palmer Drought Severity Index: Limitations and Assumptions, *J. Clim. Appl. Meteorol.* 23(1): 100-109.

Coughlan, M.J. (1987) Monitoring Drought in Australia, in D.A. Wilhite and W.E. Easterling (eds.), *Planning for Drought: Toward a Reduction of Societal Vulnerability*, Boulder, Colorado: Westview Press, 131-144.

Edwards, D.C. and McKee, T.B. (1997) *Characteristics of 20th Century Drought in the United States at Multiple Time Scales*, Climatology Report Number 97B2, Department of Atmospheric Science, Colorado State University, Fort Collins.

Gibbs, W.J. and Maher, J.V. (1967) Rainfall Deciles as Drought Indicators, *Bureau of Meteorology Bulletin No. 48*, Melbourne, Australia.

Guttman, N.B. (1998) Comparing the Palmer Drought Index and the Standardized Precipitation Index, *J. Am. Water Res. Assoc.* 34(1): 113-121.

Guttman, N.B., Wallis, J.R., and Hosking, J.R.M. (1992) Spatial Comparability to the Palmer Drought Severity Index, *Water Resources Bulletin* 28(6): 1111-1119.

Heddinghaus, T.R. and Sabol, P. (1991) A review of the Palmer Drought Severity Index and where do we go from here? Preprints, *7th Conf. On Applied Climatology*, Salt Lake City, UT, Am. Meteorol. Soc., 242-246.

Jose, A.M., Magnayon, F.O., and Hilario, F.D. (1991) Climate Impact Assessment for Agriculture in the Philippines, unpublished paper, National Workshop on Drought Planning and Management in the Philippines, Quezon City.

Karl, T.R. and Knight, R.W. (1985) Atlas of Monthly Palmer Hydrological Drought Indices (1931-1983) for the Contiguous United States, National Climatic Data Center Historical Climatology Series 3-7, Asheville, NC, 319 p.

Kogan, F.N. (1995) Droughts of the Late 1980s in the United States as Derived from NOAA Polar-Orbiting Satellite Data, *Bull. Amer. Meteorol. Soc.* 76(5): 655-668.

Lee, D.M. (1979) Australian Drought Watch System, in M.T. Hinchey (ed.), *Botswana Drought Symposium*, Gaborone, Botswana: Botswana Society, 173-187.

McKee, T.B., Doesken, N.J., and Kleist, J. (1993) The Relationship of Drought Frequency and Duration to Time Scales, *8th Conference on Applied Climatology*, Boston: American Meteorological Society.

McKee, T.B., Doesen, N.J, and Kleist, J. (1995) Drought Monitoring with Multiple Time Scales, *9th Conference on Applied Climatology*, Boston, American Meteorological Society.

McKee, T.B. (1996) Personal communication, February 29, 1996.

Palmer, W.C. (1965) Meteorological Drought, *Research Paper* No. 45, Washington, D.C.: U.S. Weather Bureau.

Palmer, W.C. (1968) Keeping Track of Crop Moisture Conditions, Nationwide: The New Crop Moisture Index, *Weatherwise* 21(4): 156-161.

Riebsame, W.E., Changnon, S.A. Jr, and Karl, T.R. (1991) *Drought and Natural Resources Management in the United States: Impacts and Implications of the 1987-89 Drought*, Boulder, Colorado: Westview Press.

Shafer, B.A. and Dezman, L.E. (1982) Development of a Surface Water Supply Index (SWSI) to Assess the Severity of Drought Conditions in Snowpack Runoff Areas, *Western Snow Conf.* 164-175.

Smith, D.I., Hutchinson, M.F., and McArthur, R.J. (1993) Australian Climate and Agricultural Drought: Payments and Policy, *Drought Network News* 5(3): 11-12.

Soulé, P.T. (1992) Spatial Patterns of Drought Frequency and Duration in the Contiguous USA Based on Multiple Drought Event Definitions, *Int. J. Clim.* 12: 11-24.

Tannehill, I.R. (1947) *Drought: Its Causes and Effects*, Princeton, N. J.: Princeton University Press.

Wilhite, D.A. and Glantz, M.H. (1985) Understanding the Drought Phenomenon: The Role of Definitions, *Water International* 10: 111-120.

Wilhite, D.A., Rosenberg, N.J., and Glantz, M.H. (1986) Improving Federal Response to Drought, *J. Clim. Appl. Meteorol.* 25: 332-342.

Wilhite, D.A. and Wood, D.A. (1994) Drought Management in a Changing West: New Directions for Water Policy. *IDIC Technical Report Series 94-1.*

Willeke, G.J., Hosking, R.M., Wallis, J.R., and Guttman, N.B. (1994) The National Drought Atlas, U.S. Army Corps of Engineers Institute for Water Resources Report 94-NDS-4, 587 p.

COMPARISON OF TWO DROUGHT INDICES FOR DROUGHT MONITORING IN HUNGARY – A CASE STUDY

S. SZALAI and Cs. SZINELL
Climate Research and Development Division
Hungarian Meteorological Service
1024 Budapest, Hungary

Abstract. This paper presents two indices developed for drought monitoring purposes. In a first step, the need for drought monitoring and analysis in Hungary is demonstrated against the climatic background of the region. In the following, the Palmer Drought Severity Index (PDSI) and the Standardized Precipitation Index (SPI) are briefly described and their monitoring capabilities are demonstrated for the drought of 1983 in south-eastern Hungary.

1. INTRODUCTION

Drought is a recurrent feature of any climate and its impacts extend over numerous sectors of the economy. Droughts might recur at different locations and with different severity in subsequent years, therefore it is important to be able to provide policy-makers with estimates of the spatial extent, severity and duration of a given drought. Several indices have been developed to quantify these characteristics. Normally, drought is defined by these indices exceeding certain thresholds.

In this paper a short overview of the climatic features of Hungary is given and the serious droughts of the 1980's and 1990's are put in a historic perspective. The performance of the Palmer Drought Severity Index (PDSI, Palmer, 1965) and the Standardized Precipitation Index (SPI, McKee *et al.*, 1993; 1995) is studied during the serious drought of 1983 in order to analyse their monitoring capabilities.

2. CLIMATIC BACKGROUND

Hungary is situated in the Carpathian Basin. Its climate is determined mainly by the large-scale circulation patterns of maritime, continental and Mediterranean air masses, modified by the topography of the basin. With respect to the basin surroundings this results in an increased sunshine duration, a decreased precipitation amount, weaker winds, greater amplitudes of the daily and yearly temperature ranges and a greater spatial variability of precipitation (the annual mean maximum is 879 mm, while the minimum is only 453 mm). The mean annual temperature is about 10°C and exhibits a zonal pattern modified by the altitude. Rainfall distribution over Hungary is uneven, as

J.V. Vogt and F. Somma (eds.), Drought and Drought Mitigation in Europe, 161–166.

it is apparent from figure 1. The most humid parts of the country in the West receive somewhat less than 900 mm of rain per year, about twice the precipitation of the driest areas in the Hungarian Plain, which is the most important agricultural area of the country. The climate of the Hungarian Plain is characterised by a tendency to dryness and often insufficient rain for agriculture during summer months. The highest monthly precipitation values are measured in June (55-80 mm), and January and March are the driest months. While on the one hand monthly precipitation can exceed 100 mm or sometimes even 200 mm in any month, on the other hand months without any rainfall may occur any time of the year. The growing season (April-September) exhibits an even larger variation of monthly precipitation sums.

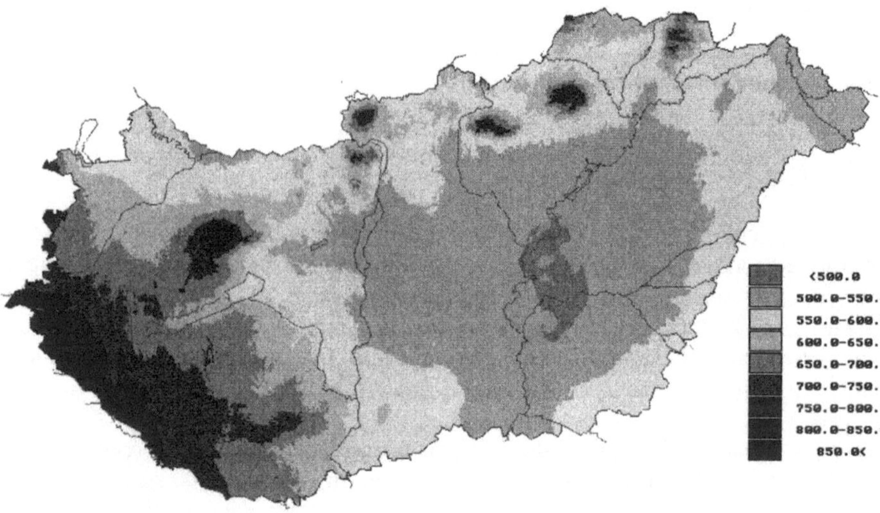

Figure 1: Mean annual precipitation in Hungary.

The occurrence and characteristics of historic droughts in Hungary have been analysed by Szinell *et al.* (1998). They found that droughts have been recurring in the past century. Historic references (e.g. Gunst, 1993) suggest that, although extremely dry periods occurred already before the beginning of regular meteorological observations (e.g. between 1779 and 1794), the series of severe droughts from 1983 to 1995 is rather exceptional. Szinell *et al.* (1998) statistically analysed PDSI series between 1881 and 1995 at 15 locations in Hungary. Both linear regression and the Mann-Kendall test revealed a significant decrease of the PDSI at most of the stations. At the same time, analyses of drought occurrence suggested an increase in the frequency of droughts in different severity classes.

3. APPLIED DROUGHT INDICES

3.1. The Palmer Drought Severity Index

The *Palmer Drought Severity Index* (PDSI) is one of the most widely used drought indices (Palmer, 1965). It is based on the concept of a hydrologic accounting system, which is a useful technique for tracking long-term moisture anomalies and for evaluating drought frequencies. It was designed to simulate the supply and demand concept of the water balance equation. A two-layer soil component has been built into the index. Therefore the PDSI is a complex index, requiring weekly or monthly inputs of precipitation and temperature. It is a standardized measure of water availability, thus allowing to compare the values at different locations. Numerous studies have been published on the PDSI (for European examples see, for example, Briffa *et al.*, 1994; Szinell *et al.*, 1998; Bradford, 2000). Many policy-related decisions regarding the need for emergency drought relief have been based on the PDSI.

3.2. The Standardized Precipitation Index

The *Standardized Precipitation Index* (SPI) is a relatively new drought index developed in 1993 by researchers at Colorado State University (McKee *et al.*, 1993; 1995). The SPI can be calculated at multiple time scales, providing the flexibility to monitor short term water supplies, important for agriculture, and longer term water supply conditions, important for reservoir and groundwater management. Calculation of the SPI is based on a long-term precipitation database. The long-term record is fitted to a probability distribution and then transformed into a normal distribution so that the mean SPI for the any location and time period is zero. Any particular precipitation total is then given an SPI value in units of standard deviation of this distribution. Thus, positive SPI values indicate greater than median precipitation, negative values less than the median precipitation. The SPI and its calculation procedure are further presented by Wilhite *et al.* (2000), for example.

4. A CASE STUDY AT SZARVAS, 1983

In 1983 an extended and severe drought caused significant damages and far-reaching economic consequences in Hungary. This drought event was particularly extreme in the south-east of Hungary, and therefore it has been chosen to exemplify the monitoring capabilities of the PDSI and SPI.

In figure 2, PDSI, 3-month and 12-month SPI time series are presented for the years 1981 to 1985. The data shown are representative of the eastern part of the Hungarian Great Plain, which is an important agricultural area and was the most affected by the drought. Szarvas, for which the indices have been calculated, is a town situated in the heart of this area.

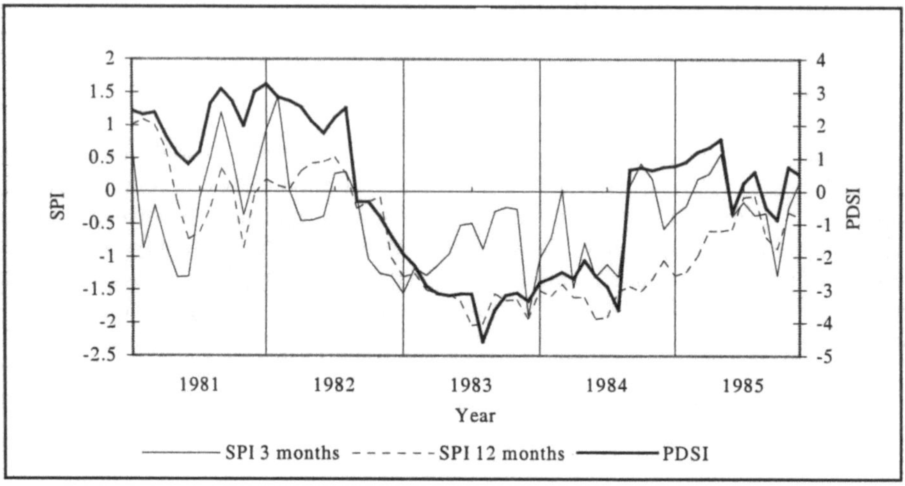

Figure 2: SPI and PDSI time series at Szarvas during the 1983 drought in Hungary.

The two SPI time series are positive by the end of 1981 and early 1982. During the spring of 1982, the 3-month SPI falls below zero, but the dryness becomes established when the 3-month SPI falls below –1.0 in September 1982.

In 1983, the 12-month SPI continues to decrease, especially in the second part of the year, while the 3-month SPI does not indicate serious dryness. It is apparent, therefore, that even though those months were not very dry, the total rainfall was not sufficient to compensate for the water deficit so far accumulated. Along with the low 12-month SPI, the evolution of a long and serious drought is indicated, which largely affected the agricultural production in the area. The 3-month SPI recovers to values above zero at the end of 1984, but the 12-month SPI only slowly improves, remaining negative until May 1991. Later the index values fall again, correctly detecting the droughts in the 90's. It achieves large positive values only during the wet autumns of 1996 and 1998.

In 1981 and during summer of 1982, the PDSI maintained its values around +2.0, indicating normal to wet conditions. In September 1982, however, PDSI fell sharply to zero and continued decreasing, falling below -2.0 (upper limit for definition of moderate drought) in January 1983, while both SPI series indicated a gradual drying during autumn of 1982. However, the course of PDSI is fairly consistent with the 12-month SPI. Once the PDSI approached its minimum, it remained consistently around -4.0 (extreme drought) until late 1984, and generally below -2.0 until September 1984. Then it rapidly increased to around zero, indicating the end of this drought event, and remained at this level. This can be associated to rainfall also appearing in the 3-month SPI, but long term dryness did not end with this rain, as can be seen from the 12-month SPI series. The described abrupt changes in the PDSI series are consistent with its distribution characteristics (e.g. Guttman, 1998).

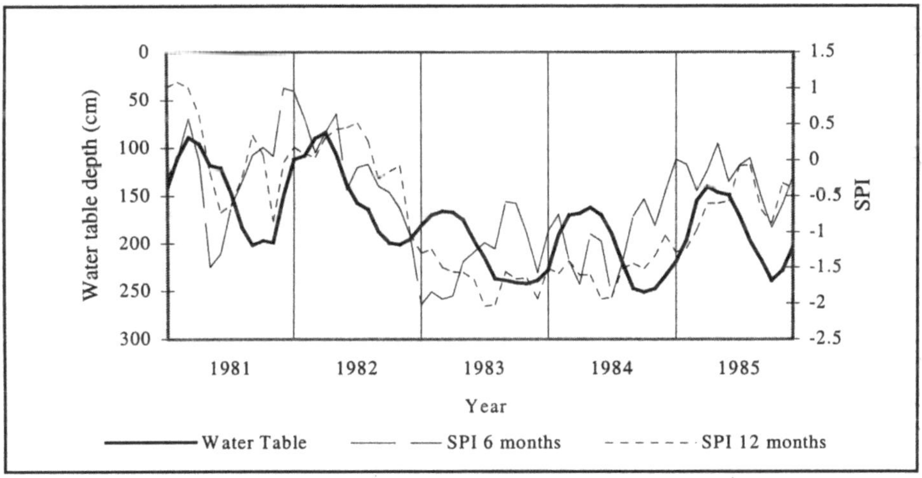

Figure 3: Water table depth and SPI time series at Szarvas during the 1983 drought in Hungary.

In addition to the poor crop yields in 1983, well data from near Szarvas also confirm the impacts of the drought. In figure 3, the 12-month SPI is again presented, this time along with series of the 6-month SPI and the water table depth measured near Szarvas. The level of the water table has a natural annual cycle in Hungary, falling during summer and reaching its lowest point in the autumn, then increasing and recovering by spring. As shown in figure 3, the SPI series indicate dryness during the autumn of 1982, which is also apparent from the decreasing level of the water table. It is also apparent that, although the 1982 autumn water table did not fall beyond its usual range, which is around 200 cm depth, the dry autumn and winter did not allow full recover to normal levels, remaining below 150 cm, which is in good agreement with the low SPI values. The usual summer and autumn fall of the water table started from a deeper level, thus decreasing to particularly low levels. Recovery started in early 1985, marked by a higher peak in 1985 than observed in 1983 and 1984. This increase was anticipated by increasing values of the 6-month SPI series.

5. CONCLUSIONS

Drought, as a result of below normal precipitation, is a recurrent phenomenon in Hungary. However, the sequence of frequent and serious droughts during the 1980's and 1990's caused significant economic losses. The performance and monitoring capability of two indices (SPI and PDSI) have been investigated in a case study for a time interval centred around 1983, one of the most severe drought years during the above dry period.

The SPI can be calculated with several time scales, which gives the index flexibility for monitoring, as different aspects of drought can be studied using different time scales. In this study, the course of the PDSI was found to be in fairly good agreement with the 12-month SPI. However, it showed a certain degree of stability during small fluctuations of moist and dry conditions, which is consistent with earlier findings. This characteristic may be misleading when trying to determine the start and end of a drought. The SPI, with its multiple time-scale capability, is a useful tool to detect and monitor droughts, although using more than one drought index can improve the results. Short term SPIs (e.g. 3 months) can detect early stages of drought development, while longer term SPIs (6 or 12 months) can characterise longer term water supply conditions or variables with a slower response to changing precipitation conditions (e.g. water table level). Therefore, a successful drought watch system should combine different drought indices with real-time monitoring of streamflow, reservoir, groundwater, and soil moisture levels.

REFERENCES

Briffa, K.R., Jones, P.D. and Hulme, M. (1994) Summer Moisture Variability Across Europe, 1892-1991: An Analysis Based on the Palmer Drought Severity Index, *Int. J. Climatol.* 14: 475-506.

Bradford, R.B. (2000) *Drought Events in Europe*, in J.V. Vogt and F. Somma (eds.), *Drought and Drought Mitigation in Europe*, Kluwer Academic Publishers, Dordrecht, 7-20.

Gunst, P. (1993) Droughts and the Hungarian State, in: C. Baráth, et al. (eds.), *Drought 1983*, University of Horticultural Sciences, Budapest, 131-159 (in Hungarian).

Guttman, N.B. (1998) Comparing the Palmer Drought Index and Standartized Precipitation Index, *J. Amer. Water Resour. Assoc.* 34: 113-121.

McKee, T.B., Doesken, N.J., and Kleist J. (1993) The Relationship of Drought Frequency and Duration to Time Scales, *8th Conference on Applied Climatology*, Boston, American Meteorological Society.

McKee, T.B. (1995) Drought Monitoring with Multiple Time Scales, *9th Conference on Applied Climatology*, Boston, American Meteorological Society.

Palmer, W.C. (1965) Meteorological Drought, *Research Paper* No. 45, Washington, D.C., U.S. Weather Bureau.

Szinell, C., Bussay, A., and Szentimrey, T. (1998) Drought Tendencies in Hungary, *Int. J. Climat.* 18: 1479-1491.

Wilhite, D.A., Hayes, M.J., and Svoboda, M.D. (2000) Drought Monitoring and Assessment: Status and Trends in the United States, in J.V. Vogt and F. Somma (eds.), *Drought and Drought Mitigation in Europe*, Kluwer Academic Publishers, Dordrecht, 149-160.

DROUGHT MONITORING FROM SPACE

J.V. VOGT, S. NIEMEYER and F. SOMMA
Space Applications Institute
EC Joint Research Centre
21020 Ispra (Va), Italy

I. BEAUDIN and A.A. VIAU
Centre de recherche en géomatique, Université Laval
Ste Foy, Québec, G1K 7P4, Canada

Abstract. In order to accurately assess the occurrence, extent and severity of a drought it is necessary to get a correct picture of the spatial and temporal distribution of a number of meteorological, hydrological and surface variables. Space observations have the potential to make a significant contribution in this field. Due to the capability to retrieve surface parameters with high spatial and temporal resolutions over large areas, a comprehensive view of the situation can be obtained. The satellite sensor measures, however, radiation fields and the conversion of these measurements into surface parameters and further into information regarding drought conditions needs sophisticated analysis tools.

This paper reviews current possibilities to use remote sensing data for drought detection and monitoring and indicates near-future developments. Examples are given for the Mediterranean regions of Andalusia in southern Spain and Sicily in southern Italy.

1. INTRODUCTION

The occurrence of relatively dry periods is an inherent phenomenon of most climates, leading to a drought when the available water is insufficient to meet the requirements of nature and/or society. Depending on the vulnerability of the affected region, such drought situations can have serious environmental and socio-economic impacts. A thorough analysis of the water needs of natural ecosystems and society, the accurate monitoring of the duration, severity and spatial extent of the water stress, and medium term weather forecasts, form the basis for adequate management of such events and for the mitigation of possible impacts.

The purpose of this paper is to discuss the possibilities for using remote sensing data to monitor and assess droughts at different spatial scales. Over the last two decades, the theoretical basis of remote sensing has rapidly evolved. Together with significant improvements in computing power and data storage capabilities, this has allowed the analysis of massive data sets covering the full globe. With the launch of a whole set of new instruments during the next two years, major advances in the derivation of

J.V. Vogt and F. Somma (eds.), Drought and Drought Mitigation in Europe, 167–183.

important surface parameters can further be expected. The conversion of these data into useful information for drought detection and monitoring, however, requires a good understanding of the physics of the measurement as well as sophisticated analysis tools.

Aside from remote sensing, different tools and indicators for drought detection and monitoring are currently in use. The preference for one or another of these methods depends on the definition of drought and on the specific application (e.g. in meteorology, hydrology, agriculture), the size of the study area, and the availability of basic input data. Examples are meteorological indicators such as the Palmer Drought Severity Index (PDSI) or the Standardized Precipitation Index (SPI), hydrological indicators such as low flow characteristics and groundwater levels, agricultural indicators such as the soil moisture reserve and expected yields, and socio-economic indicators such as shortages in water supplies (e.g. Palmer, 1965; Chang, 1991; McKee *et al.*, 1993 and 1995; Clausen and Pearson, 1995; Cancelliere *et al.*, 1995).

When studying larger areas, the limited availability and spatial density of the necessary measurements may pose serious limitations to the spatial representativity of the derived information (e.g. Rossi *et al.*, 1992). The distributed modelling of the evolution of some of the relevant parameters using Geographic Information Systems (GIS) can improve this situation. An example is the Crop Growth Monitoring System (CGMS) used by the Space Applications Institute for agrometeorological modelling at a European scale (Vossen, 1992; Supit *et al.*, 1994). While such systems are powerful tools, their results are strongly dependent on the spatial representativity of the meteorological input data and on assumptions concerning land cover type or crop type and phenology.

Given this situation, the main advantage of space-borne measurements is to be seen in the complete coverage of extended regions with a spatial resolution of the order of a few hundred meters to a few kilometres. However, since satellite sensors measure electromagnetic radiation, they can reveal water stress only through its effects on these radiation fields. In practice, this means exploiting the link between the characteristics of the vegetation cover and/or the upper soil layer and the reflected and emitted radiation.

2. APPROACHES TO SATELLITE BASED DROUGHT MONITORING

During the last decade, satellite data have been used alone or in conjunction with other data for drought monitoring from regional to continental scales. Examples are the use of vegetation indices (e.g. Gutman, 1990; Eklundh, 1996; Gosh, 1997; McVicar and Jupp, 1998; Vogt *et al.*, 1998), surface temperatures (e.g. Courault *et al.*, 1994), a combination of both (e.g. Kogan and Sullivan, 1993; Kogan, 1995; Liu and Kogan, 1996; Kogan, 1997; Kogan, 1998), or the microwave backscatter signal (e.g. Wagner, 1998). Finally, remote sensing data have been used together with meteorological and land cover data in energy balance models (e.g. Niemeyer and Vogt, 1998).

The main working agents are the reflectance pattern of vegetation in the visible and near-infrared spectral regions and the surface skin temperature as derived from measurements of emitted radiation in the thermal infrared spectral region. Whilst the former gives information on the vigour of the vegetation cover, the latter gives information on the surface energy balance. In addition, passive and active microwave data might be considered for the retrieval of the soil moisture close to the surface (Wagner, 1998).

Two approaches will be detailed in the following sections. The first one is an index-based approach, using satellite derived vegetation indices and surface temperatures as well as traditional drought indices to monitor the level of water stress of the vegetation cover. The second approach consists of a surface energy balance modelling, giving spatially distributed information on the so-called evaporative fraction. This approach exploits the difference between the satellite-measured surface skin temperature and the air temperature, interpolated from measurements at the meteorological station network.

2.1. Drought Indices Based on Reflected and Emitted Radiation

The basis for the use of such indices is the assumption that the vitality of the vegetation cover is a good indicator of the environmental conditions in an area. Since in semi-arid environments water availability is usually the limiting factor for vegetation development and vitality, the vigour of the vegetation cover should be a good indicator of the occurrence and severity of water stress. Input parameters to these indices are the reflectances in the visible and near-infrared spectral regions and the surface skin temperature (Ts). For critical reviews of different vegetation indices and algorithms to retrieve land surface temperatures see Gutman (1991), Leprieur *et al.* (1994) and Vogt (1996).

Examples for well-known indices are the Normalised Difference Vegetation Index (NDVI), the Global Environmental Monitoring Index (GEMI, Pinty and Verstraete 1992), the Vegetation Condition Index (VCI) and the Temperature Condition Index (TCI) (Kogan 1997). The NDVI has been used as the basic vegetation index in this study and is defined in equation 1, where ρ_{NIR} stands for the reflectance in the near-infrared and ρ_{RED} for the reflectance in the red spectral region.

$$NDVI = \frac{\rho_{NIR} - \rho_{RED}}{\rho_{NIR} + \rho_{RED}} \quad [--] \qquad (1)$$

The VCI is defined according to equation 2, where $NDVI_{min}$ and $NDVI_{max}$ refer to the absolute minimum and maximum NDVI measured for a specific 10-day period during several years. *NDVI* refers to the respective values for the same 10-day period in the year of interest.

$$VCI = \frac{NDVI - NDVI_{min}}{NDVI_{max} - NDVI_{min}} \quad [--] \tag{2}$$

The VCI is an indicator of the vigour of the vegetation cover as a function of the NDVI minima and maxima encountered for a given land cover type and a given 10-day period during many years. It normalises the NDVI (or any other vegetation index) according to its variability over many years and thus results in a consistent index for different land cover types. It is an attempt to separate the short-term weather-related signal from the long-term climatological signal as reflected by the vegetation. In this sense, it is a better indicator of water stress conditions than the NDVI (Kogan and Sullivan, 1993).

It is evident that the significance of the VCI depends on the strength of the relation between the vegetation index and the vitality of the vegetation cover under investigation, as well as on the number and quality of images available for the calculation of the absolute minimum and maximum.

The TCI is defined in analogy to the VCI, with Ts replacing NDVI (equation 3).

$$TCI = \frac{T_{max} - Ts}{Ts_{max} - Ts_{min}} \quad [--] \tag{3}$$

The TCI is indicator equivalent to the VCI, based on the fact that the skin temperature of a vegetation canopy or the soil surface will rise with increasing water stress due to reduced evaporative cooling.

Both, the VCI and the TCI, are dimensionless and range from zero and one. Zero indicates the worst condition encountered over the years covered by images, one indicates the best condition encountered during the same period of time. If the period covered includes dry and wet years and under the assumption that the vegetation condition is mainly related to the water availability, these indicators have a high potential for monitoring water stress. The empirical relation of these indices with drought indices (as derived from meteorological measurements or through agro-meteorological and hydrological models at controlled sites) can be used to calibrate the satellite indices for drought detection and drought severity assessment.

The high spatial information content of the satellite data allows accurate mapping of the extent of the water stress. A combination of this information with land cover information from databases such as the European CORINE[1] database leads to a more detailed assessment of the impact of drought events on the these land cover types.

---

[1] CORINE: Co-ordination of Information on the Environment

2.2. Energy Balance Modelling

This approach is based on a model of the surface energy balance and the derivation of the so-called evaporative fraction (*EF*, equation 4) with a spatial resolution equivalent to the spatial resolution of the remote sensing data (100 to 1000 meters, for example). Among the various flux ratios used to characterise the energy balance, EF has received special interest (e.g. Brutsaert and Seguita, 1992; Zhang and Lemeur, 1995; Crago, 1996). The retrieval algorithm makes use of remote sensing data from the visible, near-infrared and thermal infrared spectral regions together with data on surface cover types, elevation, and daily meteorological data (see figure 1). The resulting *EF* can be validated against measurements made at surface stations, although the difference between the measurement types (e.g. point measurement vs. integrated measurement over a large surface) makes a straightforward comparison difficult.

$$EF = \frac{\lambda ET_i}{(R_{net} - G)_i} \qquad [\text{--}] \qquad (4)$$

EF is the ratio between the latent heat flux (λET_i) and the available energy at the surface, defined as the net radiation (R_{net}) minus the ground heat flux (G_i). The suffix *i* stands for instantaneous fluxes. *EF* indicates how much of the available energy is used for evapotranspiration, that is, for transpiration of the vegetation and evaporation from the soil. As long as moisture is available, energy will be used for its evaporation and EF will have values close to one. With no moisture left, all available energy will be directed into the sensible heat flux and *EF* will approach zero. A more detailed description of the individual steps in the algorithm is given in the appendix to this paper.

EF is considered to be indicative of the moisture status of the various surface cover classes. It is calculated as an instantaneous value. Under the assumption that the early afternoon instantaneous *EF* is representative for the daily energy partitioning, estimates of daily rates of evapotranspiration (*ETa)* can further be derived.

Basic remote sensing inputs are the surface skin temperature (Ts) and the NDVI. While the former is a direct input to the energy balance algorithm, the latter serves to parameterise the yearly evolution of the vegetation cover and, therefore, together with the land cover information, the roughness of the surface. The reflectances in the visible and near infrared as well as the emitted radiation in the thermal infrared are further used in the estimation of the net available energy at the surface.

The spatial distribution of the *EF* and its temporal evolution are then used to assess the occurrence and severity of water stress. The combination of these maps with maps on land cover types, catchment limits and administrative boundaries leads to a detailed assessment of potential impacts.

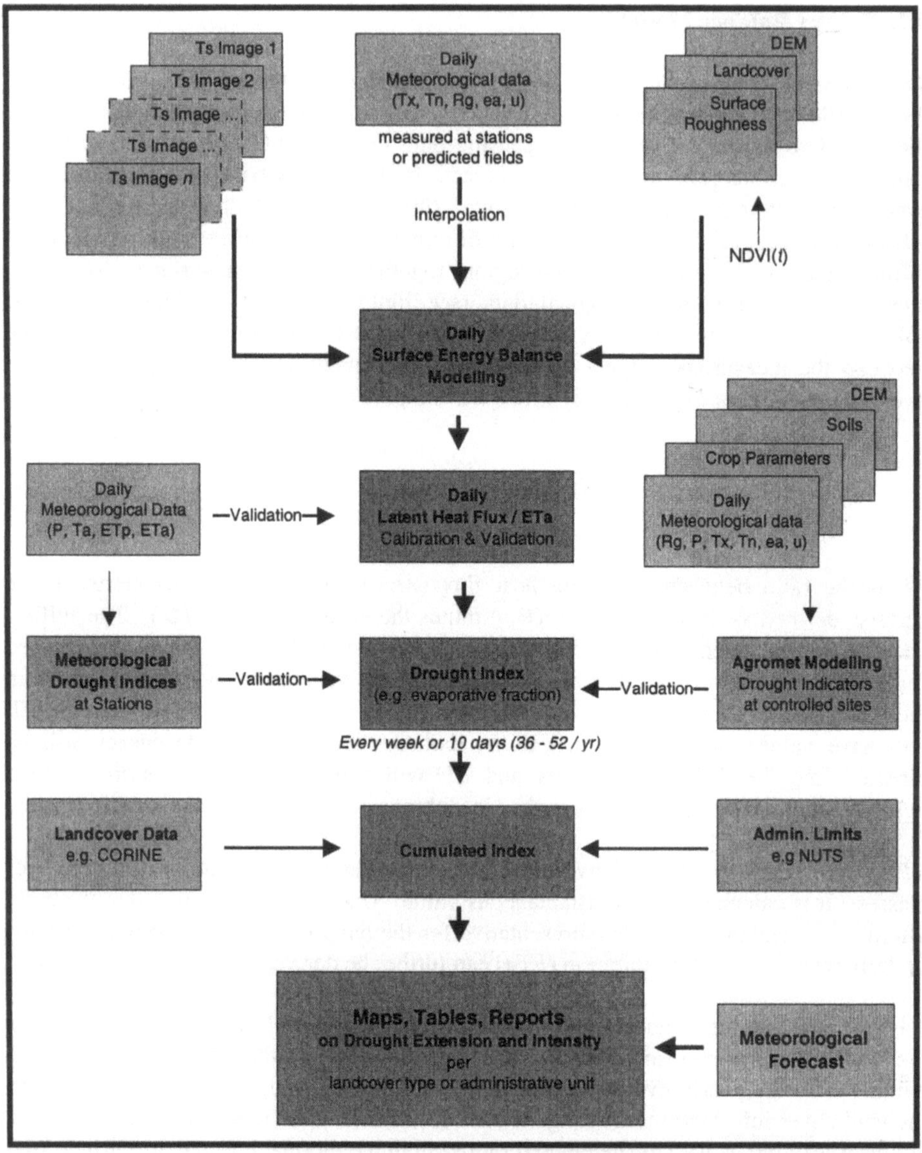

Figure 1: Flowchart of data processing for the energy balance approach to drought monitoring using remote sensing data. This paper describes the process up to the calculation of the drought index (Tx = daily maximum air temperature, Tn = daily minimum air temperature, Rg = global radiation, ea = vapour pressure, u = wind speed, P = precipitation, Ta = daily mean air temperature, ETp = potential evapo-transpiration, ETa = actual evapo-transpiration, NUTS are the European statistical units).

3. EXAMPLES OF REMOTE SENSING BASED DROUGHT MONITORING IN THE MEDITERRANEAN

In the following, the described approaches using (a) satellite-based indices and (b) an energy balance approach will be detailed for the regions of Andalusia in southern Spain and Sicily in southern Italy (Figure 2). Both regions have a pronounced relief and a typical Mediterranean climate with moist and cool winters and hot and dry summers. Mean yearly temperatures depend on the altitude and range from 17 to 19 °C for the coastal areas and from 10 to 14 °C for altitudes above 1000 meters. Yearly total precipitation ranges from about 200 to about 2000 mm in both regions.

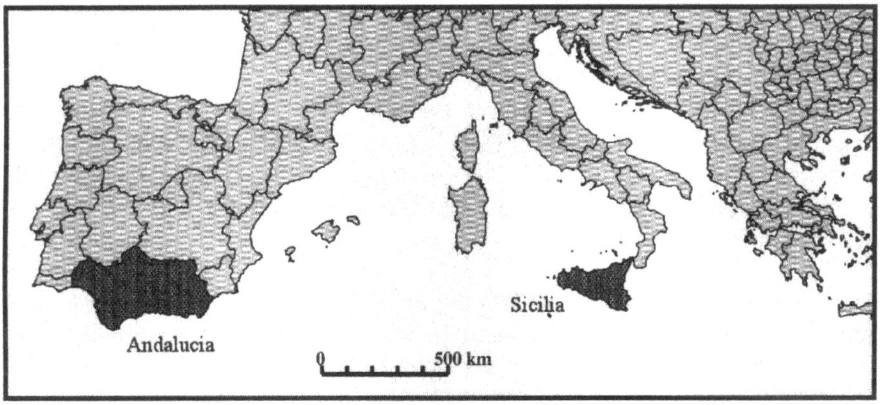

Figure 2: Location of the study areas in the western Mediterranean Basin.

3.1. Index-Based Drought Monitoring in Andalusia (Southern Spain)

From 1992 to 1995 Spain suffered from a severe drought, resulting in a significant reduction of the agricultural production and in huge economic losses (see Garrido and Gómez-Ramos, 2000). Andalusia, the most southern region of Spain suffered from the most severe impacts. This event has been used to test the applicability of the VCI, TCI and NDVI through a comparison of their temporal evolution with the SPI as calculated at selected stations of the Andalusian meteorological network.

The SPI (McKee *et al.* 1993 and 1995) has been chosen among the meteorological indicators due to various reasons: (1) it is relatively simple to calculate and needs only precipitation data as input, (2) it can be calculated for varying time scales, and (3) it has received wide-spread interest in the scientific community in recent years. The SPI is a statistical indicator, evaluating the lack or surplus of precipitation during a given period of time as a function of the long-term average precipitation and its distribution. It is calculated using a continuous, long-term (more than 30 years) series of historic monthly precipitation records. Because rainfall is not normally distributed for aggregation periods of less than 12 months a theoretical distribution is fitted to the frequency distribution. The SPI is then given in units of standard deviation from the mean for an equivalent normal distribution with a mean of zero and a standard deviation of one (see also Wilhite, 2000: 156-158).

Meteorological data were drawn from a regional database with daily measurements of temperature and precipitation for more than 100 stations in Andalusia. The image data were taken by the AVHRR[2] aboard the NOAA[3] polar orbiting satellites and pre-processed with the SPACE[4] software, developed at the Space Applications Institute, Ispra. The resulting images have a nominal spatial resolution of 1.1 km and contain calibrated and atmospherically corrected reflectances in the visible and near-infrared channels and brightness temperatures in the thermal infrared channels.

VCI and TCI have been calculated on a 10-day basis for the years 1989-1993, based on 10-day maximum value composite NDVI layers. NDVI, VCI and TCI values for individual stations were then extracted from the images as the mean value for 3.3×3.3 km windows, centred at each station. The resulting values were then compared to the SPI for the respective station. For the purpose of this paper two stations have been selected to present the results.

In figure 3, the results for the stations of Aldea de Cuenca with a mean annual rainfall of 564 mm and Galaroza with a mean annual rainfall of 1020 mm are presented. The selected period includes the relatively wet winter of 1989-1990 and the first years of the drought starting in the winter of 1991-1992.

The land use in the 3.3 km×3.3 km window centred at the Aldea de Cuenca station is mostly dry-farming (71%) and figure 3a demonstrates that the curve of the NDVI follows the vegetation dynamics throughout the growing season. Values range from 0.15 for bare soil to 0.50 for the period of maximum vegetation cover. However, the index does not show a significant relation to the increasingly unfavourable growing conditions in the years 1992 and 1993 as shown by the SPI (SPI values below –1 are considered as dry; Wilhite, 2000: 157). The VCI, on the contrary, shows significant changes in the vegetation conditions over the four-year period. It follows the SPI over the growing period with a certain time lag due to the response time of the vegetation cover. The low rainfalls in the winter of 1992/93 apparently had no worse effect on the vegetation cover than during the 1991/92 season. The TCI follows the SPI very sharply in 1993 and somewhat in 1992, but its evolution during the wetter years is not readily explainable.

---

[2] AVHRR: Advanced Very High Resolution Radiometer with daily global coverage
[3] NOAA: US National Oceanic and Atmospheric Administration
[4] SPACE: Software for the Processing of AVHRR images for the Commission of the European Communities

Figure 3b shows the situation for the station of Galaroza. The main land use types in the 3.3×3.3 km window centred at this station are forest (63%), tree crops (13%) and dry farming (20%). This gives rise to a smoother NDVI curve, although some clear reduction of the NDVI is seen during the spring time of 1992 and 1993. However, the VCI and the TCI reveal a much clearer and more significant relationship with the drought conditions as shown by the SPI.

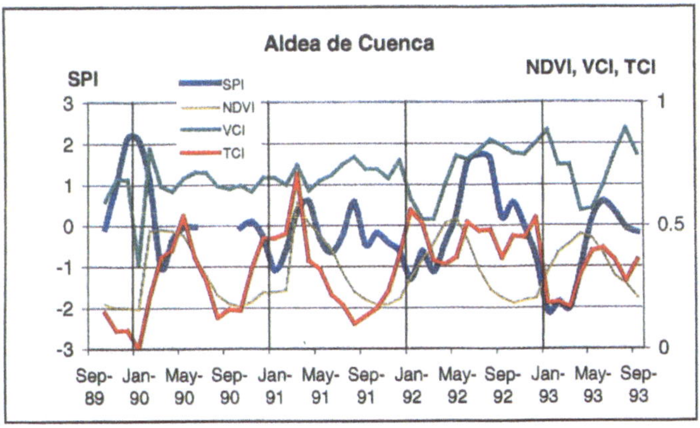

a

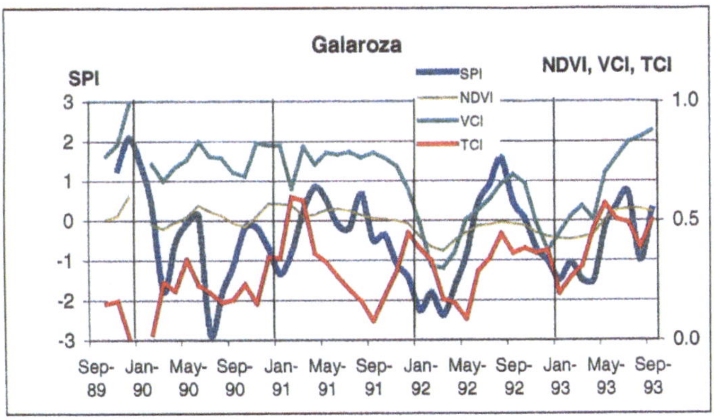

b

Figure 3: 3-month running mean of the 10-day NDVI, VCI, and TCI and the 3-month SPI for the stations of Aldea de Cuenca (a) and Galaroza (b) from Sept. 1989 to Sept. 1993.

The apparent contradiction between the SPI and the VCI in the summer months of 1990, when the SPI falls to a value of -3 whilst the VCI remains at a normal level, is indicative of the different nature of the two indices. Since the mean rainfall over the summer months is very low in this area (less than 20 mm), a standard deviation (SPI) of -3 is actually not significant in terms of drought conditions. The SPI needs careful interpretation in this sense. The VCI, on the contrary, reflects the effects of the rainfall on the vegetation cover and, therefore, indicates no significant departure from the expected conditions for this period of the year.

Figure 4 presents a comparison of VCI images for selected 10-day periods in 1989 and 1992. Values range from very bad conditions in red to favourable conditions in dark green. Clouds appear in white. The images show the difference between the relatively humid spring of 1989 and the beginning of a prolonged drought in 1992.

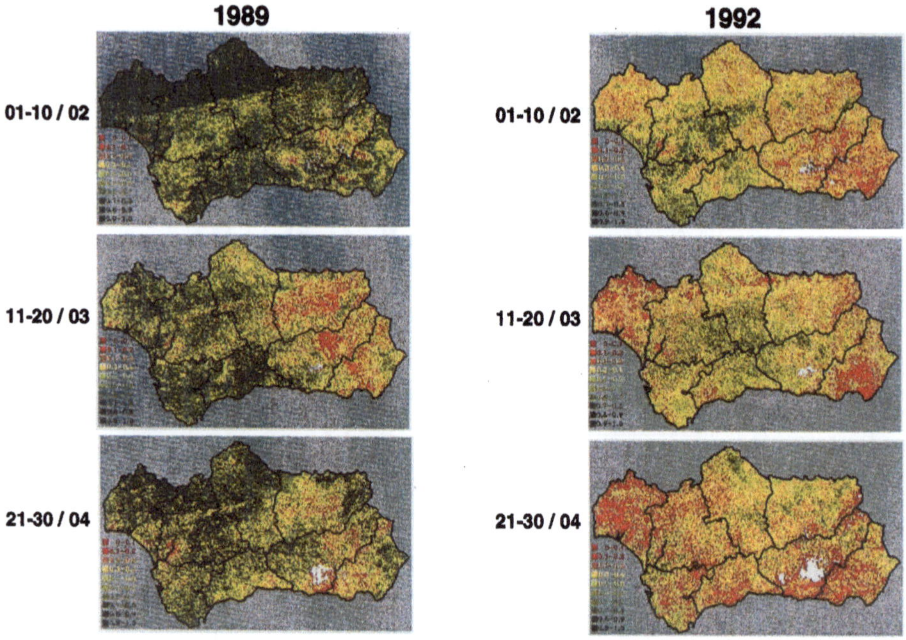

Figure 4: *VCI images of Andalusia in Spain. The images show the situation in early February, in March and in late April for the years 1989 and 1992. Colours vary from red (signifying serious water stress) through yellow, brown, and light green to dark green (signifying good vegetation conditions). Overlaid are the province boundaries of Andalusia.*

Compared to maps resulting from the interpolation of meteorological indices such as the SPI or the PDSI, the satellite data give a very detailed picture of the situation. In addition, they result from actual measurements of the vegetation cover and, therefore, represent current surface conditions, whilst the meteorological data represent a climatic situation with variable effects on the different ecosystems (forest, rainfed cereals, irrigated cereals, tree crops, etc.).

3.2. Evaporative Fraction based Drought Monitoring in Sicily (Southern Italy)

From 1988 to 1990 extended areas of the Mediterranean basin suffered from abnormally low precipitation, including large parts of Italy (Rossi and Margaritora 1994). This event, which severely affected the Italian region of Sicily, was used to test the applicability of the energy balance approach.

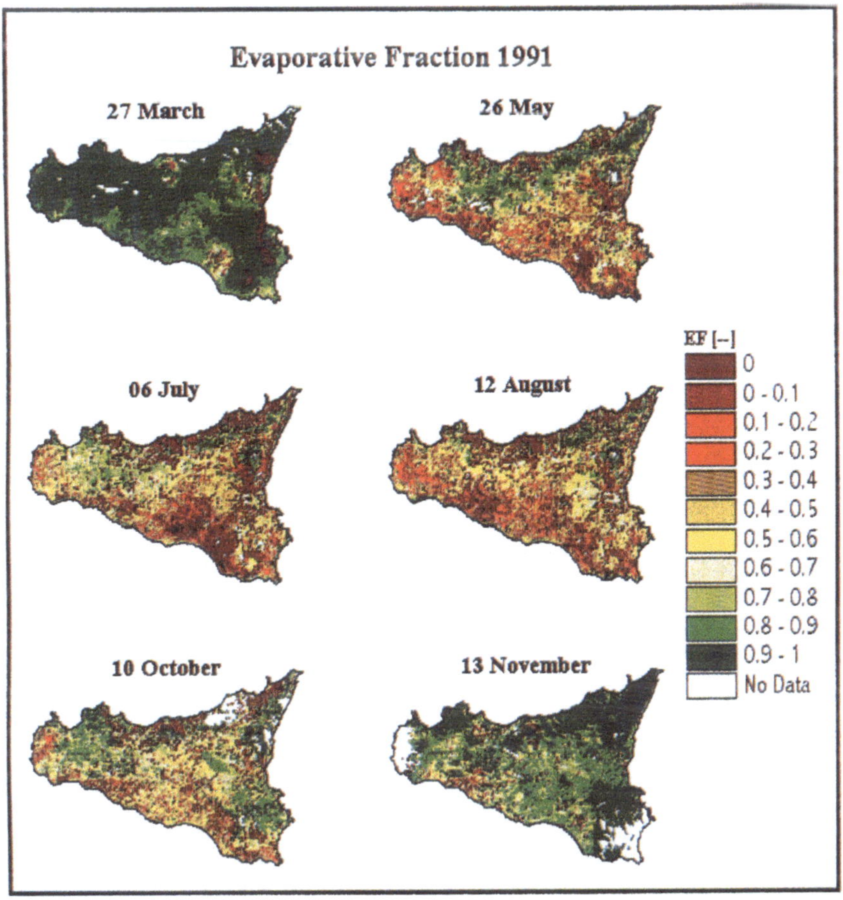

Figure 5: Examples of the evaporative fraction on selected days in 1991 for Sicily.

Figure 5 presents images of the evaporative fraction (see equations 4 and A6 in the appendix) as derived for selected days in 1991, an average year regarding precipitation and its distribution. Green colours indicate areas with ample water supply, where evapotranspiration is limited mainly by the available energy. Yellow and especially red colours indicate a lack of moisture at the surface, resulting in seriously restricted evapotranspiration.

A clear and realistic evolution of *EF* can be seen over the year. The image of 27th March 1991 shows a fairly uniform distribution of high *EF* values, which indicates that evapotranspiration throughout Sicily is unlimited by the water supply. However, by the end of May large areas exhibit yellow and orange colours with *EF* values as low as 0.1. These values represent regions where evapotranspiration is strongly limited by the available water.

During the summer (June until August) only the mountainous regions in the north and northeast of Sicily remain with sufficient water supply for evapotranspiration, whilst the extent of areas with very low *EF* reaches its maximum, covering up to 90% of Sicily. This situation changes only in October and November with the onset of the autumn and winter rains. This sequence represents a typical situation as to be expected for a Mediterranean region with hot and dry summers.

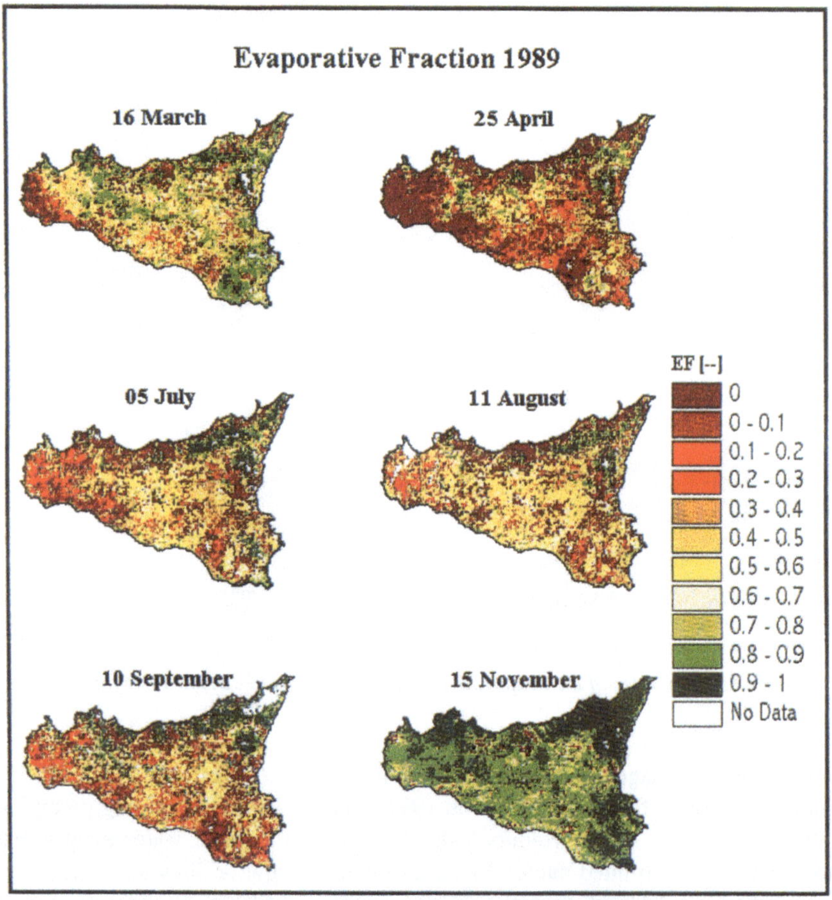

Figure 6: Examples of the evaporative fraction on selected days in 1989 for Sicily.

The year 1989, on the contrary, can be considered as a drought year. This is evident from figure 6. In contrast to the situation in 1991, evapotranspiration is restricted by limited water supply in large areas of Sicily in early spring 1989. Whilst low values of the evaporative fraction during the summer describe a normal situation in Sicily, the early decrease in 1989 indicates a large moisture deficit with severe consequences for agriculture and natural vegetation. Water stress only becomes less severe in November 1989 as shown by *EF* values ranging from 0.6 to 0.9.

These examples demonstrate the potential to map drought relevant surface parameters with a high spatial resolution when using a combination of remote sensing data, meteorological data and surface data such as land cover maps.

4. FUTURE PROSPECTS

The potential use of remote sensing data for drought monitoring will considerably increase with the recent and near-future launch of new satellites. The instruments on-board these spacecrafts will be significantly improved compared to the AVHRR used in the above examples. While the AVHRR currently is the only instrument with a long history of daily data with adequate spatial and temporal resolutions (i.e. 1 – 4 km spatial resolution, daily global coverage), it was originally designed for meteorological applications and its characteristics are not optimal for land surface applications. The new instruments will have a much improved performance in terms of signal to noise ratio, calibration, and spatial, spectral and radiometric resolution. These improvements, together with the development of advanced retrieval algorithms, will allow for the production of more precise and accurate surface information. A few of these instruments and their major characteristics are detailed in table one. For more detailed information on each of these instruments the reader is referred to Kramer (1996).

5. CONCLUSIONS

The examples given in this paper have shown that the combination of meteorological data, satellite data and land cover data gives promising results for the mapping and monitoring of drought conditions. Both approaches, using either a combination of meteorological and satellite-based indices or an energy balance model, have proven their capability to detect and monitor water stress conditions of the vegetation canopy. The approaches will have to be tested at different scales, such as the whole Mediterranean basin, and in different climates. Validation at these scales remains a challenging problem and will require the comparison to different drought indices and to the results of different types of models, including regional circulation models, hydrological models and agro-meteorological models.

The major advantage of the presented methodology is seen in the high spatial information content of the satellite data, which allows for accurate mapping of the spatial extent of water stress conditions. A combination of this information with land cover information from databases such as CORINE will enable a more detailed assessment of the impact of drought events on various land cover types.

Given the complex nature of droughts an optimal use of this type of information is, however, to be seen in a combined analysis with drought indicators based on other information sources.

Table 1: Potential sensors for drought monitoring.

Sensor	Resolution		No. of	Spectral	Local Solar	Launch
	Spatial (2)	Temporal	Channels	Range (3)	Time	
NOAA/METOP(1) AVHRR	1100 m	6 hours	5	V, N, M, T	2:00 8:00 14:00 18:00	1981 →
SPOT VEGETATION	1150 m	1 day	4	V, N	10:30	03-1998
TERRA MODIS	250 – 1000 m	1 day	36	V,N,M.,T	10:30	12-1999
EOS-PM MODIS	250 – 1000 m	1 day	36	V,N,M.,T	13:30	12-2000
ENVISAT MERIS	300 – 1200 m	3 days	15 (4)	V, N	10:00	06-2001
ENVISAT AATSR	500 - 1000 m	3 days	8	V, N, M, T	10:00	06-2001
ADEOS II GLI	1000 m	4 days	36	V, N, M, T	10:30	2001
MSG SEVIRI	>5000 m	15min.	11	V, N, M, T	---	07-2001

(1): From 2000 on the AVHRR instrument will also be flown on the European METOP spacecraft
(2): at nadir
(3): V = visible, N = near infrared, M = mid infrared, T = thermal infrared
(4): 15 bands programmable in position and width

ACKNOWLEDGEMENTS

The financial support provided by the JRC for Isabelle Beaudin and Alain A.Viau is kindly acknowledged.

REFERENCES

Brutsaert, W. and Sugita, M. (1992) Application of Self-preservation in the Diurnal Evolution of the Surface Energy Budget to Determine Daily Evaporation, *J. Geophys. Res.* 97, D17: 18377-18382.

Cancelliere, A., Ancarani, A. and Rossi, G. (1995) Identification of Drought Periods on Streamflow Series at Different Time Scales, in Tsiourtis, N.X. (ed.), *Water Resources Management Under Drought or Water Shortage Conditions*, Rotterdam, 49-57.

Chang, T.J. (1991) Investigation of Precipitation Droughts by Use of Kriging Method, *Journal of Irrigation and Drainage Engineering* 117: 935-943.

Choudhury, B.J., Idso, S.B. and Reginato, R.J. (1987) Analysis of an Empirical Model for the Soil Heat Flux Under a Growing Wheat Crop for Estimating Evaporation by an Infrared-temperature Based Energy Balance Equation, *Agric. Forest Met.* 39: 283-297.

Clausen, B. and Pearson C.P. (1995) Regional Frequency Analysis of Annual Maximum Streamflow Drought, *Journal of Hydrology* 173: 111-130.

Courault, D., Clastre, P., Guinot, J.P. and Seguin, B. (1994) Analyse des sécheresses de 1988 à 1990 en France à partir de l'analyse combinée de données satellitaires NOAA-AVHRR et d'un modèle agro-météorologique, *Agronomie* 14: 41-56.

Crago, R.D. (1996) Conservation and Variability of the Evaporative Fraction During Daytime, *Journal of Hydrology* 180: 173-194.

Eklundh, L. (1996) *AVHRR NDVI for Monitoring and Mapping of Vegetation and Drought in East African Environments*, Meddelanden fran Lunds Universitets Geografiska Institutioner, 126, Lund University Press, Lund, Sweden, 187 p.

Garrido, A. and Gómez-Ramos, A. (2000) Socio-Economic Aspects of Droughts, in J.V. Vogt and F. Somma (eds.), *Drought and Drought Mitigation in Europe*, Kluwer Academic Publishers, Dordrecht, 197-207.

Gosh, T.K. (1997) Investigation of Drought Through Digital Analysis of Satellite Data and Geographical Information Systems, *Theor. Appl. Climatol.* 58: 105-112.

Gutman, G.G. (1990) Towards Monitoring Drought from Space, *Journal of Climate* 3: 282-295.

Gutman, G.G. (1991) Vegetation Indices from AVHRR: an Update and Future Prospects, *Rem. Sens. Env.*, 35: 121-136.

Kogan, F. and Sullivan, J. (1993) Development of a Global Drought-watch System using NOAA/AVHRR Data, *Adv. Space Res.* 13: 219-222.

Kogan, F.N. (1995) Droughts of the Late 1980s in the United States as Derived from NOAA Polar-orbiting Satellite Data, *Bull. Am. Met. Soc.* 76: 655-668.

Kogan, F.N. (1997) Global Drought Watch from Space, *Bull. Am. Met. Soc.* 78: 621-636.

Kogan, F.N. (1998) A Typical Pattern of Vegetation Conditions in Southern Africa During El Niño Years Detected from AVHRR Data Using Three-channel Numerical index, *Int. J. Rem. Sens.* 19: 3689-3695.

Kramer, H.J. (1996) *Observation of the Earth and Its Environment. Survey of Missions and Sensors* (3rd enlarged edition, Springer), Berlin.

Leprieur, C., Verstraete, M.M. and Pinty, B. (1994) Evaluation of the Performance of Various Vegetation Indices to Retrieve Vegetation Cover from AVHRR Data, *Remote Sensing Reviews* 10: 265-284.

Liu, W.T. and Kogan, F.N. (1996) Monitoring Regional Drought Using the Vegetation Condition Index, *Int. J. Rem. Sens.* 17: 2761-2782.

McKee, T.B., Doesken, N.J. and Kleist, J. (1993) The Relationship of Drought Frequency and Duration to Time Scales, *Proc. 8th Conf. on Appl. Clim.*, 17-22 Jan. 1993, Anaheim, CA, 179-184.

McKee, T.B., Doesken, N.J. and Kleist, J. (1995) Drought Monitoring with Multiple Time Scales, *Proc. 9th Conf. on Appl. Clim.*, Dallas, TX, 233-236.

McVicar, T.R. and Jupp, D.L.B. (1998) The Current and Potential Operational Uses of Remote Sensing to Aid Decisions on Drought Exceptional Circumstances in Australia: a Review, *Agricultural Systems* 57: 399-468.

Niemeyer, S. and Vogt, J.V. (1998) Towards Monitoring Drought Conditions in Sicily Using an Energy Balance Approach, *Proc. 7th Int. Conf. for Comp. Techn. in Agriculture*, Florence, Italy, 8 p.

Palmer, W.C. (1965) *Meteorological Drought* (US Dept. of Commerce, Weather Bureau Research Paper, 45), Washington, D.C., 58 p.

Pinty, B. and Verstraete, M.M. (1992) GEMI, a Non-linear Index to Monitor Global Vegetation from Satellites, *Vegetatio* 101: 15-20.

Rossi, G., Benedini, M., Tsakiris, G. and Giakoumakis, S. (1992) On Regional Drought Estimation and Analysis, *Water Resources Management* 6: 249-277.
Rossi, G. and Margaritora, G. (1994) *La Siccità in Italia 1988-1990*, Presidenza del Consiglio dei Ministri, Dipartimento della Protezione Civile, Rome, Italy, 193 p.
Supit, I. (1994) *Global Radiation*, EUR 15745 EN, Luxembourg, 192 p.
Supit, I., Hooijer, A.A. and van Diepen, C.A. (1994) *System Description of the WOFOST 6.0 Crop Simulation Model Implemented in CGMS. Volume 1: Theory and Algorithms*, EUR 15956 EN, Luxembourg, 146 p.
Vogt, J.V. (1996) Land Surface Temperature Retrieval from NOAA AVHRR Data, in G. D'Souza, A.S. Belward, J.P. Malingreau (eds.), *Advances in the Use of NOAA AVHRR Data for Land Applications*, Kluwer, Dordrecht, 126-152.
Vogt, J.V., Viau, A.A., Beaudin, I., Niemeyer, S. and Somma, F. (1998) Drought Monitoring from Space Using Empirical Indices and Physical Indicators, *Proc. Int. Symp. on Satellite-Based Observation: A Tool for the Study of the Mediterranean Basin*, Tunis, Tunisis, 7 p.
Vossen, P. (1992) Forecasting national Crop yields of E.C. Countries: The Approach Developed by the Agriculture Project, *Proc. 2nd Conf. on the Application of Remote Sensing to Agricultural Statistics*, 26-27 November 1991, Belgirate, Italy, (EUR 14262), Brussels/Luxembourg, 159-176.
Wilhite, D.A., Hayes, M.J. and Svoboda, M.D. (2000) Drought Monitoring and Assessment: Status and Trends in the United States, in J.V. Vogt and F. Somma (eds.), *Drought and Drought Mitigation in Europe*, Kluwer Academic Publishers, Dordrecht, 149-160.
Wagner, W. (1998) Soil Moisture Retrieval from ERS Scatterometer Data, EUR 18670 EN, JRC Ispra, Italy.
Zhang, L. and Lemeur, R. (1995) Evaluation of Daily Evapotranspiration Estimates from Instantaneous Measurements, *Agric. Forest. Met.* 74: 139-154.

APPENDIX

Description of the algorithm to derive *EF*.

The energy balance model is based on the general energy balance equation:

$$Rn = \lambda E + H + G \quad [\mathrm{W/m^2}] \qquad \text{(A1)}$$

with: Rn = net radiation flux density; λE = latent heat flux density; H = sensible heat flux density; G = ground heat flux density

In a first step the instantaneous available energy $(R_{net}-G)_i$ and the instantaneous sensible heat flux $(H)_i$ are derived independently. The latent heat flux (λET) is then obtained in a second step as the residual of the energy budget equation (in W/m^2):

$$\lambda ET_i = (R_{net} - G)_i - H_i \qquad \text{(A2)}$$

The components of the net radiation (R_{net}) are derived as follows (in W/m^2):

$$R_{net} = R_G * (1-\alpha) + L\downarrow - L\uparrow \qquad \text{(A3)}$$

For the case of Sicily, global radiation (R_G) is taken from existing estimates for an operational agrometeorological model (Supit 1994) and the surface albedo (α) is derived from the short-wave reflectances as measured by the AVHRR sensor. The long-wave components $L\downarrow$ and $L\uparrow$ are estimated from the air temperature and humidity and from the long-wave emittance as measured by the AVHRR.

The instantaneous soil heat flux (G_i) is approximated by a fraction of the net radiation, as a function of the Normalized Difference Vegetation Index (Choudhury et al.1987).

The sensible heat flux (H_i) is estimated from the difference between radiometric surface temperature (Ts) surface-measured air temperature (Ta), and the formulation of a bulk aerodynamic resistance (r_a) (in W/m^2):

$$H_i = c_p * \rho_{air} \, {}^{1}\!/_{r_a} * (T_s - T_a)_i \tag{A4}$$

c_p being the specific heat of air, and ρ_{air} the density of air.

The difference between the known radiometric surface temperature and the unknown aerodynamic surface temperature, which theoretically should be applied, is taken into account within the formulation of r_a. This is achieved through the definition of different roughness lengths for heat (z_{0h}) and for momentum (z_{0m}) (in s/m):

$$r_a = \frac{\ln\left(\frac{z_m - d}{z_{0m}}\right) * \ln\left(\frac{z_h - d}{z_{0h}}\right)}{k^2 * u(z_m)} \tag{A5}$$

d being the displacement height, k the von Karman constant, z the observation height and u the wind speed at height z_m. Finally, the evaporative fraction (EF) is derived according to equation A6:

$$EF = \frac{\lambda ET_i}{(R_{net} - G)_i} \tag{A6}$$

SCALE ISSUES IN DROUGHT MONITORING

A.A. VIAU [1] and J.V. VOGT [2]

[1] *Centre de recherche en géomatique, Université Laval Ste Foy, Québec, G1K 7P4, Canada*

[2] *Space Applications Institute, EC Joint Research Centre 21020 Ispra (Va), Italy*

Abstract. Drought has spatial and temporal dimensions and can be defined at a variety of scales. The complex nature of the phenomenon requires the consideration of very different data for monitoring and impact assessment. Albeit their differing spatial and temporal properties, these data need to be integrated in adequate models. One of the challenges for drought monitoring, therefore, is related to the determination of appropriate scales of analysis, since scale considerations are integrally linked to the definition and objectives of drought monitoring. This paper discusses different aspects of scale with relevance to drought monitoring and indicates the potential of GIS and remote sensing for data integration.

1. INTRODUCTION

Scale is one of the most fundamental and ambiguous concepts underlying geographical and environmental information. The term 'scale' has multiple meanings, including absolute size, relative size, spatial resolution, temporal resolution and duration (see, for example, Goodchild and Quattrochi, 1997; Lam and Quattrochi, 1992, Meentemeyer, 1989). In the context of drought monitoring it is important to consider the way meteorologists, climatologists, agronomists, hydrologists and other experts managing drought situations handle scale-related issues and how varying spatial and temporal scales of analysis influence our capabilities to detect and monitor drought and its impact.

Drought as such is difficult to define. Two of the simplest yet meaningful meteorological definitions are 'a prolonged and abnormal moisture deficiency' and 'a worrisome lack of rain' (Palmer, 1965; Wilhite and Glantz, 1985). However, drought needs to be defined more specifically in terms of the impact of the moisture deficiency in agricultural, hydrological, ecological or socio-economic terms. In relation to these categories, water stress establishes at different spatial and temporal scales. An agrometeorological drought, for example, may establish over a few days or weeks due to missing rainfall combined with a strong evaporative demand, leading to rapid soil moisture depletion. Hydrological drought establishes at longer time scales. It is a function of catchment response, which depends on the size of the catchment, its climate, geology and land use and the complex relation between surface- and groundwater resources. In a meteorological sense, drought may be characterised by a long-term

J.V. Vogt and F. Somma (eds.), Drought and Drought Mitigation in Europe, 185–193.

cycle. Such a cycle, however, will be apparent only if the time series of precipitation data studied are long enough to capture the periodicity in the signal. In shorter time series, the same events may have an episodic characteristic.

The concept of drought is further complicated due to the fact that it is a meteorological *and* a socio-economic phenomenon. Its impact on society results from the interactions between the natural event (less precipitation than expected due to natural climatic variability) and societies' demand for water supply (Figure 1).

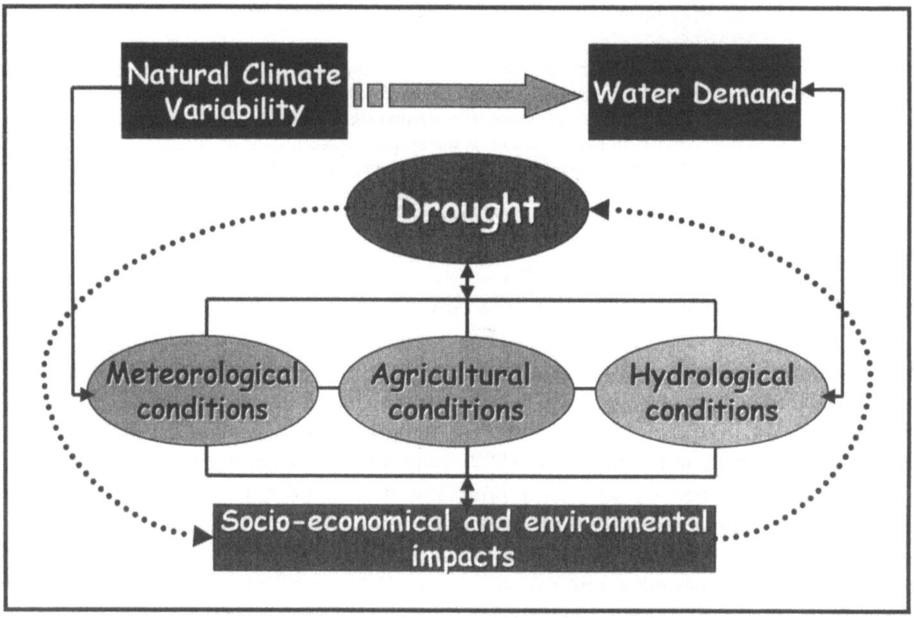

Figure 1: Natural and socio-economic factors determining drought.

As a consequence, drought means different things to different people and affects them at different spatial and temporal scales. For the farmer drought results in crop failure due to soil moisture shortages at critical times (field; week, month, season). The hydrologist will be concerned with reduced reservoir storage and streamflows over more extended periods of time (catchment; month, season, year). The ecologist will study the short and long-term impact on plant and animal communities (region; month, season, year, decade), and the economist will show interest in a drought only when it significantly affects the economy of a relatively large area (region, country; years, decade).

In summary, the spatial and temporal scales of droughts and their impact range from the field to extended regions and from a few days to several years.

2. SCALE AND DROUGHT MONITORING

As a consequence of its specific spatio-temporal characteristics, the severity and impact of a drought are very difficult to measure. A meteorological drought establishes over relatively long time periods and may affect large areas with varying local impacts, depending on the water demand of nature and society. The extent of these impacts in both space and time determines its severity. An agrometeorological drought, for example, can have a severe impact because it covers a large area for one growing season or it can be severe because it affects a smaller area for several years.

Drought monitoring requires the consideration of a large number of variables on climate and meteorology, land cover and land use patterns, pedology, and socio-economic demands for water. Depending on the application, different types of data, such as weather data, measurements of river discharge, groundwater levels and soil water content, or information on water abstraction rates are major sources of information. These variables, which have to be considered in both their spatial and temporal dimensions, have decisively different characteristics. They have different temporal dynamics and are referenced to very different units such as points, polygons or administrative entities. The vegetation cover, the underlying soil type and the surface topography, for example, are highly variable in space. Therefore, both the microclimatic conditions and the soil moisture conditions are characterised by a high spatial and temporal variability. This poses considerable problems to the interpolation or generalisation of point measurements. For the interpolation of meteorological data, for example, the general characteristics of the measurement site and the specific characteristics of the measurement station need careful consideration (Viau *et al.* 2000).

As a consequence, the following scale-related questions have to be considered for each field of application:

- Which are the most representative spatial and temporal scales of measurement?
- Which is the adequate period of reference?
- Which indices should be used (based on which parameters)?
- Which surface characteristics should be monitored?
- How can the information be spatially and temporally aggregated?
- How can the error propagation during data processing be quantified?
- How can the accuracy of drought indices be evaluated at different temporal and spatial scales?

Figure 2 schematically summarises the different domains of the Earth system involved in processes leading to a drought as well as their spatial, temporal and vertical dimensions.

2.1. Spatial Scale

Integrating climatological and meteorological data with land cover and soil data is essential for drought monitoring and for the development of indicators that can be used at different spatio-temporal scales. The level of detail or resolution of the required input data can vary depending on the goal of the investigation. Addressing the water needs for crops, for example, may require high resolution information to adequately describe specific soil characteristics and habitat requirements. Lower resolution data will be sufficient for identifying the diverse crops necessary to support agricultural biodiversity under given or changing climate conditions.

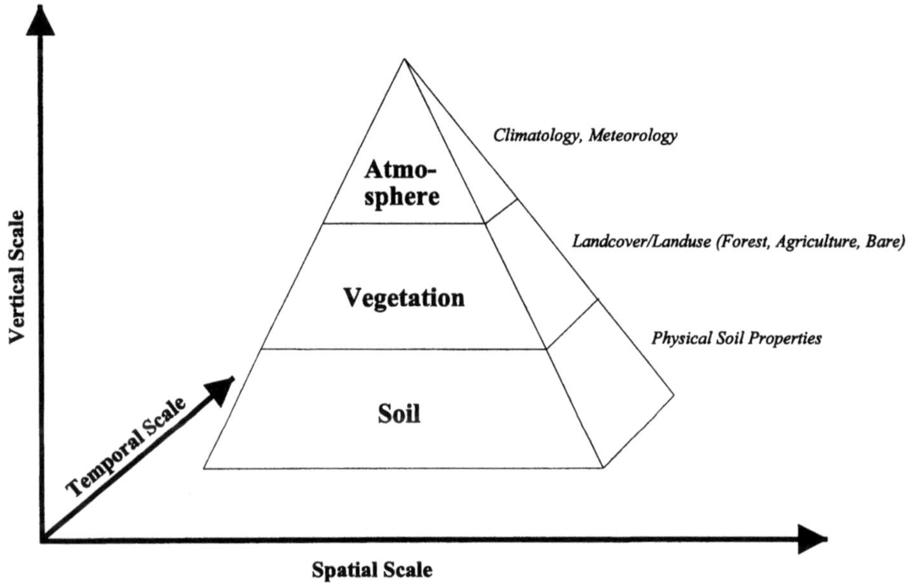

Figure 2: Scales and domains important in drought monitoring (Vogt and Viau, 1998).

Climate conditions and land cover changes are key elements in drought understanding and the measurement of its intensity. In this context, it is becoming increasingly important to model interactions between human activities and the state of the environment. Environmental processes and natural and human-induced disturbance produce patterns and cause changes in the diversity the landscape and of natural resources at a range of spatial scales. The basis for analysing the environmental impact of such changes is a detailed understanding of the relationships within the soil-vegetation-atmosphere system and their spatial and temporal variability, especially with respect to the energy and water balance.

Since the response of the vegetation to changes in the environment and climate is variable in space and time, drought analysis needs to consider three criteria of particular importance:

- behaviour as geographic scale changes,
- behaviour in relation to hierarchical changes in soil-vegetation-climate relations,
- the influence of spatial data quality.

In summary, in order to get a realistic picture of the spatial pattern of water stress, the spatial distribution of a drought indicator needs to be modelled at different scales, using environmental and/or socio-economic data with an adequate level of detail.

2.2. Temporal Scale

Temporal scales play a role in drought monitoring and management through both the consideration of planning horizons and through the identification of appropriate historical perspectives. Drought frequency is a function of the natural variability of climates and water demand and drought disturbance can last from weeks to several years. The persistence of a drought can initiate irreversible impacts such as desertification processes and may cause a severe degradation of the environment with considerable feedback on the climate. It is, therefore, important to consider potential long-term changes in climate and its effects on the precipitation regime, the vegetation distribution and the soil physical properties.

A good understanding of historical ranges of variability can provide important reference points in this endeavour. Whilst droughts have always been influencing ecosystems to varying degrees and spatial scales, human alteration of natural ecosystems has been dramatic over the last two centuries. This has been changing the impact of drought disturbance at a variety of spatial and temporal scales.

Temporal aspects of drought are difficult to define and the analysis has to be performed for different time-scales with the respective application in mind (McKee *et al.* 1993, 1995). The usual temporal sequence of agricultural, hydrological and socio-economic impacts of a drought may highlight the issue. In the course of a drought, rainfed agriculture is usually the first to be affected because of its heavy dependence on stored soil water. Soil water can be rapidly depleted during dry periods, especially when high temperatures and winds occur. Only with continuing precipitation deficiencies, sectors relying on other sources of water will be affected as well. Those sectors relying on streamflows and/or water stored in reservoirs and lakes and/or groundwater (i.e. irrigated agriculture, industries, urban supplies) are affected with a delay, if at all. A short-term drought that persists for 3 to 6 months may have little impact on these sectors, depending on the characteristics of the hydrologic system and the water requirements.

When precipitation returns to 'normal' and meteorological drought conditions have abated, the sequence is repeated for the recovery of surface and subsurface water supplies. Soil water reserves are replenished first, followed by streamflow, reservoirs and lakes, and groundwater. Drought impacts may diminish rapidly in the agricultural sector because of its reliance on soil water, but linger for months or even years in other sectors dependent on stored surface or subsurface supplies. Groundwater users, often the last to be affected by drought during its onset, may be last to experience a return to 'normal' conditions. The length of the recovery period is a function of the intensity of the drought, its duration, and the quantity of precipitation received as the episode terminates.

In summary, drought monitoring needs to consider a variety of temporal scales in order to draw a realistic picture of drought impacts in different environmental and socio-economical domains. A comprehensive drought watch system will have to include drought indices covering multiple time scales in order to serve the needs of society and decision-makers.

3. SCALE AND DATA INTEGRATION IN GEOGRAPHICAL INFORMATON SYSTEMS

The data needed for drought monitoring and analysis originate from many different sources and are referenced to very different spatial units. Geographical Information Systems (GIS) allow to integrate these data during the analysis process and to represent phenomena at multiple scales. The nature of the data (i.e. point, vector, polygon) and the size of their spatial reference unit determine the level of spatial detail and the scale of monitoring and modelling. The problems related to the interpolation of point data and the conversion of vector and polygon data to grid data are important issues in this context. It is important to recognise that data may carry different information when presented at different scales and that process models are scale dependent. Models developed on one scale may, therefore, not be applicable at another scale (Bian, 1997). The questions of up-scaling and down-scaling from the local to the global level and the integration of process models from different scales, therefore, have emerged as primary issues over the last decade. The use of remote sensing data and the integration of data and model results in GIS play a central role in this discussion (e.g. Stewart *et al.*, 1996; Quattrochi and Goodchild, 1997).

However, GIS systems still offer only limited capabilities for multi-scale analysis. Albeit these difficulties, developers of Decision Support Systems (DSS) pay little attention to scale and how scale issues influence decision processes. Particular questions to be studied include the scale dependence of landscape and land use patterns and of meteorological and environmental processes and how this is understood by users and decision-makers. Other questions relate to the difficulties in communicating issues of scale, and to more or less effective ways to accurately represent drought information at different scales.

Very little research has been performed to date to guide the development of Decision Support Systems for drought monitoring and only few examples exist. One example is a Drought Impact Information System for Spain, which has been presented by Killmayer (1993). It is based on remote sensing and covers different scales of analysis. It shows how remote sensing data of different resolution can be used in the stages of drought monitoring and impact analysis.

One of the major uses of GIS for drought monitoring is to provide an environment that facilitates the integration of very different data for distributed environmental modelling. The input fields required to drive such models are derived from a diversity of data types, such as raster data (e.g. remote sensing data), point data (e.g. agrometeorological data), and vector data (e.g. cartographic data, pedological data). Each type of data needs pre-processing to insure spatial conformity and temporal integration of all information.

Remotely sensed data and model output fields share an important property concerning their spatial variability: both are inherently scale limited. For remotely sensed observations, little information can be inferred below the resolution of the sensor. For distributed outputs, the model grid spacing determines the smallest spatial scale at which fields may be realised. The limitations imposed by resolution and grid spacing, or their interaction, can seriously hamper modelling efforts over large areas, especially when the process being modelled is spatially autocorrelated. Such processes often scale non-linearly such that the moments of the field obtained at one spatial scale may be significantly different from the ones obtained at a larger or smaller scale. The usual mode of exploration has been to realise fields at fine scales, then to aggregate and compare the results with those realised at coarser scales. The hope is that either linear or otherwise parameterisable relationships may be found that link field statistics, at least for some limited range of scales.

The increasing availability of GIS technology and decreasing restrictions in hardware and software recently led to very high demands for computerised databases. Difficulties to handle the amount of data lead to the question of the minimum level of detail necessary to achieve a maximum of information, especially at regional to continental scales.

Although more and more digital data are available, adequate data, for example on land cover, soil physical characteristics, or high resolution meteorological fields, are often missing for large-scale drought impact studies. Remote sensing can play a key role in filling this data need. It allows for the retrieval of relevant surface parameters with different spatial and temporal resolutions. The resulting gridded data are easily integrated into a GIS for further analysis in conjunction with other data layers. This information can be highly relevant for the assessment of the extent and severity of droughts and can provide a relevant contribution to a global monitoring of drought conditions (e.g. Gutman, 1990; Kogan 1997; Vogt *et al.* 2000).

4. CONCLUSION

Cost- and time-efficient methods for drought monitoring and assessment will need flexible tools, capable to integrate data from many different sources and with different characteristics in both space and time. Understanding the factors that control the spatial distribution of drought requires an intimate knowledge of the interaction between landscape composition, soil characteristics and meteorological conditions. These interactions, however, are strongly scale dependent and need to be modelled at scales ranging from the farm to the continent.

The goal of a drought watch system, therefore, is the prediction and monitoring of drought events and their impacts at different spatial scales. The analysis of the spatial and temporal variability of these impacts will lead to a better understanding of likely vegetation and surface changes. Models capable to quantify the severity of water stress over space and time and capable to correlate it with ecological processes will be of use for such impact assessment. GIS systems provide the basic tools to develop such models and to analyse the data at a variety of scales. However, this needs an improved understanding of scale inherent problems such as the relation between scale and model results, the optimum spatial and temporal resolution of input data, or the error due to data interpolation and aggregation.

REFERENCES

Bian, L. (1997) Multiscale Nature of Spatial data in Scaling Up Environmental Models, in: D.A. Quattrochi and M.F. Goodchild (eds.), *Scale in Remote Sensing and GIS*, Lewis Publishers, Boca Raton, New York, London, Tokyo, 14-26.

Goodchild, M.F. and Quattrochi, D.A. (1997) Scale, Multiscaling, Remote Sensing and GIS, in: D.A. Quattrochi and M.F. Goodchild (eds.), *Scale in Remote Sensing and GIS*, Lewis Publishers, Boca Raton, New York, London, Tokyo, 1-11.

Gutman, G.G. (1990) Towards Monitoring Droughts from Space, *Journal of Climate*, 3, 282-295.

Killmayer, A. (1993) Drought Impact Information System, *Proc. 6th AVHRR Data Users' Meeting*, Belgirate, Italy, 349-354.

Kogan, F.N. (1997) Global Drought Watch from Space, *Bulletin of the American Meteorological Society* 78(4): 621-636.

Lam, N. and Quattrochi, D.A. (1992) On the Issue of Scale, Resolution and Fractal Analysis in the Mapping Sciences, *Prof. Geogr.* 44: 88-98.

McKee, T.B., Doesken, N.J., and Kleist, J. (1993) The Relationship of Drought Frequency and Duration to Time Scales, *Proc. 8th Conference on Applied Climatology*, Anaheim / CA, 179-184.

McKee, T.B., Doesken, N.J., and Kleist, J. (1995) Drought Monitoring With Multiple Time Scales, *Proc. 9th Conference on Applied Climatology*, Dallas / TX, 233-236.

Meentemeyer, V. (1989) Geographical Perspectives of Space, Time, and Scale, *Landscape Ecology* 3(3/4): 163-173.

Palmer, W.C. (1965) *Meteorological Drought*, Research Paper No. 45, U.S. Department of Commerce Weather Bureau, Washington, D.C.

Quattrochi, D.A. and Goodchild, M.F. (eds.)(1997) *Scale in Remote Sensing and GIS*, Lewis Publishers, Boca Raton, New York, London, Tokyo, 406 p.

Stewart, J.B., Engman, E.T., Feddes, R.A. and Kerr, Y. (eds.)(1996) *Scaling Up in Hydrology Using Remote Sensing*, Wiley, Chichester, 255 p.

Viau, A.A., Vogt, J.V. and Paquet, F. (2000) Évaluation de la représentativité spatiale thermique des stations météorologiques du réseau d'Andalousie, *International Journal of Remote Sensing* (in press).

Vogt, J.V. and Viau, A.A. (1998) *DEMIAN: Drought Detection, Monitoring, and Impact Assessment. Drought Research and Development Issues*, Space Applications Institute (internal document), 12 p.
Vogt, J.V., Niemeyer, S., Somma, F., Beaudin, I., and Viau, A.A. (2000) Drought Monitoring from Space, in J.V. Vogt and F. Somma (eds.), *Drought and Drought Mitigation in Europe*, Kluwer Academic Publishers, Dordrecht, 167-183.
Wilhite, D.A. and Glantz, M.H. (1985) Understanding the Drought Phenomenon: the Role of Definitions, *Water International* 10(3): 111-120.
Wilhite, D.A. (ed.) (1993) *Drought Assessment, Management and Planning: Theory and Case Studies*, Kluwer Academic Publishers, Boston, 289 p.

Part 5
Drought Impact

SOCIO-ECONOMIC ASPECTS OF DROUGHTS

A. GARRIDO and A. GÓMEZ-RAMOS
Departamento de Economía y Ciencias Sociales Agrarias
ETSI Agrónomos, Universidad Politécnica de Madrid
28040 Madrid, Spain

Abstract. The 1992-95 Spanish drought has been denominated the worst of the century. Its severity caught water managers and politicians devoid of instruments and strategies to mitigate its serious effects. Overall water shortages ignited conflicts and forced urban suppliers to cut water services affecting millions of citizens. The agricultural losses affected rain-fed farming, but were far more remarkable in irrigated agriculture, which suffered a series of three seasons of virtually no water supply. In the South of Spain, and particularly in the Guadalquivir basin, the drought had the largest social impact.

The objective of this paper is to present a methodology to evaluate the effects of the drought in the irrigation sector. It aims to provide an *ex-post* overview of various key variables, such as farmers' net returns and utilised seasonal labour along the period 1991-1998, comprising the drought seasons. Although the modelling framework is still under development, the preliminary results seem to support the view that under drought conditions hydrological criteria might be in conflict with economic conditions. Furthermore, various economic interests linked to the supply of irrigation water are not perfectly correlated. These findings support two preliminary conclusions. Firstly, the rules to allocate scarce water under stressful conditions are subject to multiple and non-compatible objectives. And secondly, despite the previous point, there might be alternative rules to allocate water on a multi-annual basis that provide more benefits to all interested groups than the rules followed during the last drought experience.

1. INTRODUCTION

To a Spaniard the occurrence of drought is perceived as common as Russians expect freezes in winter. Society as a whole has struggled to adapt to lapses short of rain since the beginning of the ages. Depending on the severity or length of a dry period, the people of rural towns resort to various religious rites, each one with increasing importance as problems get worse, pleading God, local saints or virgins for the return of the rains.

Up until the 1992-95 severe drought, hardly any effort to reduce the probability and the severity of dry periods was not related to the construction of hydraulic works. The same homogeneity goes for the panoply of alleviating measures delivered when drought conditions set in, these mostly being economic relief packages for farmers. The well established fact that drought effects can be offset both by public policies *and* by individual agents has always been dismissed.

J.V. Vogt and F. Somma (eds.), Drought and Drought Mitigation in Europe, 197–207.

In a country so prone to droughts such as Spain — no matter how they are defined —, drought management should be a common and permanent source of concern. But drought management is primarily a forward looking task, and ultimately results in different patterns of risk distribution among social groups and individuals. Both make drought management very complex, but certainly no more than allocating public funds to the different programs and priorities of a other government branches.

In the absence of markets for drought insurance, the evaluation of the economic effects of drought is hampered with all sorts of theoretical, ethical and empirical difficulties. The closer we get to individuals to study how drought affects their lives, the more complex is any evaluating effort. But only under severe drought conditions do individuals become affected by the shortage of water. And we would contend that although severe droughts may inflict serious harm to the peoples that suffer from them — a fact that justifies emergency measures —, less severe ones constitute a research topic of enormous economic importance, particularly in regions in which water supply is the basis of a significant proportion of employment and output generation (Garrido *et al.*, 1998).

1.1. The 1992-95 Spanish Drought

The average precipitation in the Guadalquivir Basin is about 600 mm. In the four year period 1992-95, the annual average precipitation was 397 mm. The sequence of allotments to the irrigation sector that relies on the Basin's regulatory system were: 2,100 m^3/ha in 1992, 180 m^3/ha in 1993, 760 m^3/ha in 1994, and nil in 1995 (4,000 m^3/ha being the post-drought average allotment) (Saura Martínez, 1995). Although many other basins and regions were severely injured by the 1992-95 drought, the Guadalquivir Basin's represents by far the most dramatic example of society's vulnerability to water shortages.

According to various farmers' representatives, the drought affected about 500,000 hectares of irrigated land and caused economic losses of 3.0 to 4.2 billion Euros, directly or indirectly linked to the agricultural sector. Economic losses resulting from water shortages and from the reduction in water quality in many urban areas, including cities such as Seville, have not been made yet. But we must take into account that the Water Company that serves the city of Seville is considering building another dam to increase the guarantee of supply and to avoid the severe situation of the 1994-95 years.

During the 1993-95 period, the government issued various decrees aiming to solve specific problems caused by water shortages. First, urban suppliers are granted minimum volumes to meet their water demands and farmers with permanent crops and intensive livestock farms are given high priority as well. Second, farmers not included in the previous categories were granted special tax relief and the government provided job opportunities to unemployed seasonal farm workers. In addition to these measures, small-scale projects such as desalinisation plants, drilling wells or water carriers were financially promoted and contracts awarded following urgent procedures. While no in depth study has been made to evaluate the success of these last structural measures, apparently hardly any of these projects began to supply water before the drought finished in December 1995.

One of the most controversial issues that has emerged in Andalusia once the drought finished is the project to build a new dam, called *El Melonares*. Its main purpose would be to add extra security to the water supply of the city of Seville, and secondarily to serve one irrigation district, called *El Viar* (Del Moral, 1998). Despite the ecological cost of erecting such facility in a very valuable ecosystem, its construction is challenged by proponents of an agreement between the farmers and Seville's water company. Such an agreement, inspired by ideas already developed by Michelsen and Young (1993), Waters (1995) and Fisher *et al.* (1995), would contractually dictate the distribution of water amongst users under normal as well as abnormal conditions. Perhaps, the fact that the Water Law establishes clear priorities with regard to the rationing of water among different users explains the reluctance of the farmers to "sell" the water and to enter into negotiations with the city. They fear that eventually the city would invoke the Water Law to appropriate the water without compensating the farmers.

2. OBJECTIVES AND SCOPE

This paper has three objectives. First, it addresses the issue of drought and how it is perceived in Spain, highlighting the findings of the most recent literature. Secondly, it presents a methodology to evaluate the socio-economic effects of drought in the irrigated agriculture in the South of Spain. Although our modelling effort still yields preliminary results, it is halfway towards a more general and comprehensive framework to get economic drought indexes. Third, it shows that drought effects affect various social groups belonging to the rural world differently.

Although the paper's scope is limited to irrigated agriculture, it aims to look beyond net revenue downfalls that result from water shortages. Intensive agriculture demands seasonal labour and variable inputs, and investments in irrigation equipment. Whereas net revenue measurements certainly provide a clear view of how droughts affect a regional economy, of equal importance are the trend of hired labour and the indirect effects on the rural economies.

The Spanish Water Law (enacted in 1986) sets unambiguous guidelines and criteria to allocate water during drought conditions. First in the list is the supply of water to cities; and second — and above industrial users — are the agricultural users. In the Guadalquivir Basin and under normal conditions, irrigators represent more than 75% of the basin's total consumptive users. This means that water allotments are subject to reductions more often than most people realise. Now, if water is rationed among irrigators so frequently, multiple questions arise as regards to:

- whether water allocation is efficient and equitable among farmers;

- whether, given the best information at hand, water could have been allocated more efficiently and/or equitably, in full compliance with the Water Law's criteria; and,
- whether alternative allocative rules could have performed better.

In this paper, we attempt to characterise the past drought and seek to measure its effects focusing on farmers' returns, labour demand, and capital investment in irrigation equipment. By linking economic indexes of various variables with a simple index of water availability, and tracing them along the 1991-1997 period we will show the differing levels of vulnerability of farmers, farm workers and suppliers of irrigation equipment.

3. METHODOLOGY

Our modelling framework originates from a previous generation farming model whose details are given in Varela-Ortega *et al.* (1998) and Sumpsi *et al.* (1998)[1]. Essentially, the model simulated a given farmers' behaviour with regard to water use, investment in irrigation technologies, land allocation and production techniques, solving a forward looking dynamic problem with perfect information on prices, water availability and agricultural policy constraints. Whereas in the cited references, we looked at the steady-state solution that resulted from a change in the criteria, system and level of irrigation water taxation, in this paper the focus is centred on how variables change each farming season and the model is fed with the sequence of water allotments that actually took place during the 1991-95 period.

The central assumptions of the model are the following:

(1) Crop and input prices are non-stochastic.
(2) The objective value is the discounted sum of a farm's returns.
(3) Water availability is known in advance and reconstructs the sequence that took place in the simulated period.
(4) Farmers can adapt to water shortages pursuing various non-exclusive strategies:
    - Growing rain-fed crops,
    - Shifting to less water demanding crops,
    - Investing in water-saving irrigation technologies,
    - Fallowing land.

The technical parameters of the model have been obtained through in-depth interviews of the managers of the irrigation districts and technical experts knowledgeable of the region's agriculture. Both input and output prices as well as agricultural subsidies are based on the Spanish and EU official statistical sources.

---

[1] We acknowledge the joint effort of J. Sumpsi, M. Blanco, C. Varela-Ortega and E. Iglesias to develop the original model and the Spanish Ministry of Agriculture for funding the project "*Water economics and the competitiveness of the Spanish irrigation sector*", 1995-96.

The model is run on GAMS (Brooke *et al.*, 1992), using a non-linear algorithm to find model convergence and solutions. In total, seven representative farms belonging to two irrigation districts located in the Guadalquivir basin have been simulated. Each one's parameters attempt to represent various kinds of farm types and farming practices. Both cropping patterns and irrigation techniques are based on field work conducted in 1995.

4. RESULTS

Among the model's results, we are mainly interested in tracing a few key variables along the period 90-98, which comprises the 92-95 drought period. The results pertain to two irrigation districts of interest that are located in the Low Guadalquivir basin. The first, called "El Viar", uses surface water and has good soils, and favourable hydrological conditions due to the fact that it has a reservoir with solid water rights. The other, called "Bajo Guadalquivir (B-XII)", uses surface water as well, but has salinity problems and its water supply depends on the general basin's regulation, which in turn depends on the water available from six important Guadalquivir dams.

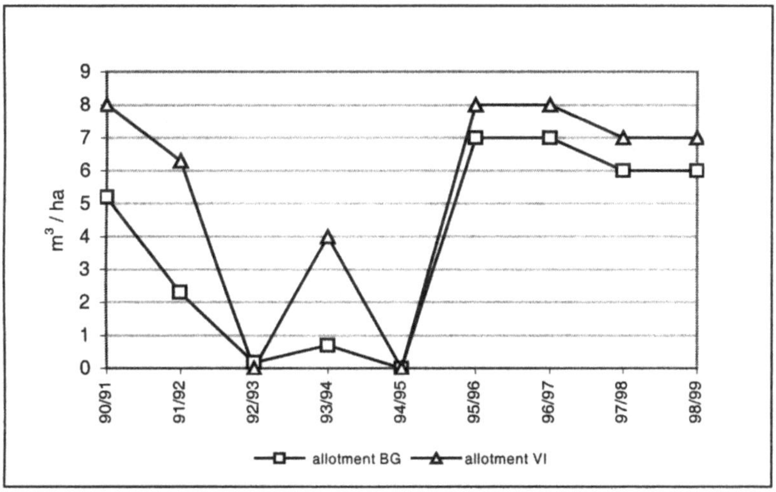

Figure 1: Water allotments in the Bajo Guadalquivir (BG) and El Viar (VI) irrigation districts.

The actual values of water allotments for both districts are depicted in figure 1. The first thing that would seem to be in conflict with the Water Law is that water was not rationed equitably. The "El Viar" (VI) water district had a reasonable good year in between the two most severe ones, namely 92/93 and 94/95, whereas the "Bajo Guadalquivir" (BG) water district was given less than 10% of its regular allotment during the 93/94 season. Several explanations might be given for this anomalous behaviour: the first one is the fact that VI's representatives are far more numerous on their dam's release committee than BG's representatives in the general basin's committees; second is the fact that BG is a tail district and water allocation under stressful conditions tends to favour to a greater extent the middle basin districts and those that are directly linked to single dams; and thirdly that precipitation in the VI catchment area was larger than in the whole Guadalquivir Basin. None of the three explanations have any support according to our water law, which states clearly that all farmers should have suffered to the same extent the impacts of a given drought. Despite this, the VI district was given priority in 1993/94 and granted water supply to make emergency water applications on tree orchards. Since in the previous and subsequent season the VI district did not supply water at all, farmers with permanent crops tapped the shallow aquifer to obtain the needed flows to insure the survival of their trees. It is noteworthy to mention that this groundwater abstraction is not permitted and that, due to its proximity to the river bed, the pumped resources reduced the main river flow.

We look now at the influence of water shortages in the representative farmers' net returns. Figures 2 and 3 depict the models' results for each of the water districts' representative farms.

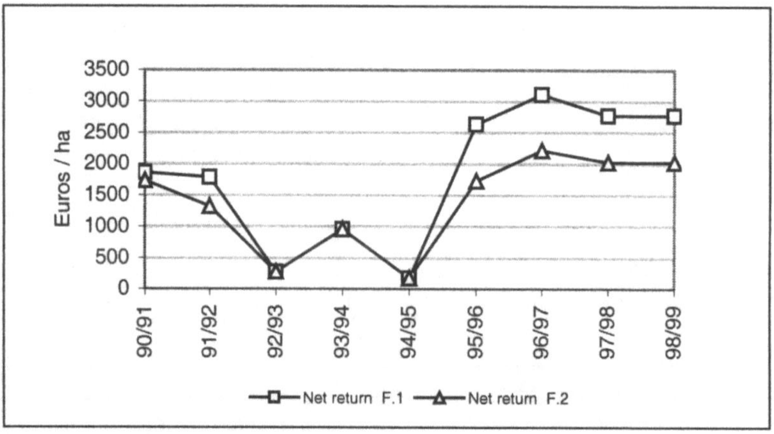

Figure 2: Net returns of two representative farms in the Bajo Guadalquivir irrigation district.

Each line represents the model's outcome with regard to per hectare net revenue for each representative farm. Differences among representative farms result from widely differing cropping patterns and attitudes towards economic risk (a fact that has already been established by Amador *et al.*, (1998)). On both districts, net returns virtually collapse to zero on the two seasons with zero allotments (92/93 and 94/95) and they increase more than proportionally in the in-between season, because the prices of cotton and other basic commodities rose as a result of short supply. In the VI district the financial results are outstanding, better than in normal years, simply because output prices were high, and the district managed to supply their farmers about 4,000 m^3 of water per hectare. In addition to that, the VI district signed a contract with the water company of Seville. With this contract the district "sold" 30% of its regular allotments at a reasonable price to the water company.

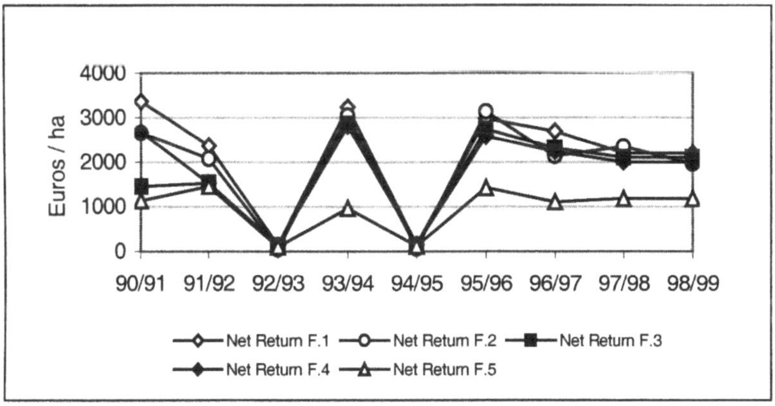

Figure 3: Net returns of five represenative farms in the El Viar irrigation district.

Another significant favourable factor is the new economic environment that the reform of the Common Agricultural Policy (CAP) gave rise to after the 1992/93 season, by which price support mechanisms were replaced by income support systems unrelated to production levels. Without the direct support system that came along after the CAP reform, both the shortage of irrigation water and the extremely low yields in rain-fed crops would have caused a wave of farms bankruptcy. Farmers could at least meet their financial obligations and finish those years with slightly positive returns.

As might be expected, the amount of hired external labour experienced a significant reduction during the drought. Figures 4 and 5 depict the simulation results with regard to the value of seasonal labour hired in each of the representative farms. Two remarks need to be made to avoid reaching misleading conclusions about the drought effects on the labour market. The figures show that under normal hydrological conditions, total labour use was reduced in VI and has risen in BG. In the VI district, the effect of the CAP reform, providing larger relative incentives to grow crops entitled to direct payments than before the 1992/93 season, has had a larger impact than in the BG district. Contrary to the farm workers' interests, the irrigated crops entitled to direct payments use less labour than those that are not irrigated, and this explains the relatively lower level of labour utilisation from 1995-96 onwards than in the in pre-drought seasons. Despite this, the effects of drought differ across water districts. For instance, the BG's representative farm coded with 1 (F.1 and L.Use 1) has recuperated significantly both its labour utilisation and its net returns as a result of its intensive farming of horticultural crops (Figures 3 and 4).

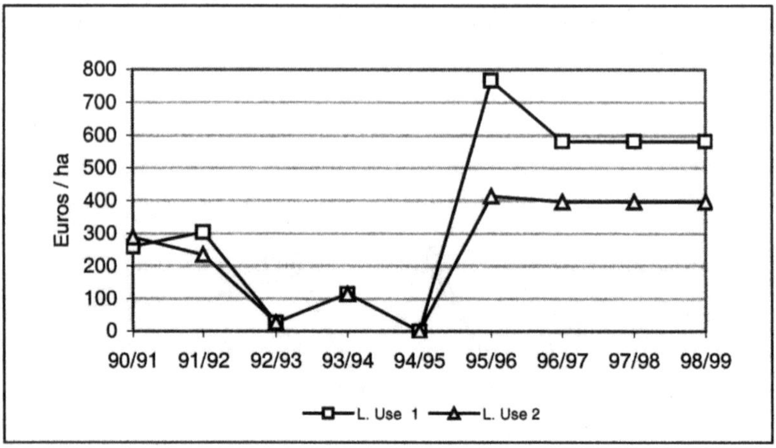

Figure 4: *Seasonal labour utilisation in the Bajo Guadalquivir irrigation district for two representative farms (L.Use 1 and L.Use 2).*

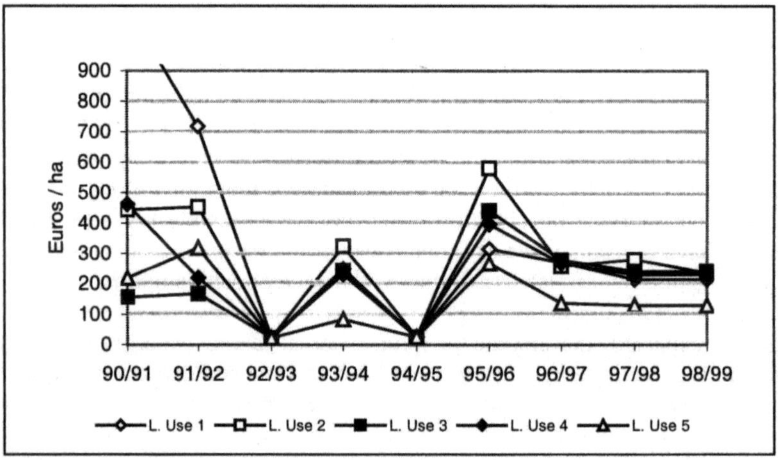

Figure 5: *Seasonal labour utilisation in the El Viar irrigation district district for five representative farms (L.Use 1 through L.Use 5).*

In the VI district, labour use increased in the 1993/94 season to a lesser extent than farmers' net returns (Figures 3 and 5). This indicates that the benefits of VI's extra water reserve accrued on farmers more than on farm workers, as a result of the larger net margins generated by the rise of output prices. In the BG district the labour utilisation and the farmers' net returns are more correlated, even in the anomalous 1993/94 season (Figures 2 and 4).

Table 1: Evolution of various key variables in the Bajo Guadalquivir and El Viar irrigation districts.[1]

Years	Bajo Guadalquivir district			Viar District		
	Subsidy /returns (%)	Non irrigated Surface/ total (%)	Investment total cost Euros/ha	Subsidy /returns (%)	Non irrigated Surface/ total (%)	Investment total cost Euros/ha.
90/91	2.24	1.88	80	1.21	12.45	70
91/92	2.98	29.64	240	1.17	0.40	170
92/93	57.31	96.50	450	83.74	100.00	0.00
93/94	28.35	86.39	30	11.12	15.15	70
94/95	70.52	100.00	0.00	81.62	100.00	370
95/96	6.25	0.00	80	6.90	12.11	60
96/97	6.85	0.00	880	10.03	11.89	60
97/98	6.85	0.00	130	8.96	11.89	60

[1] The results for each irrigation district are obtained as a weighted average of the representative farms.

Table 1 reports the simulation results with regard to the percentage of CAP subsidies over the farmers' net returns, the percentage of fallowed land or non-irrigated land over total land, and the total expenditure in investments in irrigation equipment and maintenance, measured in Euros/ha. The results show unambiguously that the subsidies of the CAP reform provided 57% to 84% of the farmers' net returns, offering a financial cushion to the landowners or tenants. This result has been confirmed by Gómez-Limón *et al.* (1996). Also the other two variables show that the indirect effects on the rural economies must have been noteworthy, although difficult to evaluate. Assuming that the variable costs of a hectare devoted to cotton or sugar beet are in the range of 1200 to 1600 Euros/ha, and that none of these crops were planted during three years, the economic losses of local business are surely remarkable. Similarly, the suppliers of irrigation equipment and maintenance services experienced low sales and activity.

5. DISCUSSION

The curves shown previously tell interesting stories about the management of drought, at least from the perspective of the irrigation sector. Some are supported by the empirical investigations performed for the purposes of this paper, whereas some others will require more sophisticated models and more realistic assumptions.

First, we summarise the points that can be made as a result of this preliminary analysis:

- When managing water under stressful conditions, it might be the case that hydrological criteria are not fully consistent with economic criteria. For instance, the value of water for a given district rises significantly if in other districts water supply is very short. Hence, in the absence of water markets, districts with ample scope and technical capacity to manage their water reserves on a multi-annual basis, might obtain significant benefits if their water supply curve is counter-cyclical to that followed by the majority of the districts.

- With regard to economic performance, there is another source of conflict among the farmers and farm workers. In the middle of a drought the speculative farmers' value of water rises more than does the farm worker's value of water. Furthermore, a farmer has in principle more financial capacity to save and defer consumption than do seasonal workers, thus it is in his/her interest to maximise the present value of the flow of benefits. Yet, for a low-income farm worker two factors might support a water objective other than maximising the present value of farmers' net revenue. First, his capacity to escape from the a steady flow of income, saving and borrowing is constrained. Second, the option to save water for subsequent seasons in order to take advantage of higher output prices does not benefit farm workers because their labour supply is very elastic, implying that farm workers' wages will not increase as a result of labour demand increases.

- The direct payments based on the CAP-reform have provided a financial cushion for irrigated agriculture during drought years. In the absence of them, irrigators would have suffered more financial losses. But, again, farm workers do not benefit at all from these subsidies.

In addition to the conclusions that are supported by this study, there are a few emerging ideas that merit attention and will be addressed in future studies:

- The fact that within the 1992 - 1995 period there were two seasons with virtually no water supply conveys a clear idea that drought management demands multi-period planning. Looking into the economics of that past drought suggests that other allocative criteria might have been used in the region. In particular, the hypothesis that water allotments could have been set accordingly with economic criteria, and not only based on hydrological procedures, should be tested. If an alternative pattern of water allocation criteria comes out from a pure forward looking methodology, then the basin's agency should modify its criteria.

- However, as this paper has shown, there is no single objective which the agency should try to optimise. During a drought, the value of water for farmers and for farm workers evolves differently, up to the point of being negatively correlated. In the region investigated — with high unemployment in the rural areas —, this translates into a political problem because the Spanish Water Law states at the outset that water resources belong to the public domain, and this implies that they ought to be managed to promote the general interest.

- Agricultural subsidies as they exist now might change in the future. Our empirical application has shown that the financial revenues from agriculture are largely dependent on the crop related magnitude of these subsidies. As the future prospects of the CAP show that both the criteria and the level of agricultural subsidisation might change, the economic management of droughts in the irrigation sector might demand fundamental changes. This reinforces the need to base decisions on allocation in the economics of farming, going beyond strict hydrological criteria.

ACKNOWLEDGEMENTS

This work is a result of an ongoing EU research project titled: *"Societal and Institutional Responses to Climate Change and Climatic Hazards: Managing Flood and Drought Risks (SIRCH)"* General Directorate XII, Science, Research and Development, European Commission Contract nº ENV4-CT97-0447, 1998-2000.

REFERENCES

Amador, F., Sumpsi, J.M., and Romero, C. (1998) A Non-Interactive Methodology to Asess Farmers' Utility Functions: An Application to Large Farms in Andalusia, Spain, *European Review of Agricultural Economics* 1998, 25(1): 92-109.

Brooke, A., Kendrick, D., and Meeraus, A. (1992) *GAMS. A User's Guide, Release 2.25*, The Scientific Press Series, Danvers, Massachusetts.

Del Moral, L. (1998) El Sistema de Abastecimiento de Agua de Sevilla: Análisis de Situación y Alternativas al Embalse de Melonares, *Nueva Cultura del Agua, Serie Informes* 5.

Fisher, A., Fullerton, D., Hatoh, N., and Rainelt, P. (1995) Alternatives for Managing Drought: a Comparative Cost-Benefit Analysis, *Journal of Environmental Economics and Management* 29: 304-320.

Garrido, A., Iglesias, E., and Calatrava, J. (1998) *Economics Applied to Drought Planning and Management for Damage Mitigation*, SIRCH Working Paper no. 2., Department of Agricultural Economics, Univiersidad Politécnica de Madrid.

Gómez-Limón, J.A., Sánchez, F.J., Rodríguez, A., and Lara, P. (1996) Evaluación del Impacto Socio-Económico de la Sequía en los Regadíos de la Campiña Baja (Córdoba). Un Aproximación Multi-Criterio, *Revista Española de Economía Agraria* 178: 163-192.

Michelsen, A. and R. Young (1993) Optioning Agricultural Water Rights for Urban Water Supplies During Drought, *American Journal of Agricultural Economics* 75: 1010-1020.

Saura Martínez, J. (1995) Recursos Superficiales y Subterráneos, in *La Gestión del Agua en Andalucía Ante las Experiencias de la Sequía*, Fundación El Monte, Sevilla.

Sumpsi J.M., Garrido, A., Blanco, M., Varela, C., and Iglesias, E. (1998) *Economía y Política de Gestión del Agua en la Agricultura*, Mundi-Prensa, Madrid.

Varela-Ortega, C., Sumpsi, J.M., Garrido, A., Blanco, M., and Iglesias, E. (1998) Water Pricing Policies, Public Decision Making and Farmers' Response: Implications for Water Policy, *Agricultural Economics* 19: 193-202.

Waters, P.A. (1995) *Efficient Pricing of Water Transfer Options: Non-Structural Solutions for Reliable Water Supplies*, Unpublished, University of California, Riverside.

ASSESSMENT OF THE VULNERABILITY OF A RIVER SYSTEM TO DROUGHT

S. DEMUTH, B. LEHNER and K. STAHL
Institute of Hydrology
University of Freiburg
79098 Freiburg, Germany

Abstract. Based on the regionalisation of recession parameters a method has been developed to estimate streamflow drought vulnerability along a river system. Drought predisposition of a river is defined by the time period T_D until the threshold for drought flow Q_D is reached under dry weather conditions. A multiple regression analysis with basin properties was carried out using the parameters of 55 gauged catchments. Recession parameters, initial discharges Q_{MS} and drought threshold discharges were regionalised. Supported by GIS functions and network topology, aggregated recession curves were derived for every 2.5 km along the river network and thus the time period T_D could be estimated. This first test of the method was carried out in the study area of southwest Germany and the results and comparisons with earlier studies are satisfying. The visualised results show the degree of vulnerability of river stretches to droughts.

1. INTRODUCTION

In recent years droughts have occurred in different regions of Europe and the demand for water is especially critical when low flow periods prevail for a long time. In these times a conflict is evident between different water users like industry and agriculture and the socio-economic and ecological demands. The water industry has to guarantee a sustainable water management both in time and space, which is especially difficult during extreme low flow periods. Many studies have been dealing with the definition of surface water stress on the one hand, and its frequency, regional and temporal extent, and the processes that lead to low flow situations on the other hand.

Tallaksen *et al.* (1997) defined streamflow droughts based on the theory of runs and compared the distributions of drought parameters such as the duration and the deficit volume (severity), depicted when different pooling methods are applied. Based on the same approach, Demuth and Külls (1997) evaluated different procedures to estimate the probability distribution of droughts (duration and deficit volumes) and investigated the regional pattern of droughts. The regional study revealed the benefits of applying the partial probability weighted moments (PPWM) in fitting the generalised extreme value (GEV) distribution to drought parameters. The years 1947 and 1991 have been selected to show the regional variation of drought duration in southwest Germany expressed as recurrence intervals. The two drought years show completely different patterns. The drought of 1947 was most pronounced in the Upper Main region with recurrence

J.V. Vogt and F. Somma (eds.), Drought and Drought Mitigation in Europe, 209–219.

intervals between 23 and 84 years. In contrast, the 1991 drought shows dry spells of shorter duration in the entire study area. They do not exceed 93 days and the respective recurrence intervals are below 27 years.

Demuth and Heinrich (1997) investigated the occurrence and the space-time structure of streamflow droughts at a regional scale in south Germany, the same area as in the present study. Pronounced winter drought events (October-March) were found to occur regularly in the Alpine regions resulting in small coefficients of variation (between 90% and 130%) of the inter-annual variability of drought severity. The Pre-Alps and the River Rhine regions show particular summer droughts with higher coefficients of variation (between 170% and 220%). The high values for the Rhine region indicate the irregular temporal occurrence.

Drought occurrence was also used in another investigation to define homogenous regions within the same study area (Stahl and Demuth, 1999). A regional index describing streamflow drought was derived and linked to atmospheric circulation pattern frequencies using a logistic regression model. The decision to use a constant 30-day time-lag to analyse the circulation patterns that cause the dry spell was based on an average time of streamflow recession from mean flow to Q90 (the threshold for drought definition). One objective for the present study was therefore to derive a spatially more detailed regionalisation of the characteristic streamflow response of a region to a prolonged atmospheric dry signal. This 'vulnerability of a river to streamflow drought' was defined by a recession curve derived parameter.

A number of authors have considered the influence and importance of basin properties, particularly geology, on low flows: e.g., Toussaint (1981) for Hessen in Germany, Pereira and Keller (1982) for the Pre-Alps, Demuth (1989) for north west Europe, Gustard *et al.* (1992) for the United Kingdom and Demuth (1993) and Demuth and Schreiber (1994) for south-west Germany. Especially in extreme low flow periods with lack of rain, the geology of a catchment dominates the runoff.

In the present study the discharge and recession parameters were regionalised for areas of equal size in order to account for scaling problems. Technically supported by GIS functions, recession curves for equally spaced points along the river network were then aggregated from the upstream areas following the idea of multi-storage recession. Hence, it could be analysed how the parameters change along the river network. The method is designed to be applicable at a pan-European scale and thus renounces resources available only for small regions and furthermore strongly considers computing time efficiency.

2. STUDY AREA

The study area is situated in the south of Germany and covers six main regions with distinct differences in climate, geology and hydrology (Alps, Pre-Alps, Black Forest,

Bavarian Forest, Swabian and Franconian Alb, and the Upper Main region) (Schreiber and Demuth, 1997). The catchments in the Upper Main region show mostly pluvial regimes with high inter-annual variation of runoff and low flows mainly in summer (between August and September). The Swabian and Franconian Alb, the Bavarian Forest and the Black Forest are characterised by nivo-pluvial and pluvial regimes with moderate inter-annual variability and low flows occurring mainly in late summer. In addition, the flow regime of the Swabian and the Franconian Alb, which consist mainly of limestone, are influenced by karstic phenomena. The Alps are characterised by a nival regime with low flows only during the winter season. The region exhibits a high degree of seasonality and little inter-annual variation of low flows. The maximum flow occurs in June and July due to snowmelt and a maximum of precipitation. The Pre-Alps shift from a nival to a pluvio-nival regime with less seasonality and low flows in winter and in summer. The weak seasonality indicates extended aquifers with high storage capacity in that region.

The hydrological data available for the study area originate from the European Water Archive (EWA), the database of the UNESCO FRIEND-Project (Flow Regimes from International Experimental and Network Data). A subset of 55 stations with daily discharge records between 1961 and 1995 was selected. The basin areas are smaller than 500 km^2. Records of four larger basins are available to validate the regionalisation model.

The river network, which was digitised from maps at a scale of 1:500 000, is used in the Hydrological Atlas of Germany. Implemented into a GIS, an automated river coding system was applied to assign the topological information of the river network to the individual river stretches (Lehner, 1998b). The coding system is platform independent and is designed especially for river network analyses. The system allows performing fast queries e.g. for all up- and downstream flow connections to a given location, even outside the GIS.

Most of the study area consists of hard rocks, in particular the crystalline bedrocks and New Red Sandstone of the Black Forest, the Jurassic of the Swabian Alb and the Keuper and Coquina of the Franconian Alb. The unconsolidated rocks of the Quaternary are located in the Rhine valley and the Pre-Alps. The areal information on petrography, land use and soil was taken from the CORINE database at a scale of 1:1 000 000 (Commission of the European Communities, 1989). The data were provided by the European Community. The petrographic properties (parent material) are divided into eleven classes. In terms of land use, the study area is covered by 26% forest, 16% grassland and 58% arable land. Elevation, slope and exposition were derived from the digital elevation model GTOPO30 (resolution: 30 Arc Seconds) of the US Geological Survey.

3. DROUGHT VULNERABILITY

To describe the vulnerability of a river system to a meteorological drought, various parameters can be used. In the present study, the vulnerability of a river to the occurrence of streamflow drought is defined by the time period T_D, which the discharge takes to decay from an initial discharge Q_{MS} below the threshold level for drought Q_D (Figure 1). The shorter the time T_D, the higher is the risk for a region to be affected by a streamflow drought, if a meteorological dry signal prevails for a long period of time. This vulnerability description is based on the presumption of the initial discharge representing a hydrological state at which the basin's natural groundwater reservoir is not anymore filled up to its highest level (Q_{MS} for mean storage), when a prolonged atmospheric dry signal occurs over the whole catchment. Besides Q_{MS} and Q_D, the typical mean recession behaviour (master recession curve) had to be parameterised and determined from gauged discharge series in order to regionalise the drought vulnerability.

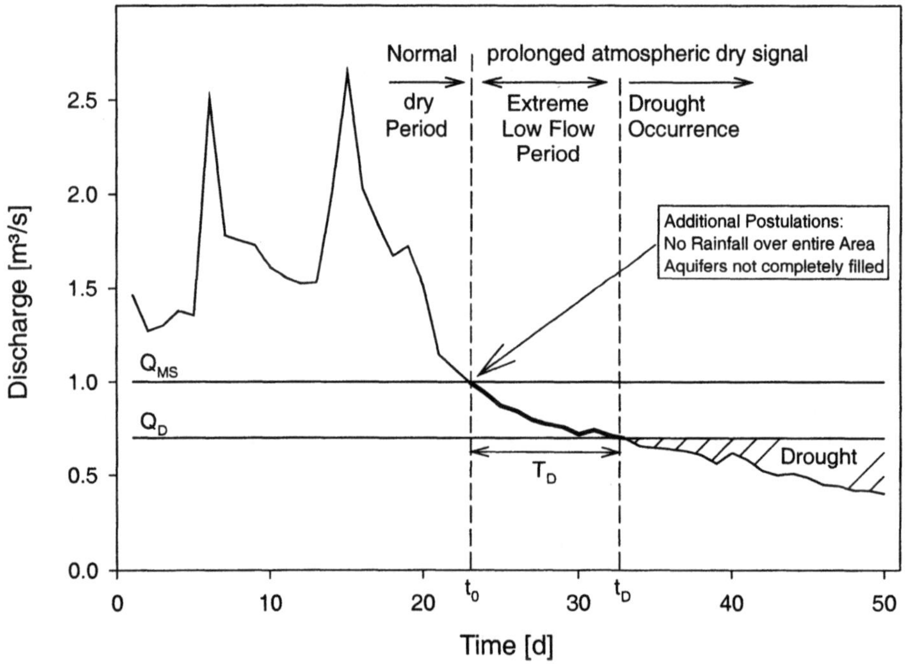

Figure 1: Approach to a definition of drought vulnerability.

In this study, attention is paid to the analysis of the characteristic streamflow behaviour during a rainless period. As the approach treats low flow as sustained groundwater flow, the outflow function Q(t) describes the outflow from groundwater storage (Gottschalk *et al.*, 1997) which describes the natural storage behaviour of a catchment in a generalised form. A simple exponential storage model has been used to depict the recession parameters. Here, the recession model is defined by the initial discharge Q_{MS} and the recession coefficient α and can be written as:

$$Q(t) = Q_{MS} \cdot e^{-\alpha t} \tag{1}$$

Hence T_D can be calculated as:

$$T_D = -\frac{\ln(Q_D) - \ln(Q_{MS})}{\alpha} \tag{2}$$

Within this first study the Q60 (flow exceeded for 60% of the time) was chosen as Q_{MS}. All streamflow recessions at a station starting at this value (±10%) and continuously decreasing for seven or more consecutive days were extracted from the 35-year streamflow series. The exponential model (1) was fitted to determine α.

The choice of the threshold level Q_D influences the drought duration and the number of droughts. In general, Q_D is an arbitrary value that depends on the objective of the drought study. The use of a statistical value (e.g. from the flow duration curve) allows the regional comparison. Tallaksen and Hisdal (1997) applied Q50, Q70 and Q90 as a threshold value for a drought investigation in Norway. Tallaksen *et al.* (1997) stress the problems of multi-year droughts using high thresholds. In contrast, low thresholds especially in regional studies with a lack of long time series lead to fewer droughts, which results in interpretation difficulties. For the region under investigation the threshold level Q90 was chosen as it has proved to be appropriate in earlier studies (Demuth and Heinrich, 1997).

4. REGIONALISATION APPROACH

4.1. Modelling Network

Thiessen Polygons are calculated for all nodes (intersection points and limiting ends of any geometric line) of the vectorised river system using GIS standard functions. Additional nodes were added to the line segments of the river network at 2.5 km distances to assure similar spatial scale of the polygons (Figure 2a). Due to their geometric nature, the Thiessen polygons delineate a middle line between neighbouring rivers, which can be interpreted as a rough estimation of the basin boundary. As the polygons are connected to the encoded river network, network queries can be applied to the polygon areas as well (Lehner, 1998a). Hence, the catchment area at any node of the river net is derived by simply summing up all polygons connected upstream (Figure 2a). In the same way, other parameters like thematic properties of the aggregated catchment upstream of any node can be automatically extracted and parameters available for the individual polygons can be visualised, cumulated or routed along the river network.

For a test region within Germany, the areas of polygon estimated catchments (PECs) have been compared with the values from digitised basins (Figure 2b). The result of the comparison shows a very good agreement between the estimated basin areas and the digitised basin areas with an increase in matching with increasing basin area. The six highlighted stations are located in the karstic range of the Swabian Alb. Here dry river valleys (no lines for rivers are digitised) lead to inaccurate estimates.

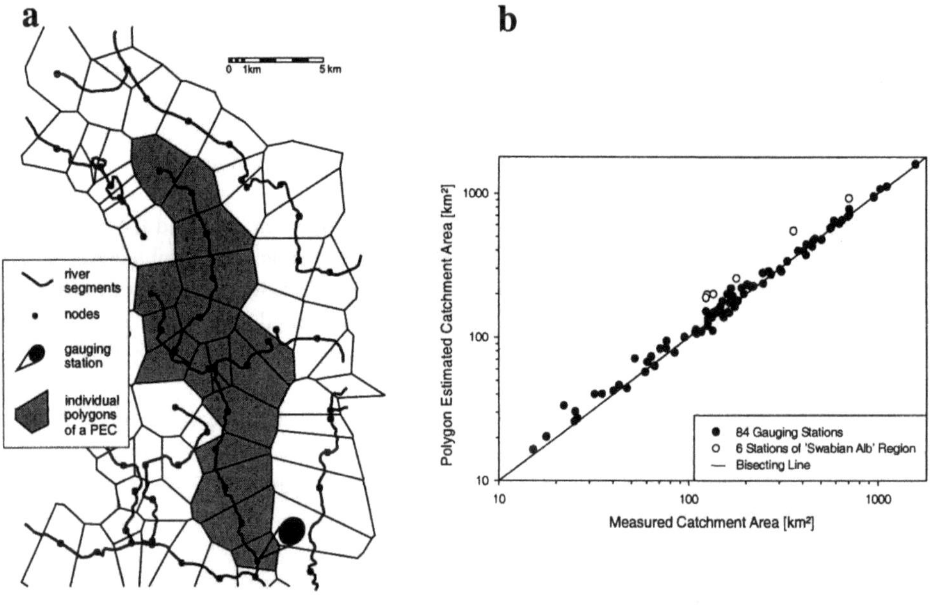

Figure 2: a) Example for a polygon estimated catchment (PEC); b) Comparison of estimated and measured catchment areas in the study area.

Besides their geometric properties, there were several practical reasons to prefer the Thiessen Polygon modelling network over the classical 'grid with DEM derived flow directions' method. The presented approach should be applicable to larger areas within Europe and for reasons of spatial accuracy be based on a quite dense river net. Therefore a DEM of very high resolution would be required to derive an adequate river network. Generally recognised difficulties with the DEM procedure are the time consuming calculation algorithms and the problems to determine the flow direction in flat areas. In the present study this has been avoided using existing vectorised river maps. Thiessen polygon functions are available in most GISs. The procedure requires only very short calculation time to build the river topology and to calculate the PECs.

4.2. Multiple Regression

Using 55 gauged catchments, a multiple regression was performed to determine a relation between the initial discharge Q_{MS}, the drought threshold level Q_D, the recession coefficient α, and catchment (PEC) properties from the thematic maps. The absolute areas of the property classes were used for the regression with Q_{MS} and Q_D; relative areas were used for α. The latter further required the introduction of the proportion of mean flow (MQ) to Q90 as an independent variable to obtain a significant result.

4.3. Regionalisation and Synthesis of Recession Curves

Individual recession coefficients, Q_{MS} and Q90 values were regionalised applying the respective regression equations to all derived Thiessen polygons without changing the scale. The results are maps, which show the spatial variation of the regionalised parameters determined for the scale of the individual polygons. However, for the identification of streamflow drought situations, the integrated runoff of the whole catchment draining to the point of interest at the river has to be considered.

Assuming that during stable low flow conditions any time delay due to runoff concentration can be neglected and the initial discharge expressed as percentile from the flow duration curve occurs simultaneously within the area, the recession curve for any particular node (e.g. node 9 in figure 3) within the river network can be calculated by summing the individual regionalised curves along the upstream river stretches. Finally, T_D, which characterises the vulnerability of the river system to droughts, can be determined from the curves at all nodes and can be visualised along the river network.

5. RESULTS AND DISCUSSION

The method has been applied to the test area in the south west of Germany. The multiple regression equations were derived estimating Q_{MS}, Q_D and α as a function of the properties of the estimated basins (PECs). The results of the multiple regression analysis are summarised in table 1.

All regionalised parameters show a high spatial variability. The maps with the distributed (for individual polygons) regionalised parameters clearly delineate the typical hydrological regions within the study area and are comparable with the patterns obtained in previous studies. However, the main result to be discussed are the parameters derived for successively growing PECs and hence the visualised development of the parameters along a river.

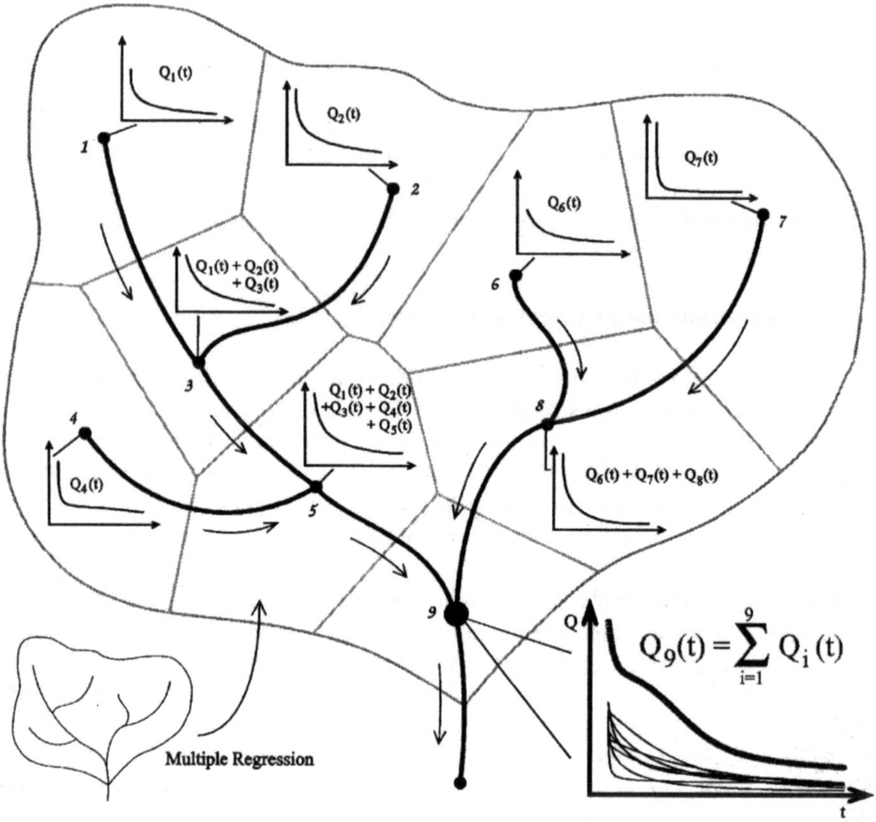

Figure 3: Aggregation of polygon regionalised recession curves to determine the characteristic streamflow recession from an initial discharge to the threshold level for streamflow drought.

Table 1: Results of the multiple regression analysis.

Parameter	Variables in the model	R^2	adjusted R^2	standard error of the estimate	Significance (F-test)
MQ	Petrography, slope, river length	0.96	0.94	0.64	>99%
Q_{MS}(Q60)	Petrography, slope, river length	0.96	0.95	0.37	>99%
Q_D(Q90)	Petrography, slope, river length	0.95	0.93	0.25	>99%
α	Petrography, slope, MQ/Q90	0.83	0.77	0.007	>99%

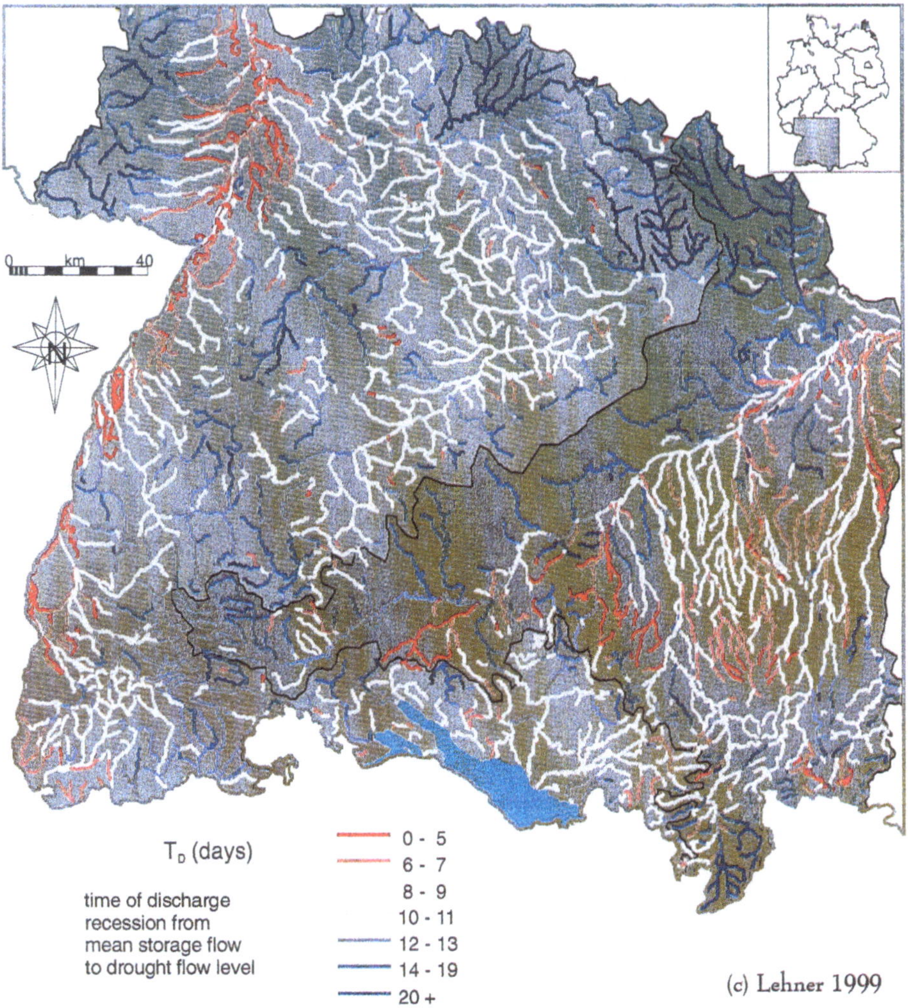

Figure 4: Map of T_D, derived from the recession curve of every node's PEC and visualised along the river network.

The map showing the drought vulnerability in terms of the time T_D is shown in figure 4. Most of the study area shows medium T_D values around 10 days and there is a tendency for the parameter to increase with growing catchment area and distance to the spring. The shortest times T_D were found in an area characterised by sandy petrographic characteristics. Neither very low Q90 flows nor a very fast recession behaviour have been regionalised for the polygons in this area. The reason for the short T_D is a stable runoff regime and thus a narrow $Q_{MS}/Q90$ relationship. The opposite is found in the karstic limestone area of the Swabian Alb, where streamflow recession is generally fast, but because of the wide relationship of $Q_{MS}/Q90$ a long T_D results. In the latter case it can be assumed that both ecosystem and surface water management concepts are adapted to high discharge variability, whereas in the former, stable regime case, a dry period may soon cause unexpected water shortage.

The results reflect the approach to define the vulnerability of a river towards streamflow drought as anomaly from the normal (statistical) discharge variation regarding strength and fastness of the reaction during an extended dry weather period. Drought vulnerability as defined in this study is a complex combination of recession behaviour and the flow duration curve of a river basin. The chosen streamflow parameters and method for determination of the master recession curve are to be re-discussed with regard to these results.

The regionalised parameters were validated comparing the parameters with measured data of a few larger basins, which were not used to derive the regression equations. However a more thorough validation is in progress. The derived values now allow for a number of further investigations for instance by plotting the development of a parameter versus the PEC area or distance from the spring of a river.

ACKNOWLEDGEMENTS

The research presented in this paper was carried out in the framework of the ARIDE Project (Assessment of the Impact of Droughts in Europe). Support is given by the European Commission under Contract no. ENV4–CT97–0553. The authors gratefully acknowledge the groups and organisations that have contributed data for the analysis presented in this paper (FRIEND-European Water Archive and CORINE). The flow data originated from State Water Authority of Baden-Württemberg (LfU). The company Terraplan, Schallstadt, kindly provided the GIS software SPANS.

REFERENCES

Commission of the European Communities (1989) CORINE (Co-ordinated Information on the European Environment), Data Base Manual, Version 2.2, compiled by Whimbrel Consultants and CORINE CENTRAL TEAM, GDXI, CEC, Brussels, edited by the Laboratory of Land Management, KUL University, Belgium.

Demuth, S. (1989) The Application of West German IHP Recommendations for the Analysis of Data from Small Research Basins, FRIENDS in Hydrology, *International Association of Hydrological Sciences, Publication no. 187*: 47-60.

Demuth, S. (1993) Untersuchungen zum Niedrigwasser in West-Europa, *Freiburger Schriften zur Hydrologie*, Band 1, Universität Freiburg, 202 p.

Demuth, S. and Schreiber, P. (1994) Studying Storage Behaviour Using an Operational Recession Method, *Proc. 2nd International Conference on FRIEND*, 11-15 October 1993, Braunschweig, *International Association of Hydrological Sciences, Publication no. 221*: 51-59.

Demuth, S. and Külls, Ch. (1997) Probability Analysis and Regional Aspects of Droughts in Southern Germany, *Proc. Symp. Sustainable Water Resources under Increasing Uncertainty*, 23 April – 3 May 1997, Rabat, Morocco, *International Association of Hydrological Sciences, Publication no. 240*: 97-104.

Demuth, S. and Heinrich, B. (1997) Temporal and Spatial Behaviour of Drought in South Germany, *Proc. Symp. Regional Hydrology: Concepts and Models for Sustainable Water Resource Management* (FRIEND '97), 30 September – 3 October 1997, Postojna, Slowenia, *International Association of Hydrological Sciences, Publication no. 246*: 151-157.

Gottschalk, L., Tallaksen, L. M., and Perzyna, G. (1997) Derivation of Low Flow Distribution Functions Using Recession Curves, *Journal of Hydrology* 194: 239-262.

Gustard, A., Bullock, A., and Dixon, J.M. (1992) Low Flow Estimation in the United Kingdom, *IH Report No. 108*, Institute of Hydrology, Wallingford, UK.

Lehner, B. (1998a) Automated Generation of a Topologic River Network System Within a Geographic Information System (GIS) as a Basis for Catchment Planning and River Management, Project Study on behalf of the Gesellschaft für Technische Zusammenarbeit (G.T.Z.), Germany, for the Department of Water Resources, Ministry of Water, Zimbabwe, G.T.Z. Report.

Lehner, B. (1998b) Automated Generation of a Topologic River Network System Within a GIS, in J. Strobl and F. Dollinger (eds.) *Angewandte Geographische Informationsverarbeitung. Beiträge zum AGIT-Symposium Salzburg '98*, Wichmann Verlag, Heidelberg: 195-201.

Pereira, L.S. and Keller, H.M. (1982) Recession Characterization of Small Mountain Basins, Derivation of Master Recession Curves and Optimisation of Recession Parameters, *Proc. Symp. Hydrological Aspects of Alpine and High Mountain Areas, International Association of Hydrological Sciences, Publication no. 183*: 243-255.

Schreiber, P. and Demuth, S. (1997) Regionalisation of Low Flows in South West Germany, *Hydrological Sciences Journal*, 42(6): 845-858.

Stahl, K. and Demuth, S. (1999) Linking Streamflow Drought to the Occurrence of Atmospheric Circulation Pattern, *Hydrological Sciences Journal*, 44(3): 467-482.

Tallaksen, L.M. and Hisdal, H. (1997) Regional Analysis of Extreme Streamflow Drought Duration and Deficit Volume, *Proc. Symp. Regional Hydrology: Concepts and Models for Sustainable Water Resource Management* (FRIEND '97), 30 September – 3 October 1997, Postojna, Slowenia, *International Association of Hydrological Sciences, Publication no. 246*: 141-150.

Tallaksen, L.M., Madsen, H. and Clausen, B. (1997) On the Definition and Modelling of Streamflow Drought Duration and Deficit Volume, *Hydrological Sciences Journal* 42(1): 15-33.

Toussaint, B. (1981) Ermittlung der Leerlaufkoeffizienten nach MAILLET und des effektiv nutzbaren Gesteinshohlraums in hessischen Flußgebieten durch Auswertung der Abflüssse im Trockenjahr 1976, *Deutsche Gewässerkundliche Mitteilungen*, 25, H. 3-4: 70-84.

KEY ELEMENTS OF REMOTE SENSING BASED SYSTEMS FOR ASSESSING DROUGHT IMPACT ON EU AGRICULTURE

W. KÜHBAUCH and I. RADEMACHER
Lehrstuhl für Allgemeinen Pflanzenbau,
Universität Bonn
53115 Bonn, Germany

Abstract. The paper presents principle effects of drought on growth and yield performance of agricultural crops. Based on drought related phenomena a retrospective learning process is suggested using satellite images either with low or high spatial resolution. In case of low spatial resolution images, the learning process is based on (1) the appearance of the crop vegetation cover in the image, (2) yield data of major crops from agrostatistics, (3) rotation patterns of cultivated crops, (4) phenological stages of cultivated crops, (5) local distribution of soil types, and (6) regional weather conditions. Parameters 3 to 6 are classified into categories of different susceptibility to drought. Combining all parameters over time will allow to reveal both periods and areas of high drought impact on crop growth. Moreover, at the end of a learning and calibration process a quantitative measure of drought impact can be related to yield. High resolution satellite images enable a different approach based on individual farms and field sites that are representative for a major cropping area with respect to soils, climate, cropping pattern and yield. In retrospective, signatures from satellite images of identical farms or field sites have to be related to crop yield in both normal and dry years.

The operational use of remote sensing for assessing drought impact on crop growth requires the availability of satellite data with different spatial resolutions and at a reasonable price.

1. INTRODUCTION

World-wide the most limiting factors to expansion and distribution of productive plant growth and farming are low temperatures and water shortage. Each year farmers suffer from losses of yield and income due to drought events in different or even the same agricultural zones. For instance, world-wide large portions of wheat are grown in semiarid zones where the most yield-limiting factor is water. Although farmers cultivate the well-adapted dwarf type varieties they still suffer from a dramatic loss of income due to drought in some years.

There exist several options to mitigate the impact of drought. The use of adapted crop types, which most farmers respect, has already been mentioned. Another option is irrigation. However, from the ecological and the economical point of view, large-scale irrigation might not be the right choice. Quite often, and especially in industrialised economies like the EU, losses of income are compensated by public payments. The amount of compensation depends on the type of crop and the magnitude of drought damage in terms of yield losses. However, in this case objective criteria are needed to verify yield loss from large areas down to districts or even individual farms. Remote

J.V. Vogt and F. Somma (eds.), Drought and Drought Mitigation in Europe, 221–229.

sensing seems to be the appropriate instrument to meet these requirements. At present, a real operational system that is based on remote sensing, and that quantitatively records and verifies drought impact on crop yield is not available.

Following an overview on soil-plant interactions during water shortage, two different concepts will be introduced to access drought impact on crops by means of remote sensing. The first concept is based on remote sensing data that are low in spatial resolution but frequently available. Examples are vegetation indices derived from Advanced Very High Resolution Radiometer (AVHRR) data aboard the NOAA-satellites. The second concept depends on remote sensing data of high spatial resolution from predefined calibration sites in drought prone areas. The two different approaches will be outlined in detail, and perspectives to use remote sensing as an operational instrument to quantify drought will be discussed.

2. PRINCIPLES OF CROP GROWTH UNDER WATER SHORTAGE

Under conditions of freely available soil water, crops evaporate water at the potential rate ET^* [mm d^{-1} or kg H_2O m^{-2} d^{-1}]. As the soil dries following rainfall or irrigation, the actual evapotranspiration (ET_a) will fall below ET^*. Thus, relative crop evapotranspiration (ET_a/ET^*) is a function of plant-available soil water. For a number of full-cover field crops it has been shown that over a wide range of soil textures ET_a/ET^* is maintained at a level of 1.0 until the available soil water content falls below 30 percent of the moisture-holding capacity (see figure 1).

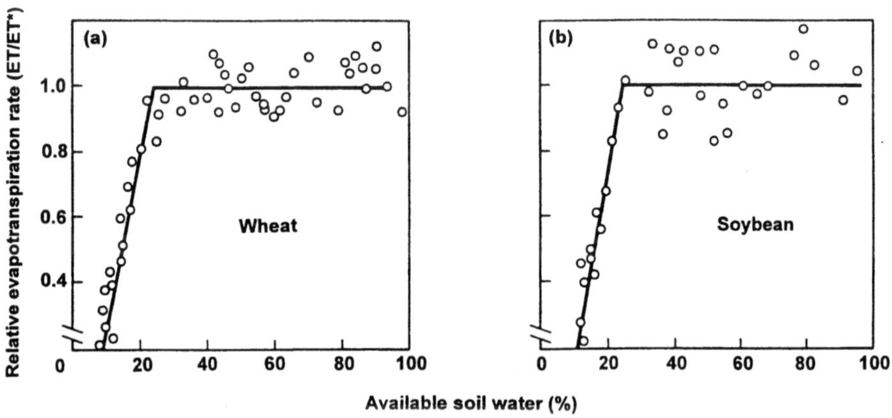

*Figure 1: Relative crop evapotranspiration rate (ET_a/ET *) as a function of available soil water content for (a) wheat, and (b) soybean (after Meyer and Green, 1981, from Loomis and Connor, 1992). Reprinted with the permission of Cambridge University Press.*

The soil water content [g H_2O/g soil] available to plants, is much higher in loamy or fine soils as compared to sandy or coarse soils, and for most field crops no more soil water is available beyond a soil matrix potential (ψ_m) of -1,5 MPa, i.e. the permanent wilting point (PWP). And, of course, in deep soils the amount of water available for transpiration is much higher than in shallow ones. Plant-available soil water and transpiration of crops are closely related. That dry matter production and transpiration of crops are directly related is generally accepted (see figure 2), since water shortage does not only dramatically reduce leaf expansion growth (Durand *et al.*, 1995) but is also associated with reduced photosynthesis per unit leaf area. The latter is due to an increased stomatal resistance against uptake of CO_2 and an overall reduced growth.

Soils do not only serve as a source of water, they also serve as a source of plant nutrients that improve plant growth. For instance, potassium decreases the plant water potential beyond that of the soil (in its extremes down to –2 to –3 MPa). Thus better soils, with more volume accessible for root growth and high amounts of available nutrients, enable plants to overcome water shortage better than shallow and poor soils will do.

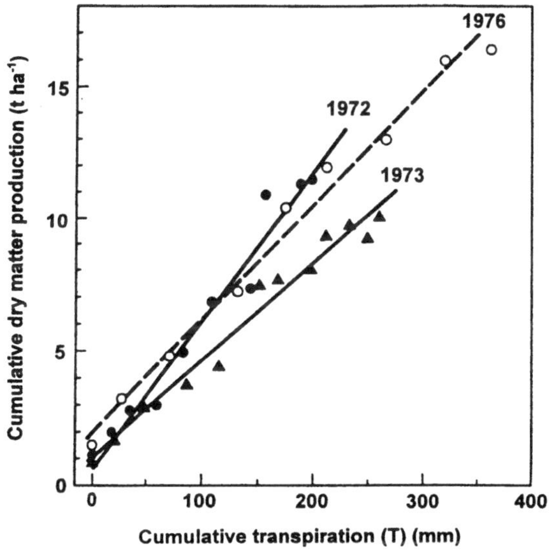

Figure 2: Relation of total dry matter production in potato crops during the period of tuber growth with cumulative transpiration T (after Tanner 1981, from Loomis and Connor, 1992). Reprinted with the permission of Cambridge University Press.

Further, there are huge differences between crop types regarding their water use efficiency (WUE) and their adaptation to water shortage. In this context, the dwarf type varieties of wheat have already been mentioned. However, with respect to water use efficiency and heat tolerance, world's agriculture probably takes most advantage from C-4 plants. In general, these plants produce much more plant dry matter per unit water transpired than C-3 plants do. The high WUE of C-4 plants is mainly due to the high affinity of their CO_2–binding enzyme and the negligible photo-oxidation under high stomatal resistance, (e.g. Mohr and Schopfer, 1978). Moreover, sensitivity against drought changes with the developmental stages of the crop. In cereals, for instance, yield may be strongly effected if water shortage occurs during the double ridge stage, i.e. the time when the positions of earlets and single kernels are established, or again during anthesis (e.g. Loomis and Connor, 1992).

In satellite images, longer periods of water shortage are often reflected by a sequence of spectral reflection patterns of the vegetation cover. A similar sequence of spectral reflection patterns is shown in figure 3, representing the canopy of wheat from early tillering to harvest.

However, neither the soil type, nor the crop type and growth stage alone provoke the impact of drought on plant growth and yield. It is rather the coincidence of all these factors, together with the prevailing weather conditions, which determine the impact.

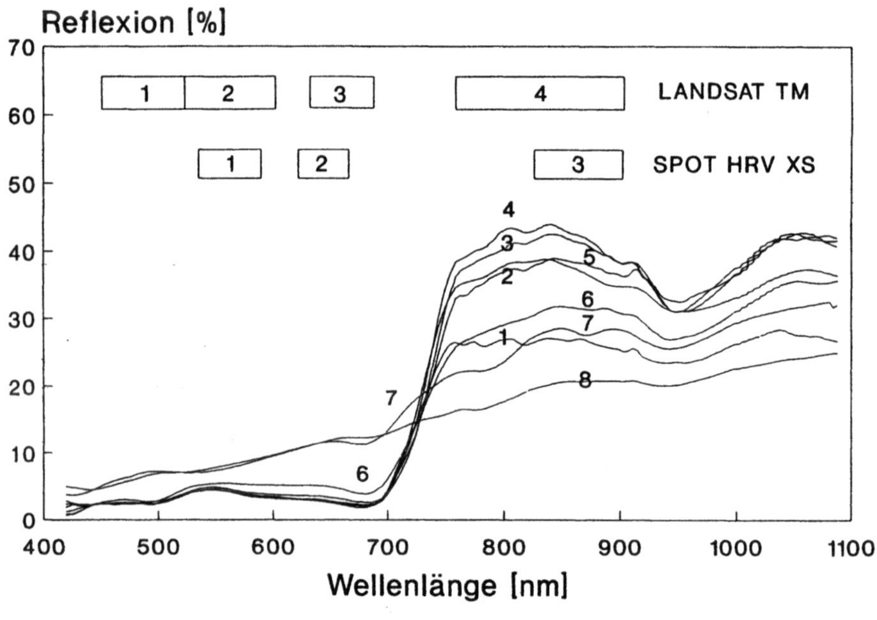

Figure 3: Spectral reflection of wheat during its growing cycle form early shooting (1) to harvest stage (8); spectral windows of LANDSAT-TM and SPOT satellite systems are indicated above the curves (Schellberg and Kühbauch, 1992).

3. KEY ELEMENTS TO VERIFY DROUGHT IMPACT BY REMOTE SENSING

Assessment of drought impact on crop growth and yield using remote sensing (RS) depends on the availability of RS-images in due time and their spectral and spatial resolution. NOAA satellite images are frequently available at high spectral but low spatial resolution, whereas systems like LANDSAT or SPOT have high spatial resolution but are less frequently available. In the following chapters two different approaches are suggested to assess drought impact based on satellite images either with low or high ground resolution. In both cases it is assumed that areas with frequent drought events are geographically well defined and that there is access to relevant information about agricultural practices, weather and soil conditions. Another requirement is the distinction between farm land and areas which might disturb the crop inventory (e.g. forest or urban regions) within the satellite image.

3.1. Satellite Images of Low Ground Resolution

In most agricultural sites of the EU, the 1×1 km pixels of the NOAA satellite images cover a number of different fields and soil types. Therefore characteristics of single fields are more or less hidden in each of these pixels. As small sub-units cannot be recognised individually, weighted categories of the drought related factors have to be applied to the 1×1 km pixels that are displayed in satellite images. With the exception of the satellite data, all other factors have to be transformed step by step into categories of susceptibility to drought in an area-weighted form (Table 1). In a first approach the number of susceptibility-categories has to be as small as possible.

Table 1: Assessment of drought impact on crop yield by satellite images with low ground resolution.

Step 1:	General classification of the susceptibility of different soils to drought ('risk factor soil').
Step 2:	General classification of the susceptibility of different crop species to drought ('risk factor crop species').
Step 3:	Evaluation of the rotation pattern of crop species according to their relative proportion within the area of interest.
Step 4:	Calculation of the weighted average of susceptibility to drought for each crop type and rotation in the area of interest.
Step 5:	Combination of both rotation and soil susceptibility to drought within the area of interest.
Step 6:	Determination of drought sensitive phenological stages for each crop in rotation within the observed area, e.g. in a sugar beet-wheat-corn rotation ('risk factor phenology').
Step 7:	Integration of weather conditions essential to soil-plant water relations ('risk factor weather') and calculation of the weighted average of phenological sensitivity to drought and weather risk at a given time.
Step 8:	Combination of steps 5, 6, and 7 as time series over the area of interest.
Step 9:	Comparison of accumulated 'drought risk' versus yield, accumulated NDVI-data versus yield, and calibration of RS-data like NDVI (Normalised Difference Vegetation Index) versus drought risk data.

In step 1 different soils are classified into drought susceptibility categories, as 'risk factor soil'. This information can be derived from soil maps. To each pixel of the satellite image, e.g. an area of 1×1 km, the related proportion of different soil risk categories is then attributed, resulting in an average soil risk factor per pixel. Similarly, in step 2 a general 'risk factor crop species' is established by classifying crops into low, medium and high susceptibility to drought. In step 3 the typical rotation patterns are evaluated and the relative proportion of each crop in rotation is attributed to the area of interest; for instance, information about typical rotation patterns can be derived from local agrostatistics. It can happen that the same soil categories are cultivated with different rotations. In this case, sub-units have to be determined and evaluated separately. In step 4 a weighted mean of drought susceptibility is calculated from the area proportion of different crop types in the area of interest. Step 5 combines the local 'drought risk factor soil' with the 'drought risk factor rotation', i.e. the weighted risk factor of the crops grown in rotation. Up to this step the drought risk is still independent of phenology and weather conditions. In step 6 the drought susceptible growth stages are estimated for major crops in the area of interest. This is done in a separate module that is based on growing degree days and/or meteorological observations. Step 7 adds a simplified 'drought risk factor weather' that can be obtained by an integration of weather conditions essential to soil-plant water relations, like climatic water balances. Since both weather conditions and crop phenology change over time, the weighted drought risk for both of these factors has to be calculated separately for each date of observation. Finally, in step 8 the single or weighted risk factors from the preceding steps are combined to give the evolution of 'total drought risk' over time and with respect to spatial differences in soil type.

The impact of drought on yield is a function of both, its time of appearance during crop development and its duration. Thus, losses in yield have to be described by the integrated 'total drought risk' over the whole vegetation period. Finally, in step 9 the 'total drought risk' is integrated over time, i.e. either over the whole growing season or only over the drought sensitive periods. The result is then calibrated against the yield of major crops (or rotations) and against vegetation indices like NDVI (Normalised Difference Vegetation Index) or other parameters derived from remote sensing data.

3.2. Satellite Images With High Ground Resolution

Satellite images with high ground resolution allow for a different approach, which is shown in figure 4. According to soil type, crop rotation and weather conditions homogenous areas can be determined, and drought impact can then be evaluated from calibration sites within these homogenous areas. However, it is essential that the selected sites or fields are representative for the whole cropping area. This procedure concentrates on a small number of fields bearing the representative crop types of the area of interest. Crop yield and drought impact on yield is recorded for each of the crop grown at the calibration sites.

Conventional yield data have to be collected with respect to soil-plant water relations over several years, including periods with or without water deficit during sensitive phenological stages of major crops.

Remote sensing data from calibration sites and training fields are compared with yields from the same sites over several years. Relations between RS-signatures and crop yield in the test sites serve as calibration standard to evaluate drought impact in the whole cropping area that is represented by the calibration sites. Using the RS-signature-yield relations from preceding years as a database, actual yield may be predicted by means of RS-data from a few sites and training fields only.

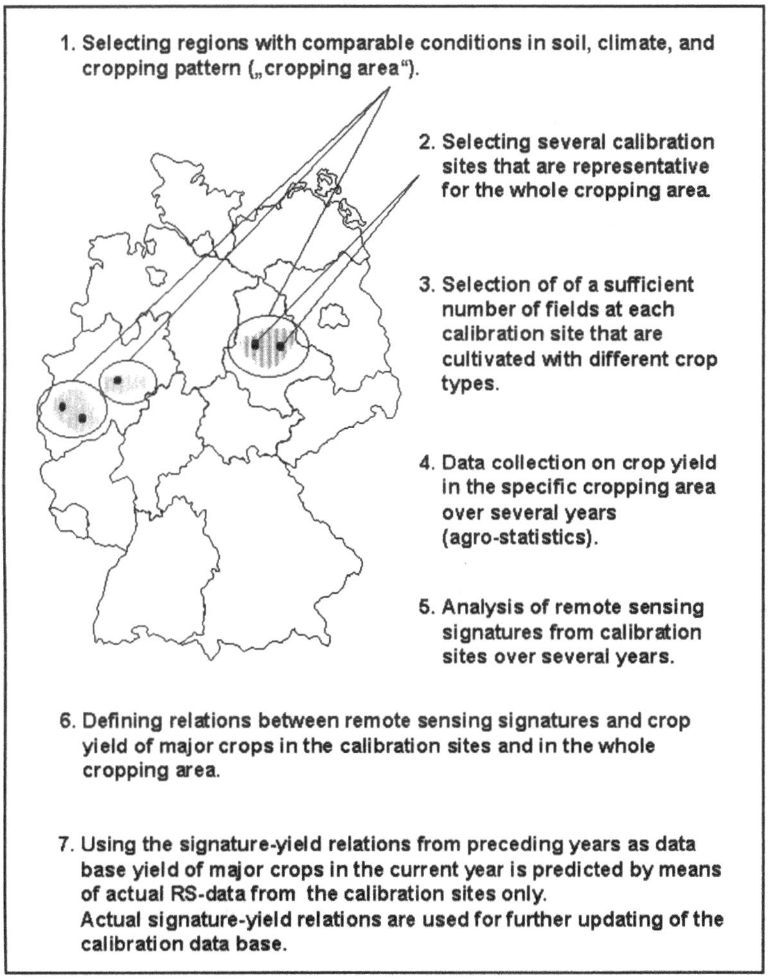

Figure 4: Assessment of drought impact on crop yield by satellite images with high ground resolution.

4. DISCUSSION

Two approaches have been suggested to assess drought impact by remote sensing (RS). The first method uses RS-data with high temporal resolution and low spatial resolution, while the second method is based on images with high ground resolution but poor repetition rates. Both procedures use field data and corresponding satellite images for a training and calibration procedure that may be further adjusted by regular updates. In principle, a combination of both procedures will enhance the capabilities of drought impact assessment by RS but each procedure has its own advantages and disadvantages.

In the procedure using low spatial resolution data simplified categories of weighted risk factors are introduced as key elements, since in most cases single fields cannot be detected in the 1×1 km pixels of NOAA-AVHRR images. Still it has to be examined if the risks are weighted adequately. Another key element in this approach is the availability of a number of information layers to be handled in a geographic information system (GIS). These data include soil conditions, crop rotation practices, proportions of major crops, and requires data on weather conditions during critical phenological stages of the crops. Finally, the yield performance of these crops needs to be analysed from the regional and national agrostatistics for the area of interest.

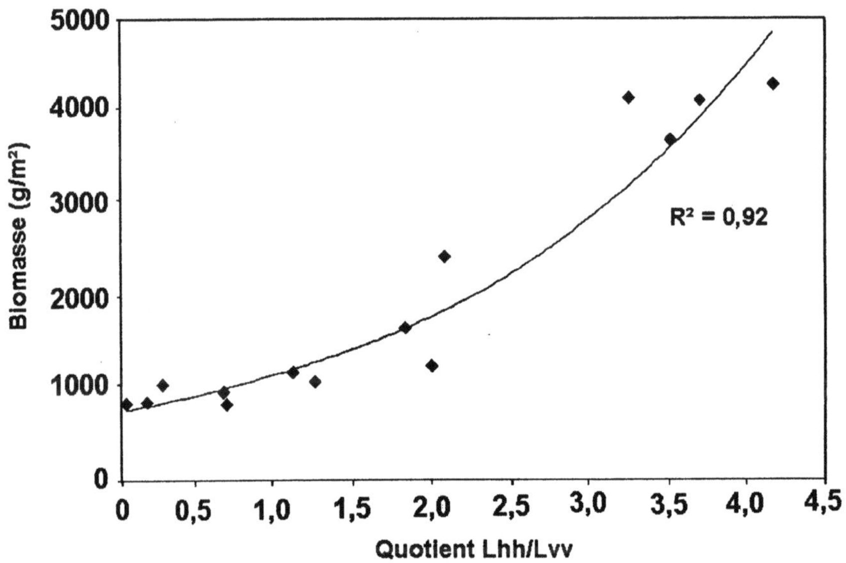

Figure 5: Fresh above ground biomass of winter barley related to microwave backscatter at wavelength of 23 cm at polarisation HH versus VV (Steingiesser and Kühbauch 1998).

In the procedure using high spatial resolution images, sites and fields representative for larger areas have to be selected. The continuos observation of these calibration sites over many years will be essential. One of the major deficits of this approach is the fact that the number of observations is likely to be too small for a substantial learning process by using optical sensors. Therefore, future options have to be considered such as new optical satellite systems with high repetition rates and/or multiband and multipolarization radar-systems. During the X-SAR/SIR-C mission in 1994 such a multiband radar system was used for Earth observation. Using these radar data, we found a close relation between the quotient of horizontal and vertical backscatter (HH/VV) in the L-band (19,35 - 76,9 cm) and fresh biomass of barley (Steingiesser and Kühbauch, 1998) as shown in figure 5.

In the drought prone areas of the EU remote sensing observations have to be performed continuously in order to improve the application of remote sensing towards an evaluation of drought impact and in order to have sufficient RS-data available in case the information is needed. This means that a network of calibration sites has to be established, observed, and evaluated continuously, not over all of the EU, but in some of the most drought prone areas. Once an acceptable relation between RS-data and crop status and yield has been established (knowing that the update of calibration will never end), it will be essential to produce commercial products with various categories of price, precision, and availability in order to create the market for these products. At the present time we do have a great demand for operational RS-systems, but we still do not have the relevant commercial products.

REFERENCES

Becker. G. (1961) Vierjährige Bodenfeuchte – Untersuchungen auf dem Dikopshof Dauerdüngungsversuch, *Zeitschrift für Acker- und Pflanzenbau*, 112: 279-291.

Durand, J.L., Onillion, B., Schnyder, H., and Rademacher, I. (1995) Drought Effects on Cellular and Spatial Parameters of Leaf Growth in Tall Fescue, *Journal of Experimental Botany* 46: 1147-1155.

Gebbing, T., Schnyder, H., and Kühbauch, W. (1998) Carbon Mobilization in Shoot Parts and Roots of Wheat During Grain Filling: Assessment by $^{13}C/^{12}C$ Steady-State Labelling, Growth Analysis and Balance Sheets of Reserves, *Plant Cell and Environment* 21: 301-313.

Loomis, R.S. and Connor, D.J. (1992) Crop Ecology, Cambridge Univ. Press, GB, 538 p.

Mohr, H. and Schopfer, P. (1978) Pflanzenphysiologie, Springer-Verlag, Berlin, Heidelberg, 608 p.

Meyer, W.S. and Green, G.C., (1981) Plant Indicators of Wheat and Soybean Crop Water Stress, *Irrig. Sci.* 2: 167-176.

Schachtschabel, P., Blume, H.P., Brümmer, G., Hartge, K.H., and Schwertmann, U. (1989) Lehrbuch der Bodenkunde, Enke-Verlag, Stuttgart, 491 p.

Schellberg, J. and Kühbauch, W. (1992) Reflexionsmessungen von Weizenbeständen zum Zwecke der fernerkundlichen Zustandsbeschreibung und Ertragsschätzung, *Berichte der GIL* 6: 61-68.

Steingiesser, R. and Kühbauch, W. (1998) Erfassung der frischen und trockenen Biomasse und Schätzung der frischen Biomasse von Wintergerste in verschiedenen Regionen Europas mit einem raumgleiter-getragenen Radarsensor, *Journal of Agronomy and Crop Science* 181: 145-152.

Tanner, C.B. (1981) Transpiration Efficiency of Potato, *Agronomy Journal* 73: 59-64.

Part 6
Drought Mitigation Strategies

DROUGHT MITIGATION MEASURES: A COMPREHENSIVE FRAMEWORK

G. ROSSI

Department of Hydraulics, Hydrology and Water Management
University of Catania
95125 Catania, Italy

Abstract. Drought mitigation measures include a large number of actions, which can be grouped into three broad categories: supply-increase, demand reduction, and drought impact minimisation. Each category affects the physical, economic and societal impacts of drought in different ways. In this paper the necessity of moving from a reactive approach to a proactive approach is discussed. The latter approach requires advance preparation of both long-term actions oriented to reduce vulnerability of water supply systems, and short-term actions to be taken during the drought period. The role of a drought monitoring system and an appropriate institutional framework for effective drought management is also highlighted. The use of multicriteria methods to compare different drought mitigation measures and choose the appropriate combination of actions is recommended.

1. INTRODUCTION

The severe drought events that have occurred world-wide during the past decades have increased awareness to the seriousness of the impact of this hydrometeorological hazard and have induced many researchers and decision-makers to develop drought mitigation measures. In particular, the necessity of an urgent response to drought problems is strongly recognised in arid and semiarid regions, where the risk of severe water shortages is growing, due to the pressure of increasing demands on scarce water resources presenting a high natural variability.

The perception of the priority actions necessary to cope effectively with drought can be very different, according to the scientific discipline of the involved researchers and to the institutional responsibility of the organisation called to mitigate adverse impacts of drought. However, a large consensus exists on the necessity of a comprehensive approach (see Yevjevich *et al.*, 1978; Wilhite *et al.*, 1987; Easterling, 1988; Rossi *et al.*, 1992; Pereira, 1992; Rossi *et al.*, 1995; Wilhite, 1997), which should include:

- a better knowledge of the drought phenomenon (in terms of general meteorological causes as well as of continuous monitoring of drought conditions);
- an accurate assessment of drought impacts on environmental, economic and social systems;

J.V. Vogt and F. Somma (eds.), Drought and Drought Mitigation in Europe, 233–246.

- the definition of a set of actions aimed to mitigate drought impacts and to reduce the vulnerability of water systems to droughts.

The objectives of this paper are to review the measures for mitigation of drought impacts with particular focus on water management strategies in Mediterranean regions of Europe, to present the main lines of a proactive approach for drought response planning, and to discuss some critical points for a successful strategy.

2. DEFINITIONS OF DROUGHT

Many definitions of "drought" are adopted in various fields, with reference to the investigated hydrometeorological variables. Some authors emphasise the necessity of considering the interrelationship between physical and socio-economic aspects. Yevjevich (1967) suggested that a drought should be defined on the basis of the difference between water supply and water demand series, and proposed an objective identification of drought characteristics based on the concept of deficit run. This is determined as the number of consecutive intervals where the selected hydrologic variable presents lower values than a truncation level, describing the assumed level of demand.

In any case, a general consensus exists in distinguishing among drought, aridity and desertification concepts (Rossi *et al.*, 1992). Unlike *aridity*, which refers to a permanent climatic condition with very low annual or seasonal rainfall, *drought* refers to a casual (random) condition of "severe reduction" of water availability compared to the normal values extending along a "significant period" of time over a "large region". Of course the criteria to define "severe reduction", "significant period" and "large region" are affected by subjectivity, as they derive from the perception of negative impacts of the water deficit as well as from the demand level. The term *desertification* is used to indicate a long-term and somehow irreversible process of decrease or destruction of biological soil potential fostered by several factors (climate, soil properties, and human activities) where drought might accelerate the process.

Drought is a complex hydrometeorological phenomenon, caused by meteorological anomalies reducing precipitation, but also affected by the state of the various components of the hydrologic cycle. In fact, precipitation deficits are modified and delayed in surface runoff, soil moisture, streamflow and groundwater flow. Recent studies suggest that, beside the direct process from the atmosphere to the land, a feedback mechanism between soil moisture anomaly and precipitation deficit should be considered in the investigation of physical characteristics of drought.

Drought is a natural hazard but can also be a man-affected phenomenon. First of all a drought is perceived as a disaster or as an adverse event when it affects the life, the economic interests or the social well being of a human community. In any case the impacts of a drought of a certain severity can largely differ according to the used percentage of the available water resources as well as the adopted strategies for drought management.

The lack of a general accepted definition of drought has been often indicated as one of the main obstacles to an effective investigation on drought. Such a lack can also partially explain the fact that appropriate drought management strategies are far from being yet largely applied in several countries.

3. DROUGHT MITIGATION MEASURES

The measures to be implemented to mitigate drought impacts can be classified in several ways. A first classification (Yevjevich *et al.*,1978) refers to three main categories: i) water-supply oriented measures, ii) water-demand oriented measures, iii) drought impact minimisation measures. A simplified sketch of the complex process, streaming from a meteorological drought event to its economic and intangible impacts through the filtering of the water bodies, the water supply system and the socio-economic structure, is presented in figure 1. In particular the role of the three mentioned categories of drought mitigation measures is highlighted. It is apparent that the first two categories of measures, by modifying supply or demand, aim to reduce the risk of water shortage due to a drought event, while the third category is oriented to minimise the environmental, economic and social impacts of drought.

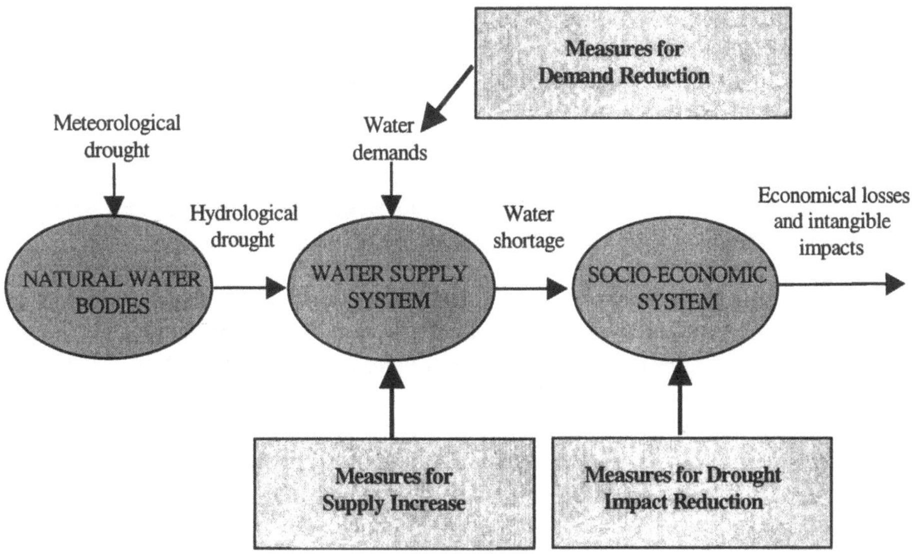

Figure 1: Drought impact process and role of mitigation measures.

A second classification focuses on the type of response to drought problems, distinguishing between a reactive and a proactive approach. The *reactive* approach consists of measures adopted once a drought occurs and its impacts are perceived. It includes the measures taken during the drought period and after the drought to minimise the impacts of the drought itself. It can be indicated as the "crisis management" approach because it is not based on plans prepared in advance. Although the reactive approach still represents the common response to drought events, there is an increasing awareness that such approach implies last minute decisions and leads to expensive actions, often with unsustainable environmental and social impacts. The *proactive* approach consists of measures conceived and prepared according to a planning strategy rather than improvised measures (Yevjevich *et al.*, 1983), before the drought event occurs. The proactive measures are devised and implemented before, during and after the drought event. In particular, measures taken prior to the initiation of a drought event and prior to the forecasting of a drought aim to reduce the vulnerability of the system to droughts and/or to improve preparedness to face drought effectively.

According to Werick (1993) and Whipple (1994), water supply planning incorporates two types of response to drought: tactical and strategic. *Tactical* measures are actions to be implemented to meet water deficiency problems after a drought has started when it is too late to build new facilities. *Strategic* measures are actions planned in advance, which involve modification of infrastructures, of existing laws and institutional arrangements.

Here we prefer to distinguish between an unplanned reactive approach and a proactive approach, which includes two main categories of measures, both planned in advance (Cancelliere *et al.*, 1998):

- long-term actions, oriented to reduce the vulnerability of water supply systems to droughts, i.e. to improve the reliability of each system to meet future demands under drought conditions by a set of appropriate structural and institutional measures;
- short-term actions, which try to face an incoming particular drought event within the existing framework of infrastructures and management policies.

For a more specific analysis of the various measures, besides the category-based and approach-based classification above indicated, the identification of the affected societal sector results to be necessary. Measures regarding urban, agricultural, industrial, recreational, energy, wildlife sectors should also be distinguished. Thus, a specific drought mitigation measure can be classified according to a three-dimensional positioning, as shown in figure 2. In table 1 a list of short-term and long-term measures is presented, subdivided into the three categories of water supply increase, water demand reduction and drought impact minimisation, pointing out the affected sectors.

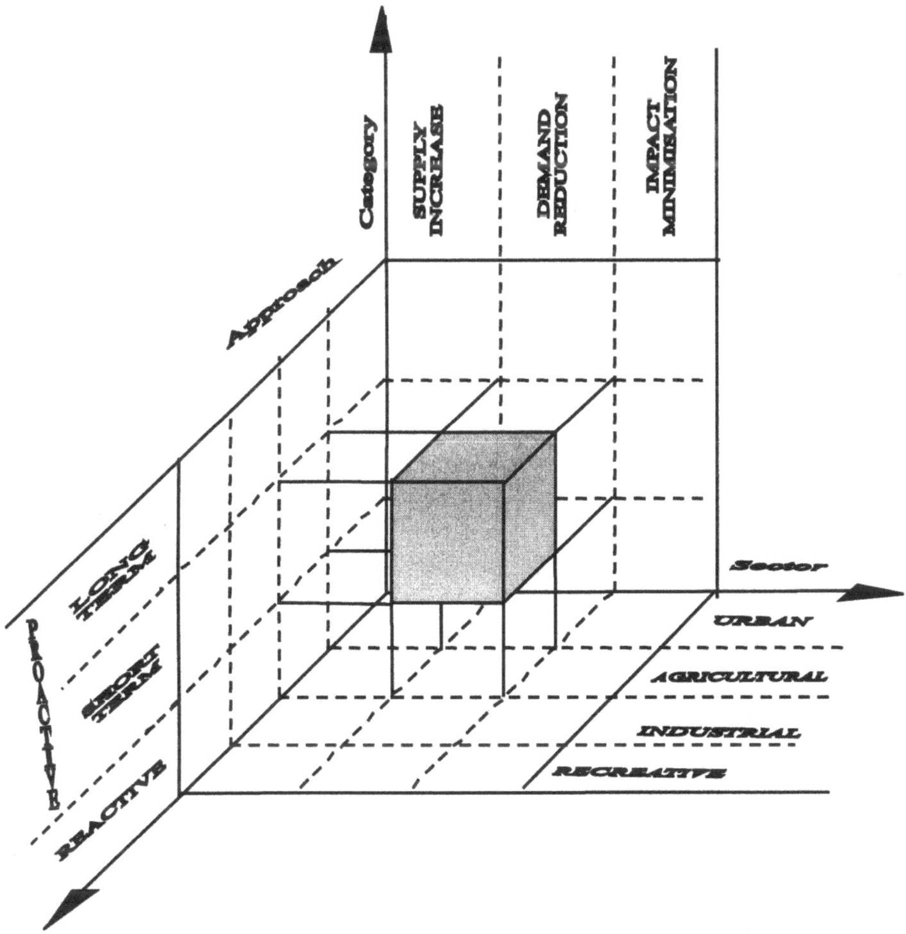

Figure 2: Three-dimensional representation of the potential drought mitigation measures.

4. IMPLEMENTATION OF A PRO-ACTIVE APPROACH

The implementation of the drought mitigation measures is difficult for several reasons, such as:

- inadequate understanding of drought problems and in particular of natural and man-induced aspects of water shortages;
- difficulty in forecasting drought characteristics (duration, severity) and time and space variability over a region;
- lack of an appropriate network for monitoring hydrometeorological variables and water resources conditions;

Table 1: Classification of drought mitigation measures.

	Short-terms actions	A.S.*	Long-term actions	A.S.*
Supply increase	- Use of additional sources of low quality water and/or high exploitation costs - Over-exploitation of aquifers - Increase diverted waters by relaxing ecological or recreational use constraints (e.g. minimum instream flow, minimum lake level) - Improvement of existing water systems efficiency trough leak detection programs, modified operation rules, etc.	U,A,I,R U,A,I U,A,I U,A,I	Augmentation of available resources through: - new surface reservoirs - inter-basin and within-basin water transfers - conveyance network for bi-directional exchanges - reuse of treated wastewater - desalination of brackish or saline waters - management of snow pack - control of evaporation losses - use of aquifers as groundwater reserve - rainfall augmentation	U,A,I,R U,A,I,R U,A,I A,I U A U,A,I U,A,I U,A,I
Demand reduction	- Restriction of some municipal uses (car washing, gardening) - Restriction of the irrigation of some crops (e.g. annual) - Pricing - Public information campaign for voluntary water saving - Mandatory rationing	U A U,A,I U,A.I U,A,I	- Dual distribution network for municipal use - Water recycle in industries - Use of less water consumptive crops - Agronomic techniques for reducing water consumption - Sprinkler or drip irrigation - Shift from irrigated to dry crops - Economic incentives for private investments in water conservation	U I A A A A U,A,I
Impact minimisation	- Temporary reallocation of water resources (on the basis of assigned use priority) - Public aid to compensate loss of revenue - Tax relief (reduction or delay of payment deadline) - Rehabilitation programs	U,A,I U,A,I U,A,I U,A,I	- Development of an early warning system - Reallocation of water resources on the basis of water quality requirements - Use of drought resistant plants - Development of a drought contingency plan - Mitigation of economic and social impacts trough voluntary insurance, pricing and economic incentives - Education activities for improving preparedness to drought	U,A,I U,A,I A U,A,I,R U,A,I U,A,I

*Affected Sector: U =urban; A = agricultural; I = industrial; R = recreation

- difficulty in quantifying the impacts of drought on different sectors of economy and on society;
- presence of strong conflicts among different users groups (farmers, municipalities, ecologists, etc.) pushing political decision-makers;
- low priority assigned to the problems of preparedness to natural hazard with respect to other investment sectors;
- legal and institutional constraints to the application of several measures (e.g. changes in water allocation, reuse of treated wastewater, etc.);

- inadequate development of decision support systems aimed to help the managers of water resources systems and/or the government officials to analyse in advance the consequences of the selected measures;
- lack of appreciation of a proactive approach to face drought which requires an attitude to plan in advance measures for an event of uncertain occurrence;
- inadequate capability and authority to manage drought problems in particular to assure horizontal co-ordination among water supply organisations and vertical co-ordination among local, regional, national and international levels.

In order to face effectively these difficulties, several courses of actions are necessary. The first priority is to shift from a reactive to a proactive approach, which should include a strategic water resources planning, taking explicitly into account drought risk, and a drought contingency plan. The latter, including the short-term actions to be adopted after the onset of a drought event, would be enhanced strongly by an early warning of potential water shortages in the different use sectors (Barbagallo *et al.*, 1992). The main steps of a proactive approach of this type is compared with the steps of a reactive approach in figure 3.

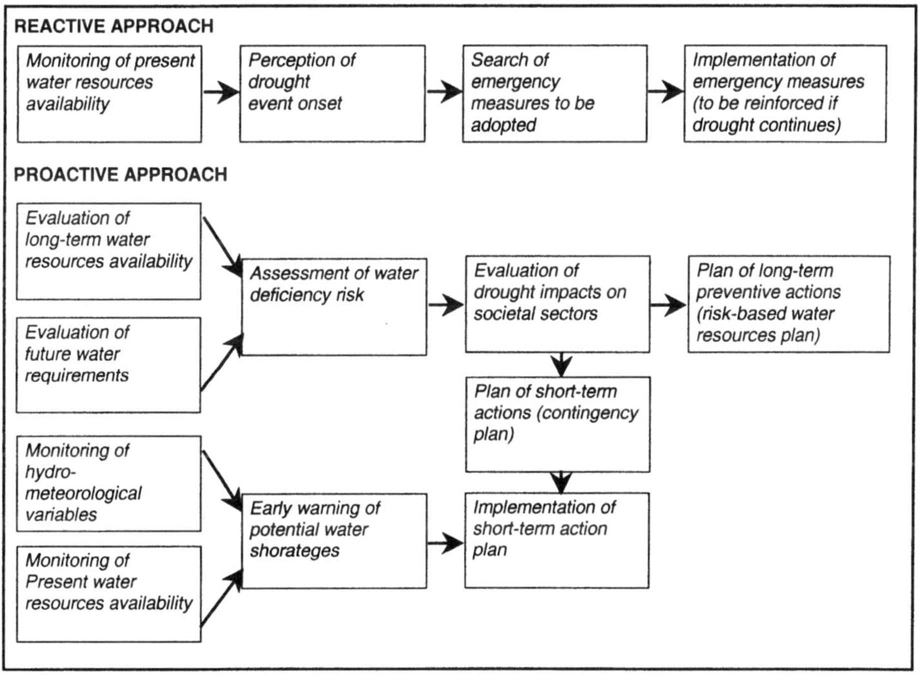

Figure 3: Main steps of a proactive approach versus a reactive approach in drought response.

As pointed out by Wilhite (1991), Grigg and Vlachos (1993), Fontane and Frevert (1995), a correct definition of the roles of the different levels of government in planning and co-ordination is a primary need in drought management processes. Such level (local, regional, national or international) can be seen as a forth dimension of the classification of the measures. Therefore, the basic tools of the proactive approach, i.e. the water resource plan and the drought contingency plan, should be developed at various decision levels or should indicate explicitly the various roles of the involved organisations.

5. ROLE OF A DROUGHT MONITORING SYSTEM

A key factor for an effective response to drought is the capability of a drought watch system to provide an early perception of the incoming drought and to monitor the evolution of meteorological, hydrological and water resources conditions. A drought monitoring system aims to help decision-makers to recognise the "drought alert" conditions and identify the appropriate drought mitigation measures on the basis of a continuous check of the drought evolution. Assessment of drought conditions requires the collection, transmission, storage, treatment and dissemination of a very large amount of information. Therefore, the choice of the appropriate data and the co-ordination of the different sources of information are critical issues for developing an effective system.

In particular the development of the watch system requires (Rossi, 1994):

- drought indicators aiming to evaluate the most significant elements enabling the assessment of the risk of severe water shortages;
- a minimal network for monitoring climate and water resources systems conditions;
- effective links with existing information systems on hydrometeorology and water resources;
- an organisational structure able to support a timely information transfer to decision makers responsible for emergency actions and prevention plans.

In spite of the fact that the role of a permanent monitoring system is generally recognised as a support to the measures for coping with a drought, a very small number of such systems are actually in operation. For example, the Drought Watch System which has been operating in Australia for the last three decades is based on a simple index of rainfall deficiency, computed using monthly rainfall observed in 600 stations and based upon the comparison of the rainfall for a period of three or more months with values of assigned percentiles derived from cumulative frequency curves (fifth-percentile for a severe deficiency, tenth percentile for serious deficiency). In several States of USA one or more drought indices are used to trigger the activation of a Drought Plan: the most popular indices are the Palmer Drought Index, based on precipitation, temperature and soil moisture data - and the Surface Water Supply Index based on precipitation, streamflow, reservoir storage and snowpack. For an analysis of the applicability of the Palmer index in Mediterranean regions see Rossi *et al.* (1996).

Although the Mediterranean regions of European countries do not have any permanent drought monitoring systems, much experience has been gained in the field of water resources monitoring during severe droughts. For example, such activities have been carried out in France and in Italy by special Work Groups for Drought, appointed in the two countries when a severe drought, perceived as a natural disaster, occurred (see table 2). In France, the Work Group appointed for the 1988-92 drought was established at "Mission interministerielle de l'eau" to collect information on the hydro-meteorological phenomenon and on the water reserves conditions and to give a synthesis to decision makers at national, regional and local levels as well as to water supply organisations and to the press (Mérillon and Scherer, 1993). In Italy, the Work Group on Drought was appointed during the 1988-90 drought at the Department of Civil Protection with the purpose of providing a description of meteorological and hydrological characteristics of the drought event, the effects on economy and environment and a synthesis of the main actions undertaken (Rossi and Margaritora, 1994).

Table 2: Comparison of the activities carried out in France and Italy by a Drought Work Group.

	FRANCE (*)	ITALY (')
Drought period	1988-92	1988-90
Responsible Organisation	- "Mission interministerielle de l'eau"	- Department of Civil Protection
Organisations represented in the Work Group	- Regional Direction of Environment - Ministry of Agriculture - Meteorological Service - BRGM - EDF (Hydroelectric Service)	- Meteorological Service - Hydrographic Service - ENEL (Hydroelectric Service) - Forestry Service - Meteomont Service (Army) - Ministry of Agriculture - Ministry of Health
Monitored Hydrometeorological Variables	- Rainfall - Streamflow - Groundwater level	- Rainfall - Streamflow - Groundwater level - Snow-pack
Monitored Water Resources	- Reservoirs storage - Irrigation supply - Municipal supply - Hydroelectric production	- Reservoirs storage - Irrigation supply - Municipal supply - Hydroelectric production
Other Monitored Conditions	- Forest fire - Navigation	
Output of the Work Group	- Monthly bulletin on situation	- Reports to the Ministry of Civil Protection - Final report on drought event

(*) Merillon and Scherer, 1993; (') Rossi and Margaritora, 1994.

These experiences were positive, as the problem of the vulnerability of water supply systems to drought in countries with abundant water resources called for a more effective co-ordination among hydrometeorological and water management agencies. Also, the revision of general water policy in order to reduce water demand and to improve the reliability of water systems was recognised as necessary. However no permanent drought monitoring system has been yet developed in these countries.

6. INSTITUTIONAL FRAMEWORK OF THE DROUGHT MITIGATION PLANNING

Another critical factor for an effective response to drought is the institutional framework of the drought mitigation planning. In the past drought received less attention than any other natural hazard in most European countries. Nowadays this situation is changing. For example in Italy the law 255/1992, organising public response to natural disasters, includes actions to face drought. The tasks and roles of governmental departments and committees at different levels in the implementation of necessary actions are indicated in table 3. In particular two main planning tools are indicated: the first one is the prediction and prevention program (to be prepared at national, regional and local level), the second one is a plan for emergency actions (contingency plan) at national and local levels (provinces and municipalities). Moreover, following the Law 36/1994 on municipal water services, a special Decree gives the guidelines for the investigations to be carried out in order to prevent water crisis due to drought.

Table 3: Organisation of public response to natural disasters in Italy according to the Law 225/1992.

LEVEL	ROLE	RESPONSIBLE ORGANIZATION	TASK
National	Decision-making	Head of Government or Civil Protection Ministry	Declaration of emergency
	Co-ordination	Department of Civil Protection	Formulation of national program of prediction and prevention
	Consultancy	National Council for Civil Protection Commission for Risk Prevention	Formulation of national program of assistance and contingency plan
	Operational	National Committee for Civil Protection	Direction and co-ordination of all institution activities Assessment of Prefect's contingency plans
Regional	Operational	Regional Committee for Civil Protection	Formulation of regional program of prediction and prevention
Local	Operational	Provincial Committee for Civil Protection	Formulation of provincial program of prediction and prevention
	Decision making and co-ordination	Prefect	Formulation of contingency plan and co-ordination of actions at provincial level
	Operational	Municipal civil protection structure	Formulation of local emergency plan and implementation of actions
	Co-ordination	Major	Direction and co-ordination of actions in municipality

Although these guidelines are oriented to assure the water supply in urban areas during drought assuming that the municipal use has priority on the other uses, the assessment of the risk of water deficiency and the definition of the measures to be implemented require the analysis of all elements of the water system as a whole. Then, the required investigations can represent the basis for the formulation of a more general drought mitigation strategy including both long-term and short-term measures.

A proposal of the necessary investigations for i) assessment of the water deficiency risk on the basis of the analysis of available water resources and the evaluation of present water consumption and future requirements and ii) identification of the long-term and short-time mitigation measures is presented in figure 4. In particular the long-term measures aim to increase robustness and resilience of the water system (*robustness* is used here to indicate the ability to avoid very adverse effects under a large range of negative conditions while *resilience* expresses the capacity to recover from a failure in meeting demands). These measures, including new facilities and legal and institutional changes in water allocation or co-ordination among water organisations, should be included in the general plan of water resources systems. As a consequence any water resources plan which did not explicitly consider the drought risk should be revised.

Short-term measures to be studied in advance to cope with potential drought events, within the existing infrastructure and management policies framework, should be included in the contingency plan to be developed according to civil protection laws. Such a plan should include actions necessary to increase water supply by using additional sources, to reduce water demand through voluntary or mandatory use restrictions, to minimise drought impacts through economic or social relief tools.

7. COMPARISON OF DROUGHT MITIGATION MEASURES BY MULTICRITERIA METHODS

The choice of drought mitigation measures to be adopted requires the assessment of the impacts of all actions, to be compared according to the different points of view of the society. Consequently, multicriteria methods represent an appropriate tool, taking into account conflicting values and giving a sound base to the decision-making process. Key aspects of the multicriteria approach are the possibility of including qualitative criteria besides the quantitative ones and the capability of providing the decision-makers with a tool to aid the identification of a compromise option in case of conflicting preferences from interested social groups. In general, the use of a multicriteria method implies:

i) the definition of criteria which measure the different impacts of each alternative action or strategy according with the different goals of various social groups;
ii) the identification of the alternative options to be compared;
iii) the application of a multicriteria technique to obtain a ranking of the alternative;
iv) the use of a sensitivity analysis to check the robustness of the resulting ranks.

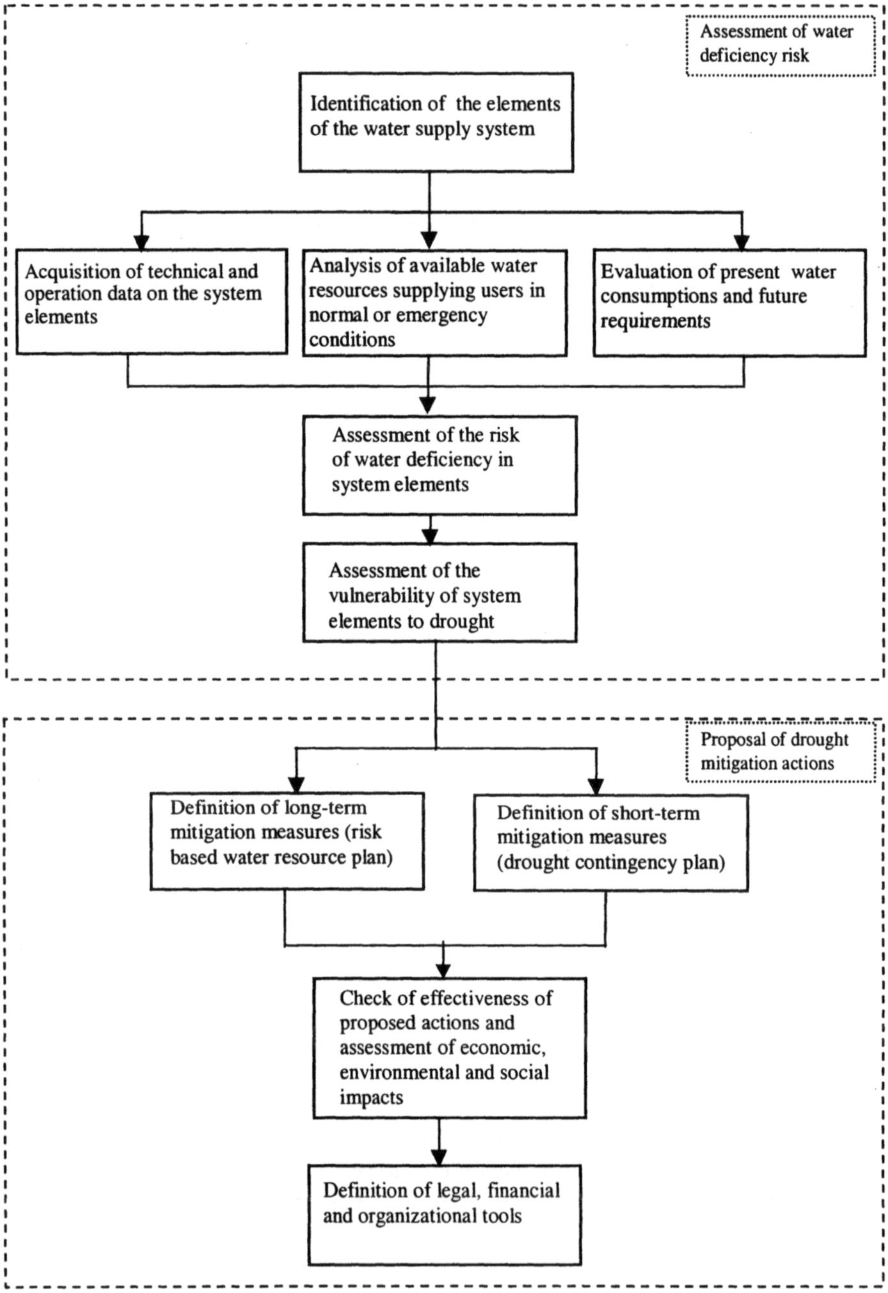

Figure 4: Proposed steps of a study for prevention of water crisis in municipal water supply systems in Italy according to guidelines of the Law 36/1994 and Prime Minister Decree 47/1996.

The criteria to be used to assess the drought mitigation measures can be distinguished into three main categories (economic, environmental and social) all of which include significant impacts resulting from the unsatisfied water demands during a drought and from the structural or non-structural actions. The alternative actions to face drought in many cases include different combinations of various measures, namely supply-oriented, demand-oriented or impact mitigation oriented. Moreover, in the framework of the general water resources planning, a trade-off exists between long-term and short-term actions. Therefore, each drought mitigation alternative could include a combination of different actions, covering explicit emergency measures to be implemented during a drought and other actions to be taken in the framework of the general water resources management, in order to minimise the vulnerability of the water systems to drought. An example of the use of the multicriteria method (NAIADE) for comparison of drought mitigation alternatives in a complex water system is presented by Munda *et al.* (1998).

8. CONCLUSIONS

Mitigating drought impacts is one of the major water management challenges in Mediterranean regions of Europe. A successful strategy should be based on a proactive approach, which requires to prepare in advance the response to drought hazard in terms of long-term and short-term planning. The definition of long-term measures to cope with droughts should be based on strategic water resources planning where the drought risk is explicitly taken into account in order to analyse the present vulnerability of the water supply systems to the drought and to define the new facilities, the operation rules and the legal and institutional arrangements necessary to improve the reliability of the supply system in meeting water demands. The definition of short-term measures aiming to mitigate the impacts of water shortages due to a drought should be developed, as for any other natural disasters, in terms of a contingency plan in the framework of civil protection responsibilities at different levels. The choice of the preferable combination of different drought mitigation measures can be supported by a multicriteria method, allowing to take into accounts different points of view.

ACKNOWLEDGEMENTS

This paper is based upon the results of the researches supported by the Italian National Research Council, Group for Prevention from Hydro-geological Disasters, and by a Contract of the European Commission, DGXII within the INCO-DC Programme on "A DSS for mitigation of drought impacts in the Mediterranean regions".

REFERENCES

Barbagallo, S., Rossi, G., and Tamburino, V. (1992) Emergency measures in water supply systems for alleviating the effects of the 1987-90 drought in Sicily, *ICID XVI European Regional Conference*, June 1992, Budapest, Hungary, 7-15.

Cancelliere, A., Ancarani, A., and Rossi, G. (1998) Susceptibility of water supply reservoirs to drought conditions, *Journal of Hydrologic Engineering* 3(2): 140-148.

Easterling, W.E. (1988) Coping with drought hazard: recent progress and research priorities, in F. Siccardi and R. L Bras (eds.), *Natural Disasters in European Mediterranean Countries*, NSF and NRC, Perugia, 231-270.

Fontane, D.G. and Frevert, D.K. (1995) Water management under drought conditions: overview of practices, *Journal of Irrigation and Drainage Engineering* 121(2): 199-206.

Grigg, N.S. and Vlachos, E.C. (1993) Drought and water supply management: roles and responsibilities, *Journal of Water Resources Planning and Management*, 119(5).

Mérillon, Y.J. and Scherer, C. (1993) La sécheresse de 1992, *La Houille Blanche* 8: 559-569.

Munda, G., Parruccini, M., and Rossi, G. (1998) Multi-criteria evaluation methods in renewable resources management: integrated water management under drought conditions, in E. Beinat and P. Nijkamp (eds.), *Multicriteria analysis in land use management*, Kluwer Academic Publishers, Dordrecht.

Pereira, L.S. (1992) Drought challenge, *Idrotecnica* 5: 283-287.

Rossi, G., Benedini, M., Tsakiris, G., and Giakoumakis, S. (1992) On regional drought estimation and analysis, *Water Resources Management* 6: 249-277.

Rossi, G. (1994) Role and requisites of a drought watch system, *Proc. of Workshop on Climate Change and Hydrogeological Hazards in the Mediterranean Area*, Colombella, Perugia, June 1994, 95-102.

Rossi, G. and Margaritora, G. (1994) *La siccità in Italia 1988-1990*, Presidenza del Consiglio dei Ministri, Dipartimento della Protezione Civile, Roma, 193 p.

Rossi, G., Ancarani, A., and Cancelliere, A. (1995) *Gestione dei sistemi idrici durante i periodi di siccità: il ruolo dei modelli*, Tipografia Leone, Catania, 91 p.

Rossi, G., Ancarani, A., and Cancelliere, A. (1996) Use of Palmer index as drought indicator in Mediterranean regions, *Proc. of IAHR Congress on "From Flood to Drought"*, Sun City, South Africa, August 1996.

Yevjevich, V. (1967) An objective approach to definitions and investigations of continental hydrologic droughts, *Hydrology Paper* no. 23, Colorado State University, Fort Collins, CO.

Yevjevich, V., Hall, W.A., and Salas, J.D. (1978) *Proc. Conference on Drought Research Needs*, Water Resources Publications, Fort Collins, 276 p.

Yevjevich, V., Da Cunha, L., and Vlachos, E. (1983) *Coping with Droughts*, Water Resources Publications, Littleton, Colorado, 417 p.

Werick, W.J. (1993) National study of water management during drought: results-oriented water resources management, in *Water Management in the 90's*, ASCE, New York, 445-450.

Whipple, W. (1994) *New Perspectives in Water Supply*, Lewis Publ., Boca Raton, Florida, 218 p.

Wilhite, D.A., Easterling, W.E., and Wood D.A. (1987) *Planning for Drought*, Westview Press, Boulder, CO, 597 p.

Wilhite, D.A. (1991) Planning for drought: a process for state government, in D.A. Wilhite D.A Wood and P.A Kay (eds.), *Drought Management and Planning*, International Drought Information Center, University of Nebraska, Lincoln, 167-178.

Wilhite, D.A. (1997) State actions to mitigate drought: lesson learned, *Journal of the American Water Resources Association* 33(5): 961-968.

DROUGHT MITIGATION STRATEGIES - A DYNAMIC APPROACH

J. LOMAS
WMO Regional Meteorological Training Centre
for Postgraduate Training in Applied Meteorology
Meteorological Service, P.O. Box 25
50250 Bet Dagan, Israel

Abstract. The mitigation of environmental and economic impacts of drought events requires the development of adequate management strategies and the education and training of the relevant decision-makers and managers as well as of the general society. This paper highlights the specific nature of drought as compared to other natural disasters and discusses educational and institutional strategies for coping with droughts. It concludes with a set of recommendations for the development of an efficient drought policy in Europe.

1. INTRODUCTION

Of the many climatic events that influence the earth's environmental fabric, drought is an important one, affecting society and natural resources of land, water and vegetation. Drought is the consequence of a natural reduction in the amount of precipitation over an extended period of time, usually associated with high temperatures, high winds and low humidity that aggravate the severity of the event. Because drought affects so many economic and social sectors, numerous definitions have been developed by the various disciplines.

Drought is a widespread misery, especially if recurrent as in the Sahel in the 1970's and 1980's, and extensive droughts recur in Asia, especially China, India, the Middle East and in Australia, North America and Mexico. In early 1997, exceptionally dry weather persisted in south-western Europe and less than 10 mm of precipitation were received between 26th January and 23rd March in much of the central and eastern Iberian Peninsula, France, much of northern Italy and parts of Corsica and Sardinia.

There is a strong correlation between rainfall and plant biomass production of both crops and pastures, especially in the agroclimatically marginal areas of production and in areas with lighter soil types. Thus, both farming and grazing activities are affected. Whereas annual field crops are affected during their growth cycle, the negative effects on pasture may linger for a number of years due to regeneration difficulties, especially when overstocking renders vegetation regeneration problematic. Under extreme conditions of water stress vulnerable dryland soils can be exposed to erosion by both

J.V. Vogt and F. Somma (eds.), Drought and Drought Mitigation in Europe, 247–252.

wind and rain. The sparser the plant cover, the more severe are the impacts of these events.

From the hydrological point of view, extreme rainfall variability forces a corresponding flow variability in rivers and consequently in runoff and soil moisture. Thus, the effect may be compounded, not only by the amount and seasonal distribution of rainfall, but also by the ability of soils to absorb and store water and by moisture losses through evapotranspiration.

Very few countries have a comprehensive drought policy and the political response to drought has mostly been on an ad-hoc basis so far. Even in the U.S. the movement to a National Drought Policy has taken a long time. First recommended in 1979, it is only recently that the U.S. inacted a comprehensive law – the National Drought Policy Act that was approved in July 16, 1998. Australia too has a National Drought Policy ratified by the state and the Federal Governments in 1992. Comprehensive institutional drought policy seems to be of recent origin.

2. CLIMATIC CHANGE AND DROUGHT

The large year to year variations in precipitation levels are inter-related with natural and external variations occurring within the entire global scale climatic system. Inter-related changes in sea-surface temperatures, land surface conditions, general atmospheric circulation patterns and atmospheric concentration of various greenhouse gases, have all been proposed to explain at least some of the variance in the observed precipitation levels. The Inter-governmental Panel on Climatic Change (IPCC) models project an increase in global surface temperature relative to 1990 of about 2°C by the year 2100. Predictions of precipitation changes are far less certain. Should the assessments be correct, the changes will most likely result in an increase of potential evapotranspiration rates. In the absence of any increase in precipitation this will increase the probability for drought events.

However, it has also been suggested that the observed and predicted changes are not different from the natural, internal variability of the climate system and that normal or even humid conditions will return in due course to the drought stricken areas as the atmospheric circulation shifts into another pattern.

Yet antropogenic pressures of overgrazing, over-cultivation, deforestation and poor land management will cause a decrease in vegetation cover, exposing vulnerable soils to erosion and degradation and thus increasing the chances for future droughts.

3. DROUGHT PREPAREDNESS AND MANAGEMENT

Drought severity is dependent on the duration, intensity and geographical extent of precipitation deficiencies on the one hand, and on the water demands by human activities and vegetation on the other hand. Two specific characteristics of drought make it different from other natural hazards: Firstly, the effects of drought often accumulate slowly over an extended period of time, so that the onset and ending of a drought are difficult to assess. Secondly, the impacts of drought are less obvious and are spread over relatively large geographical areas and consequently, damage is difficult to determine.

The economic impacts of drought range from direct losses to agriculture to losses in the recreation, transportation, banking and energy sectors. Environmental losses are due to damages of plant and animal species, damages of wildlife habitats, occurrence of forest and range fires, degradation of air and water qualities, and soil erosion. Social impacts involve public safety, health issues, conflicts between water users, losses of life, rural migration and a general reduction in the quality of life.

Planning an appropriate management program in order to mitigate drought effects is, therefore, essential. The best time to cope with the problems of drought is when no drought is apparent. The following steps are deemed necessary:

(1) Accurate appraisals of the drought probabilities on regional, district and even village basis.

(2) Development and implementation of technological practices for drought mitigation and drought recovery, divided into

    (a) short-term measures, and
    (b) long-term measures.

(3) Efficiency of government agencies in coping with drought and drought recovery.

4. EDUCATION AND TRAINING FOR DROUGHT PREPAREDNESS

Drought is a fact of life, especially in the lower rainfall regions with larger inter-annual rainfall variability. It is consequently necessary to present to the responsible government agencies the probabilities of the various types of drought and their physical and economic/financial implications. The reduction of risk and uncertainty involves costs in terms of hedges secured or opportunities foregone. There is, therefore, a need for an accepted response policy on drought mitigation measures for the farming community for the re-examination of the agronomic techniques for more efficient utilization of scarce rainfall, reducing water consumption of crops, and indeed a possible change in the farming system. At the same time it is necessary to consider the water supply sources, including the use of lower quality water, economic aspects of desalination, water re-use and the improvement of water system efficiency.

(a) Adaptation of farming systems

The awareness of the costs and benefits associated with inflexible as compared to flexible management practices needs continued educational and training efforts. This educational effort has to be directed towards the various levels of society: policy makers (legislature), administrators (local government), extension officers (agricultural advisory service), and especially the farmer. Flexibility of the farm as a business unit to meet the inevitable drought as well as the relatively favorable year takes two major forms: agronomic/physical flexibility and economic financial flexibility.

(b) Specialization versus diversification

The effect of drought on the agricultural sector will depend on the timing, severity and duration of soil moisture deficits on the one hand and the agricultural system on the other hand. There is little scientific evidence on the long-term rewards from specialization or from diversification under drought conditions. Is it economically sound to sacrifice increased returns from specialized outputs in return for some reduction in risk by diversification? Do such economic considerations vary from northern Europe where agriculture accounts for less than 10% of total water demand to the Mediterranean region where agriculture accounts for 80% of total water demand?

Optimum agronomic flexibility to meet drought has not been determined and in the final analysis the answer will be partially personal.

(c) Physical flexibility

Physical flexibility as a means by which farmers can minimize the effects of drought may take several forms: The selection of crop varieties better able to withstand stress from limited water supplies. Such consideration includes an economic analysis of varietal performance under a range of rainfall conditions. The application of water conservation measures, tillage practices, contour farming and possible summer fallowing. Part of the farming unit can be devoted to feed crops instead of utilizing all of the land to cash crops and finally consideration of irrigation.

Livestock production practices should be designed to permit flexibility to adjust to forage availability and marketing practices accordingly. While the range of such physical adjustments is great, not all of them are practical on a particular farm and some are limited by economic considerations.

(d) Financial flexibility

In drought prone regions the farmer must give consideration to the building-up of financial reserves during good production years.

Care and prudence must, therefore, be exercised in the commitment of financial resources to production expansion. Equally important are the credit facilities, the ability to borrow needed funds. It is essential for the farming sector to be able to adjust repayment of borrowed money to income availability, i.e. flexible repayment plans.

Overall flexible financial strategy will allow expansion during favorable years and does not commit the farm enterprises to an unbearable financial outlay when conditions are bad during drought years. Such financial tactics depend on the ability and willingness of the financial and business institutions to participate in them. The ultimate aim is to evolve a farming system better suited to the natural and economic conditions of the region in the face of the certitude of future drought events.

5. INSTITUTIONAL STRATEGIES

Institutional policies are advantageously best developed before a drought begins. One of the most important means is an Early Warning System and a Drought Appraisal and Information Service. To keep such a service active and efficient during periods of plentiful rainfall is administratively a difficult task.

Institutions for crop insurance need to be developed in advance of drought periods so as to spread the risk of financial collapse and to determine a premium policy. Who should pay for the losses caused by natural disasters and for the related compensations? Serious consideration must be given to financial repayments during and maybe following a drought year.

Regional and indeed national agencies should critically re-examine their programs for dealing with drought and drought recovery. This re-examination should include questions on whether the institutional infrastructure worked efficiently and where it was less successful; on what was the "time lapse" between economic loss and drought relief; and on any undesirable side effects.

6. RECOMMENDATIONS

From the above considerations there is a general need for the following steps in order to prepare a strategy for drought management and post drought recovery. Such measures should be implemented in the European Union in order to reduce the economic and environmental risks associated with drought events.

(1) Preparation of a comprehensive National and E.U. Drought Policy to respond to drought emergencies.

(2) Consideration of regional differences in the frequency and scale of drought damage.

(3) Improvement of public awareness of the need for drought prevention, mitigation, and response.

(4) Initiation of a clear and authoritative analysis of past droughts: their climatological causes, the agronomic vulnerabilities and technological responses.

(5) Implementation of a thorough study of the effectiveness of drought mitigation programs in agronomy and other water dependent sectors as well as of their socio-economic impacts.

(6) Establishment of an effective drought monitoring and early-warning system, as well as of a drought information system for the rapid flow of accurate information about the intensity and scale of drought. Primary recipients of such information should be the farmers, the decision-makers and institutions for drought response.

(7) Consideration of the adoption of a flexible technology of action as drought develops and during drought recovery. Such technology "packages" have to be verified under field conditions before being implemented.

7. CONCLUSIONS

Drought is a major natural hazard especially in the lower rainfall regions of Europe with great inter-annual seasonal rainfall variability. As drought is basically an unbalance between water supply and water demand, the Mediterranean region of Europe will be the area that is most affected by drought. The Intergovernmental Panel on Climate Change assumes an increase in the probability of drought events that will be enhanced by antropogenic pollution pressure.

Because of the slow onset nature of drought, monitoring and early warning is an essential part of a drought management plan. Yet, in spite of considerable agronomic research activities there are no national or regional response plans in the event of drought at the farming, local government or national level. An integrated response approach and educational effort are urgently required.

SOME LESSONS LEARNT DURING THE DROUGHT OF 1991-1995 IN SPAIN

M.R. LLAMAS
Department of Geodynamics
Complutense University
28040 Madrid, Spain

Abstract. In fora where water policy is discussed, policy makers from the Mediterranean countries of Europe often emphasise the great economic and social impact that droughts have in these countries. Based on the experience gained from 1990 to 1995 drought in Spain, this paper highlights potential strategies to mitigate drought effects. Some lessons learnt during recent droughts in Spain are the following: (1) droughts are not predictable but their adverse effects can be substantially mitigated; (2) dry spells in the Mediterranean countries are not a random hazard but a "normal climatic feature". Consequently European "disaster relief funds" should only be given in case of extreme droughts; (3) development of groundwater can, in many cases, play a relevant role in mitigating drought effects, both in urban and in irrigation water supply; (4) up to now, water policy to mitigate droughts has been mainly based on the construction of classical hydraulic structures paid with public funds. It seems, however, necessary to change to a water demand management based on the "full cost" or "user pays" principle; (5) it is about time to avoid situations that are prone to improvisation and/or misappropriation of public funds.

1. INTRODUCTION

In fora where water policy is discussed, policy makers from the Mediterranean countries of Europe often emphasise the major economic and social impact that droughts have in these countries in comparison with countries of central and northern Europe. Generally, just as frequently, the solutions advocated to solve these problems call on the traditional policy of increasing water supply – or, rather, the guarantee of water supply. Almost without exception, this increase in supply and its guarantee is met by constructing new dams and aqueducts and some times with interbasin transfer of water. As a corollary to this approach, appeals are often made to structural funds to help in solving these problems.

This paper sets out the economic and environmental disadvantages of these traditional options. It will frequently refer to the experience gained from the analysis of droughts in Spain over the last fifteen years.

J.V. Vogt and F. Somma (eds.), Drought and Drought Mitigation in Europe, 253–264.

In the summer of 1995, drought once again filled the headlines of the main communication media across broad areas of the centre and the south of Spain. The situation was particularly critical in the Cadiz Bay area, where for four consecutive years a number of urban centres suffered serious water restrictions. It seems that an almost equally serious situation was faced in the same years in other cities of the Mediterranean, for example in Athens.

From 1991 to 1995, the Spanish Government adopted various emergency measures at a cost to the Treasury of about 100 billion pesetas (600 million Euros). The Minister for Public Works announced that in 1995 about eight million Spaniards had suffered from water restrictions in urban areas; some newspapers raised this figure to twelve million. In addition, in the summer of 1995 not only did water transfers through the Tajo-Segura aqueduct become the cause of a renewal of "water hostilities" between several Autonomous Communities, but there was also an extension of the "water wars" to Portugal, where the draft of Spain's national hydrological plan was one of the principal issues of the programmes of the Portuguese political parties during the general elections of October 1995 (Llamas, 1997a).

However, it would be wrong to think that Spain suffered an unusual situation over this five-year period. Drought is a recurrent component of the climate of most of Spain, just as in all other Mediterranean countries. The main problem is that, when this physical fact is forgotten, economic and social tensions ensue.

Droughts are generally the "touchstone" for analysing the quality of the water policy of a country, especially if it is arid or semi-arid. In Spain, water has always been a political priority. A new Water Law was approved in 1985 with almost no societal debate. Its main new features were the declaration of groundwater as public dominion and the consideration of water planning as the key or pivot to water management. However, neither of these principles has worked in practice. Most groundwater is still *de iure* in private hands and public control *de facto* is minimal. Spain's National Water Plan has not been completed yet, although its preparation really began in 1979. A draft has been proposed in 1993 (MOPU, 1993), based primarily on constructing more dams and making major future water transfers from the rivers Duero and Ebro to Castilla-La Mancha and the Mediterranean areas. This approach was probably a strategic and tactical error, leading to an impasse in water policy and providing a catalyst for the various "water wars" at local, regional and international level (Llamas, 1997b and 1998). The new Government coming to power in 1996 has in fact followed a similar policy. The 17 Basin Water Plans were approved by the Government in 1998, and in December of the same year the Ministry of Environment presented the long ago promised White Paper on water (MIMAM, 1998). Public discussion of this thousand-page document has just begun.

Conflicts over water are not an exclusively Spanish affair, but in semi-arid countries such as Spain these conflicts are generally more serious and the solutions more complex. This is why the Spanish experience can be of special interest. In addition, these conflicts have increased in many other countries, now that water is regarded as one of the most important factors in protecting the environment (Vlachos, 1995).

2. TYPES OF DROUGHT

There is no single, universal definition of drought (Vlachos, 1995). Perhaps the simplest and most general ones consider drought as a period of time when rainfall is lower than normal, giving rise to certain undesired effects. As to the minimum time to officially declare a drought emergency and the procedures for measuring the undesired effects, opinions and regulations are logically enough varied. However, it is useful to distinguish among several kinds of drought according to their impact. In a previous work, the author has distinguished, in some detail, up to six types of drought: (a) meteorological drought; (b) agricultural drought; (c) hydrological drought (rivers and groundwater); (d) ecological drought; (e) technological drought; (f) psychological drought (Llamas, 1997b). These drought types are going to be analysed in detail in other chapters of this volume; therefore, they will not be dealt with here.

3. CURRENT INABILITY TO PREDICT DROUGHTS

At present we cannot predict with acceptable certainty if in thirty or fifty years time it will rain in a different way than it does now. The quantitative weather data available on precipitation and temperatures in Spain only date back about a hundred and fifty years, and show that (a) as in all the Mediterranean countries, the climate of the Iberian Peninsula is subject to major year-to-year variability; (b) it is not possible to identify periodic or cyclic patterns; (c) the dry periods have been more frequent than the wet periods, although the latter have generally been more intense (put in statistical terms, the median annual precipitation is lower than the mean one); (d) the Iberian Peninsula has undergone dry spells lasting for more than seven years (Almarza and López, 1994; National Meteorological Services of Europe, 1995).

Drought is an inherent climatic feature of the Mediterranean countries. The main effort to address at the moment is to mitigate the adverse effects, which this unpredictable natural event generally produces. Generally pessimistic views about climate change are one more reason to increase our efforts to improve the methods for mitigating the effects of droughts. Nevertheless, some recent papers seem to present a less pessimistic outlook on possible climate change in the southern Mediterranean countries.

4. EVALUATION OF THE SEVERITY OF DROUGHTS IN SPAIN

The social groups affected by drought generally tend to exaggerate the damage caused and treat the drought as a scapegoat to blame for their difficulties or economic losses. In view of the growing and general interference of the public authorities in the everyday life of Europe's citizen's and in view of the fact that the public authorities usually have a clear responsibility for water, it is to be expected that these attacks will generally be directed towards politicians.

There is no unique way to evaluate the severity of a drought. First because, as we saw earlier, there are several types of drought. Thus, there can be a drought which is really catastrophic for crops on non-irrigated land but which is economically beneficial to irrigated crops with available groundwater. This happened, for example, during the serious six-year drought hitting California in 1987-92, when the agricultural production of the State did not decline because of the large extent of crops being irrigated using groundwater (Llamas, 1995). There is general agreement that irrigation in Spain, covering only 17% of cultivable land, produces about 60% of its agricultural output (in monetary terms). This is equivalent to saying that the average productivity of one hectare of irrigated land is about seven times higher than the average productivity of one hectare of non-irrigated land. On the other hand, it is systematically overlooked or ignored that the economic productivity of irrigated land using surface waters (some 20 km^3 a year for the irrigation of 2.4 million hectares) is lower than the productivity of land irrigated using groundwater (4 km^3 a year to irrigate one million hectares). We can say that the economic productivity of a cubic meter of groundwater used in irrigation is four to five times higher than the productivity of a cubic meter of surface water. Among the reasons for this greater productivity of irrigation with groundwater one can mention: (1) groundwater schemes are generally more recent than irrigation using surface waters; (2) generally they have been financed on private initiative, with these agricultural transformations being implemented in areas with good soil and climate; this is not always the case for major official surface water irrigation schemes; (3) farmers irrigating with groundwater have risked their own money and have generally shown more entrepreneurship.

Obviously, during drought periods hydropower generation decreases. Nevertheless, in Spain the "economic losses" that electric utilities may attribute to drought are usually significantly lower than the parallel economic losses in agriculture (Llamas, 1997b).

To sum it up, all citizens in regions prone to drought should be told that drought is a natural event in their country and, consequently, we have to adopt measures to mitigate its effects as a normal process. It is not necessary to label it a disaster and take very costly and often ineffective emergency measures paid for by the state, i.e. impose a burden on all taxpayers.

5. METHODS TO MITIGATE THE ADVERSE EFFECTS OF DROUGHT

It is generally agreed that there are three types of measures for coping with the adverse effects of long dry periods.

The first type involves increasing the supply of water or, rather, guaranteeing the quantity of water normally required during periods of low rainfall. This is essentially the solution adopted in Spain through the construction of surface reservoirs or transferring water from one basin to another. In other words, the idea is to offset the temporary imbalance of availability of surface water in a region by storing it in surface reservoirs or by bringing it from other sites, which "theoretically" have a surplus. This was the basic philosophy behind the first draft National Water Plan (MOPU, 1993) and seems to be the same under the just presented White Paper on Water (MIMAM, 1998). For example, since the end of 1994, there has been increasing attention in the media on the possibility of bringing water from the wetter countries of Europe to semi-arid Spain. More specifically, there has been talk of transfers from the Rhone to Barcelona (Vilaró, 1998). This solution was the subject of contradictory assessments by the previous Spanish Government. The current Government does not appear to have come to any conclusions on this issue. Another way of guaranteeing an adequate supply is by tapping aquifers during drought periods, but this solution has not been much used in Spain, in part because of the systematic under-estimation of this option by the Water Authorities (Llamas, 1992 and 1998).

The second type of operation to mitigate droughts is usually described as water demand management. When a water shortage occurs in an area, it is a good idea to check to what degree the shortage is due to a poor management of water demand or to its wasteful use. This aspect of the problem has been dealt with to a slight extent in Spanish water policy. By contrast, in the latest two updates of California's water plan, containing forecasts up to 2020, it is believed that water consumption itself will scarcely rise although the population is expected to increase from 30 million to 49 million. This remarkable reduction in overall per-capita consumption is to be ensured mainly by economy measures and by joint management of surface water and groundwater (California Department of Water Resources, 1994 and 1998).

Economic incentives represent the third group of actions to mitigate drought. These may be applied since, even though using the best technology for the management of water supply and demand, there are still going to be extreme situations, genuine disasters, which are unavoidable or where the cost of preventing them would imply investments out of proportion to the gain. For example, it would be possible to guarantee the maintenance of 99% irrigation to an area but the cost would be prohibitive. To deal with this risk, there are three basic solutions: (a) make farmers aware that every so many years they should count on not achieving their expected harvest due to lack of water, and should include this factor in their economic planning; (b) set up an adequate insurance scheme; (c) introduce the idea of "disaster risk" covered by compensation from public funds (regional, national, European level).

5.1. Managing of Water Supply

Water policy in Spain in recent years has continued to be based, essentially, on traditional options, i.e. structural engineering in the form of the construction of large facilities for surface water supply (dams, pipelines, transfer channels). This is the approach proposed with the first draft of the National Water Plan (MOPU, 1993). The serious criticism that the draft received led to some slight changes of attitude. However, the sensitive point in almost all the approaches to water policy continues to be the need to transfer water to achieve a "water balance" between the drier and wetter parts of Spain. This emphasis on transfers opened up a "Pandora's box" and gave rise to a series of conflicts. Not only has there been strong opposition to such approach from experts in ecology and hydrogeology as well as conservationist groups, but it has also caused or exacerbated the so-called "water wars", first of all between a number of Spanish regions but also, as was said earlier, with Portugal (Correia, 1995).

The drought lasting from 1991 to 1995 helped persuade the Spanish Directorate-General for hydrological works to start thinking, timidly at first, that the exploitation of groundwater could solve some problems related to drought. This attitude contrasts in particular with the experience of developed countries with similar climatic conditions. The Corps of Engineers in the United States analysed the lessons to be learned from the California drought from 1987 to 1992 (Brumbaugh *et al.*, 1994; Dziegielewski 1994). In both reports, in discussing the "lessons from previous droughts confirmed in the drought of 1987-92", they state that: "Groundwater use continues to be the most effective single response against drought".

Throughout the 1991-1995 period, approved funds for drought emergency relief reached about 100 billion pesetas (600 million Euros). It happens that a good part of this money was intended for accessing groundwater. One can wonder why these works were started so late and under an emergency procedure. According to official data, the investments in studies on and facilities for groundwater access from 1983 to 1993 were between 300 million and 400 million pesetas (about 2 million Euros). In the first eight months of 1995 alone this figure came to about 15 billion pesetas (about 90 million Euros), i.e. 40 to 50 times larger (Llamas, 1997b).

In the last months of the 1991-95 drought period, there was a heated debate over the idea of desalinating sea water as a solution to water shortage in coastal areas. It is probable that the technology for desalinating sea water has advanced a great deal in the last two decades and it is also probable that it will continue to progress. But an elementary economic calculation does not suggest that in the short run this can be a significant solution to the drought problems in Spain.

A good part of all these problems can only be solved if the excessive participation of public money in financing engineering structures relating to water supplies is dramatically reduced. It is well known that this situation of aid and subsidies for hydraulic works was generally followed in all countries until relatively recent times. However, today there is more and more a tendency to introduce the principle of subsidiarity; in other words the responsibility for water management - including economic terms - should be borne mainly by the stakeholders or the beneficiaries of the water supply and not by central or regional government (Krámer, 1995; Llamas, 1995). The idea that the user should pay "the full cost" is the basis for the proposed new Framework Directive on Water from the European Commission.

5.2. Managing of Water Demand

For many years this author has argued (Llamas, 1991 and 1992) that it seems ethically indefensible to ask for the permanent transfer of water from one region to another if there is a major waste of water in the receiving region as a result of bad technical management or inadequate water pricing. This is all the more the case if these major works to provide the transfers are carried out as public works paid out of the national budget. It is worth recalling (Llamas 1998; Buckland and Zabel 1995; Krámer 1995) that there is a strong trend in all developed countries today to charge the real or full costs of hydraulic works to the operators or even better to the users themselves. Otherwise it can easily happen that the only - or the main - beneficiaries of such hydraulic works are those who design and/or build them. Moreover, we cannot ignore the potential existence of well-organised groups who legally or illegally batten onto the profits from large-scale works carried out with public money.

Over the last few years, it seems that finally some groups are acquiring the habit of being careful with water both in urban supply and, particularly, in irrigation. It is worth to mention some examples of this change in attitude. In the first draft of the National Water Plan (MOPU, 1993) mention to water saving was almost symbolic, being considered at 5% of total use in agriculture. Given that the use of water in agriculture accounts for no less than 80% of all water demand in Spain, one of the important factors in encouraging a more careful use of water was probably a parliamentary motion passed in 1994 to the effect that the National Water Plan should not proceed to the Congress of Deputies for debate until is was accompanied, inter alia, by a national irrigation plan (Plan Nacional de Regadíos). This motion was approved unanimously by all political parties and was reaffirmed and extended in February 1995, again obtaining unanimous support from all the political parties. An early version of the national irrigation plan was presented by the Ministry of Agriculture in the autumn of 1995. This Plan has, however, not yet been finalised.

With regard to irrigation using groundwater - accounting for at least half of agricultural production from irrigated land in Spain (Llamas, 1998) - better management of demand and production is linked to two main factors: reliable inventory of the groundwater wells and the formation of associations of groundwater users (Aragonés *et al.*, 1996). Remarkably, the first of these - an inventory of groundwater development- was almost left out of the White Paper on groundwater (MOPTMA and MINER, 1994). In September 1995 the Spanish Government informed the Congress of Deputies that, in order to try to resolve the major delays in making the inventory of groundwater wells, it was going to set up a programme called ARYCA, worth about 7 billion pesetas (about 40 million Euros) over three to four years. Three years or so have now gone by and the implementation of the ARYCA programme appears to be considerably behind schedule. The second goal, setting up groundwater users associations is not even mentioned in the White Paper on groundwater of 1994. The efforts being made by the current Government to encourage the set-up of groundwater users associations are negligible (Llamas, 1998).

From a political point of view, the overall priority is urban supply and modernisation, as this directly affects practically all forty million Spaniards, even though it accounts for only 10% to 12% of the national water consumption. By contrast, water for irrigation involves only a part of the farming community, which in total does not even amount to one million persons but which uses 80% to 90% of all the water supplied in Spain. However, the issue of urban supplies is primarily an economic problem given the very low urban water charges in almost all cities (Llamas, 1996). In an article published in "Zeit Magazine", a well-known German weekly newspaper supplement, a good part of our problems with urban supply was attributed to the fact that the cost of water in arid Spain is about one eight of the cost in wetter Germany (Zimmer and García Rodrígo, 1995). Buckland and Zabel (1995) analyse water rates in 12 countries of the European Union: Spain is at the tenth place, followed only by Greece and Italy.

As early as 1990, the author (Llamas, 1991) drew attention to the lack of awareness about the need to use water sparingly in Spain and the lack of regulations requiring the use of household devices (showers, flush tanks of toilets, dishwashers, etc.) designed to economise on water. Little seems to have changed on this side.

5.3. Economic Incentives

In the previous paragraph we looked specifically at some economic measures - such as changing urban water tariffs - which could be crucial to managing water demand and avoiding water wastage. A radical measure was also mentioned - going to the root of the problem - which is to reduce to a minimum the amount public hydraulic works, i.e. those financed from the national budget (or in some cases the European Union budget through the Cohesion Fund and similar aid). The reality is that these works, in the best cases, are seldom of public interest but usually benefit only very narrow interest groups.

There is, however, one relatively recent initiative, tested with apparently good results during the Californian drought of 1987-92, called "water bank". Very briefly, the water bank consists of an offer by the Governor of California to farmers who, instead of irrigating their crops with water that they had legally obtained, were prepared to "sell" the water to the Government for transfer to industrial or urban uses. The initiative was a success: in the first year the Government of California bought more water from farmers then it was later able to sell. Accordingly, in the second year, when the same operation was repeated, only the volumes of water were purchased for which there was already a buyer. A very significant fact in this whole process was that the overall annual volume of water passing via the water bank was at maximum 1.5% of the total water used for irrigation in California (Howit, 1993; Llamas, 1998). This once more seems to confirm that "noise" in the communications media (the drought was in the news all the time in California throughout those years) is representing only a small part of what is actually going on.

The water bank approach has started to be mentioned in Spain frequently, usually in connection with the need for flexibility in what is a fairly rigid legal system of water permits. Trying to transfer the Californian "water bank experiment" to Spain without sensible adaptations to the social and legal systems of Spain, however, might turn out to be a mistake.

5.4. Education and Water Awareness

Little has been done in Spain during the last few years to create water awareness in primary and secondary schools and to educate the general public. Some government institutions have invested significant amounts of money in making "publicity" for the National Water Plan and its water-transfer approach as well as for other water policies. A few examples are following. The publication of the White Paper on groundwater (MOPTMA and MINER, 1994) was announced via paid advertisements in a large number of local and national periodicals; quite unusual for official scientific or technological documents. There were a large number of programmes about the drought in state-owned and private television channels, but public debates on the subject of water - with the participation of "dissident" views - were few.

No significant effort has been made to teach water awareness to primary and secondary students. This has not only been the case at the national level but even in places such as the La Mancha Occidental aquifer (Ciudad Real), where, according to the Government, the shortage of water is a serious problem because in these same areas there has been excessive pumping of groundwater for more than fifteen years. In very few of these areas, if any, have there been mass distributions of posters, booklets and similar material to explain in simple terms what their water problems are. In these areas the relevant Water Authorities have neither developed such materials nor set up an office to provide information about the drought and water management. Nevertheless, activities of this kind would appear essential in order to secure the involvement of local people and their collaboration in managing water (Sudman, 1994).

But it is not just a matter of simple negligence over education in water awareness. It has been and still is remarkably hard to obtain information from the civil servants about environmental data in general and water issues in particular. This is in clear breach of the European Community's Directive 90/313/EEC regulating the easy access of the general public to environmental data. Its transposition into Spanish legislation was resisted at first by the parliamentary coalition, then by the government. In the face of numerous complaints to Brussels and despite the fact that Spain had held the presidency of the EU from July to December 1995, it was only at the end of June that year that the draft legislation for the transposition of Directive 90/313/EEC went to the Congress of Deputies and the Senate (Llamas, 1995; García Ureta, 1995) and that it was finally approved at the end of December 1995. However, the implementation of this Directive does not seem to be pushed ahead by the current Ministry of the Environment (Llamas, 1998).

6. CONCLUSIONS

The following five conclusions summarise the lessons which have - or should have - been learnt during the last severe drought in Spain.

One: In the current state of the art, droughts are not predictable but the effects can be mitigated by appropriate policy steps.

Two: In the Mediterranean countries, drought should not be regarded as a random hazard but as a normal climatic feature. Consequently, "disaster relief funds" should be kept to a minimum. It would be very useful to carry out an analysis of the actual use made of the national budgets and the European Commission cohesion funds which have been invested in emergency relief plans related to droughts and floods since 1990. It will probably be found that, in not a few cases, there has been much improvisation and/or corrupt diversion of funds.

Three: Groundwater can, in many cases, play a technically decisive and economically advantageous role in resolving problems due to drought. This has been experienced in other countries. The mentioned report of the US Corps of Engineers is especially significant. The main obstacles to an appropriate use of groundwater during droughts are twofold: (a) ignorance on the subject matter among a majority of engineers in relevant organisations; (b) the generalised "subsidy culture", meaning that practically all water-related facilities are declared to be of public interest and are paid for by all taxpayers and not by the final beneficiaries, who are sometimes almost the only ones designing and building them.

Four: Obviously, throughout the Mediterranean region the adverse effects of droughts cannot be solved with the use of groundwater only. Among other measures, there needs to be an analysis of the separate problems of urban supply, industrial use and agriculture. There also needs to be a change from an almost exclusive policy of supply management (more dams, interbasin transfers and exploitation of new aquifers) to a greater emphasis on a policy of demand management (reducing losses in distribution networks, appropriate tariffs, mechanisms to avoid wastage of water, "xeriscaping", etc.).

Five: It is time to create the conditions for avoiding a policy on water that seems mainly based on improvisation and/or misappropriation of public funds.

REFERENCES

Almarza, C. and López, J.A. (1994) Rachas húmedas y secas durante el período instrumental de observaciones, El caso de S. Fernando (Cádiz), in *Cambios y variaciones climáticas en España*, Universidad Internacional de La Rábida, 147-168.

Aragonés, J.M., Codina, J., and Llamas, M.R. (1996) Importancia de las comunidades de usuarios de aguas subterráneas (CUAS), *Revista de Obras Públicas*, 3355: 77-78.

Brumbaugh, R., Weiick, W., Teiz, W., and Lund, J. (1994), *Lessons Learned from the California Drought, Executive Summary*, IWR Report 94-NDS-6, U.S. Army Corps of Engineers, 36 p.

Buckland, J. and Zabel, T. (1995) Economic Instrumentation of Water Management and Financing of Infrastructures, *Conference on Institutional Dimensions of Water Resources Management*, Lisbon 10-12 July 1995, Document 9, 82 p.

California Department of Water Resources (1994) *California Water Plan Update*, Executive Summary, 39 p.

California Department of Water Resources (1998) *Proposal for the Update of the California Water Resources Plan*, 2 vol., 250 p.

Correia, M.L. (1995) La gestión de los recursos hídricos de las cuencas hidrogeológicas luso-españolas, *Eguipamientos y Servicios Municipales*, Julio-Agosto, 19-20.

Dziegielewski, B. (1994) Predictive Assessment of Severity of Agricultural Drought based in Agroclimatic Factors, *Journal of the American Water Resources Association* 33(6): 1255-1266.

García Ureta, A. (1995) Perspectivas sobre el derecho de acceso a la información sobre el Medio Ambiente, *Revista de Derecho Ambiental* 13: 119-164.

Gleick, P. (1998) *The World's Water, The Biennial Report on Freshwater Resources*, Inland Press, Washington D.C., 307 p.

Howit, R.E. (1993) *Resolving Conflicting Water Demands: A Market Approach*, Proceedings Symposium "La Economía del Agua", Sociedad General de Aguas de Barcelona, 131-164.

Krámer, R.A. (1995) *Subsidiarity and Water Policies*, Conference on Institutional Dimensions of Water Resources Management, Lisbon 10-12 July 1995, Document 16, 74 p.

Llamas, M.R. (1991) Aspectos generales de las sequías en España y posibles acciones para mitigar sus efectos, *Revista de la Real Academia de Ciencias*, Madrid 85(2-3): 193-204.

Llamas, M.R. (1992) A água - escassez ou mau uso?, *Coloquio/Ciencias*, Revista de Cultura Científica, Fundaçao Calouste Gulbenkian - Lisboa 4(12): 52-68.

Llamas, M.R. (1995) Las aguas continentales en España: Problemas y soluciones, *Revista Técnica de Medio Ambiente*, July-August: 33-45.

Llamas, M.R. (1996) El agua urbana en el Siglo XM: Futuros problemas y posibles soluciones, *Revista de Obras Públicas*, n. 3.358, October: 7-18.

Llamas, M.R. (1997a) Transboundary Water Resources in the Iberian Peninsula, in Gledish et al. (eds.) *Conflict and the Environment*, Kluwer, Dordrecht: 335-353.

Llamas, M.R. (1997b) Consideraciones sobre la sequía de 1991 a 1995 en España, *Ingeniería del Agua* 4(1): 39-50.

Llamas, M.R. (1998), Ministerio de Medio Ambiente o Ministerio de Obras Públicas, *La Ley - Actualidad*, Madrid, 19 January 1998: 1-7.

Ministerio de Medio Ambiente (MIMAM) (1998) *El Libro Blanco del Agua en España*, Secretaría de Estado para Aguas y Costas, draft mimeo, approx. 1000 p.

Ministerio de Obras Públicas y Urbanismo (MOPU) (1993) *Memoria del Plan Hidrológico Nacional*, 230 p.

Ministerio de Obras Públicas, Transporte y Medio Ambiente y Ministerio de Industria y Energía (MOPTMA y MINER) (1994) *Libro Blanco de las Aguas Subterráneas*, Centro de Publicaciones del MOPTMA, 135 p.

National Meteorological Services (1995) *Climate of Europe, First European Climate Assessment*, 74 p.

Sudman, R. (1994) *California Water Issues and Public Education*, Proceedings of the Symposium "La Economía del Agua", Sociedad General de Aguas de Barcelona: 163-178.

Vilaró, F. (1998) El acueducto del siglo XXI: Languedoc-Rosellón-Cataluña, Comunicación en las Jornadas Hidro'98 sobre "El Mercado del Agua", *Diario Expansión*, Madrid, 6-9 October 1998. Preprint 10 pp.

Vlachos, E. (1995) *Comparative Perspectives on Water Resources: Building Agendas for the 21st Century*, Conference on Institutional Dimensions of Water Resources Management, Lisbon, July 10-12 1995, Document 17, 11 p.

Zimmer, D.E. and García Rodrígo, C. (1995) "Dürre hausgemacht", *Zeit-Magazin*, Die Zeit, 21 July 1995: 8-12.

GLOBAL SYSTEMS AMELIORATE LOCAL DROUGHTS: WATER, FOOD AND TRADE

J. ANTHONY ALLAN
School of Oriental and African Studies
University of London,
London WC1H 0XG, UK

"The term 'drought' needs to incorporate a measure of the effectiveness of the climate for a specific purpose, be it the maintenance of an ecosystem, or of livestock herds. One should be able to define a 'pastoral drought' or an 'ecosystem drought' or a 'millet drought'. Drought would then be defined in terms of land use, and it follows that if the land use were to change, the frequency of drought would also change, without a necessary change in climate." (Warren and Agnew, 1988)

Abstract. The paper will show that solutions to water shortages can be found in 'problemsheds' when they cannot be found in 'watersheds'. This concept is relevant with respect to short-term drought as well as to long-term water stress. The volumes of water transferred across the world via trade, embedded in water intensive commodities such as grain, are massive in terms of the occasional and accumulating water deficits experienced in water stressed regions. Such strategic water is relatively easily mobilised; very easily mobilised compared with the problems that engineers would face in shifting such high water volumes. The reason such secure and globally available 'virtual water' is significant is that politicians can defer dealing with the impacts of a high proportion of the drought events because these reserves of accessible 'water' exist.

The study will discuss water resource deficits in Mediterranean countries in the MENA region (Middle East and North Africa region) as these economies have encountered the extreme water shortage and drought events in the first region to run out of water. Only Spain of European countries has experienced water stress on the scale experienced by the MENA economies. The MENA economies, apart from Turkey and Lebanon, moved into water deficit in the two decades before 1970. The experience of these political economies has highlighted the remedies mobilised by the political leaderships of water stressed economies. It will also be explained why environmentally and economically sound water policy reforms are difficult to implement.

1. INTRODUCTION

In discussing drought it is important to define which type of drought is being addressed. Droughts may impact water supplies for drinking water, for domestic water, for municipal water, industrial water and/or agricultural water. Interruptions to the first three types of consumption are very evident when they occur as individuals, communities and even members of the government in metropolitan centres may suffer breaks in supply. Such interruptions have immediate political significance. They draw attention to progressive declines in water availability as well as to drought events. Water shortages that impact livelihoods in industry and agriculture can also be politically as well as economically significant, but they tend to generate less political

J.V. Vogt and F. Somma (eds.), Drought and Drought Mitigation in Europe, 265–277.

heat than interruptions to drinking and domestic supplies.

The per-person use of water in these different sectors varies greatly. Very tiny amounts of water, as measured in annual consumption per head, are used by an individual for drinking - one cubic metre per year per person (m^3/p/y). About 100 m^3/p/y are used for domestic purposes in an industrialised political economy in the North; although only five m^3/p/y in a poor African country south of the Sahara. About 1000 m^3/p/y are needed to raise the food consumed annually by an individual whether living in the North or the South. They may of course consume much more in affluent economies. Meanwhile water for livelihoods in non-agricultural industrial sectors can be negligible, for example at a research institute in Ispra. On very rare occasions use per head in industry may be higher than those encountered in agriculture.

Social factors drive the levels of water used in association with food consumption. Decisions made within the family determine the overall population. Individuals in families decide how many children they will have and demographic theory has shown that these decisions are greatly affected by socio-economic circumstances. The latter also impact decisions on whether to consume a diet which involves modest inputs of water, for example that of a vegetarian, or an animal protein rich diet which requires very large volumes of water to produce animal products. It only requires about 1000 tonnes (m^3) of water to raise a tonne of wheat. It requires about 16 times that amount to raise a tonne of meat.

Water for food is the 'big' water in any national water budget. Happily it has been found that there is a global water, food and trade nexus (McCalla, 1997) which operates extremely efficiently to move water across the world from water surplus regions to those in permanent or occasional deficit. This 'virtual water', embedded in water intensive commodities such as wheat, is moved into the Middle East and North African (MENA) region to make good the 25 per cent of water which would be needed if the region was to be water and food self-sufficient.

2. APPROACH AND METHODS OF ANALYSIS

The methods of the economist, political scientist and social theorist will be the main methods deployed to explain why water policies evolve as they do in the face of progressive and periodic water shortages. It will be shown that the MENA economies that have run out of water have taken decades to reform water using policies to accord with sound environmental and economic principles. It took between two and three decades for a political economy well favoured with economic capacity and social adaptive capacity (Ohlsson, 1998), namely Israel (Allan, 1996b and 1996c; Feitelson, 1996). Other economies, both rich and poor, which have not had the social adaptive capacity to install water policy reforms are only just beginning on the process of such reforms thirty years after the water deficits could be identified.

Three sets of economic concepts are helpful. First, the notion of comparative advantage underpins the concept of 'virtual water'. Secondly, principles of water use efficiency provide foundation ideas for supply and demand management approaches to water management (Lonergan and Brooks, 1993; Brooks, 1994; Merrett, 1997; Winpenny, 1994). Thirdly, models linking environmental resource use and development are very helpful in highlighting the socio-economic contexts that enable or impede water policy reform (Karshenas, 1994; Karshenas and Allan, 1996).

Even more important are five concepts developed in the fields of political and social theory. First, the notion of the 'interest' of the community or nation in ensuring beneficial outcomes for them of any water management intervention. Secondly, the concept of discursive analysis (McHoul and Grace, 1993) which helps show how users of water contend with those advocating water policy reform. The process of contention and the achievement of 'consensus' takes time and outcomes never accord exactly with the goals of any of the contending parties; certainly not the alien scientist and consultant. A version of this explanation is that of 'sanctioned discourse' (Tripp, 1996) which is a process especially evident in political economies which have a very long tradition of water management, for example Egypt. Thirdly, the concept of strong societies and weak states (Migdal, 1988) captures the situation in many political economies where the size of rural water using communities is so large and vested with such long traditions of water management that their views prevail over those of the state. Fourthly the concept of 'windows of opportunity' (Kingdon, 1984) is especially relevant to water management in water scarce circumstances. The occurrence of a drought brings a short period of convergence in the ideas held by government officials and ministers, as well as by the public and the media (Allan, 1996a). Such convergence provides an opportunity for those wishing to introduce water policy reforms. Fifthly, in social theory the notion of 'belief systems', 'information gaps' and 'mutual knowledge' (Giddens, 1984: 334-343) provide profound insights into the processes which alien innovators and local politicians and communities encounter when 'new knowledge' is being recommended by outside professionals and scientists.

3. ADAPTING TO PROGRESSIVE AND OCCASIONAL DROUGHT: REMEDIES AND HOW THEY ARE RESISTED

3.1. Economic Theory

The concept of 'virtual water' is powerful in two senses. First it provides the remedy for progressive and occasional local 'agricultural' drought. Agricultural drought is the form of drought that is most common as deficiencies in soil water are evident in the failure of crops. The much smaller quantities of water needed for domestic use almost always come from other parts of the hydrological cycle than the soil water that supports crops. When the same surface source is used for all sectors, as in Egypt, then the volumes needed for domestic use are so small compared with those of irrigated agriculture that the prioritisation of domestic supplies is not a major political problem. Where groundwater is the resource the exhaustion of a regional water resource by a high water

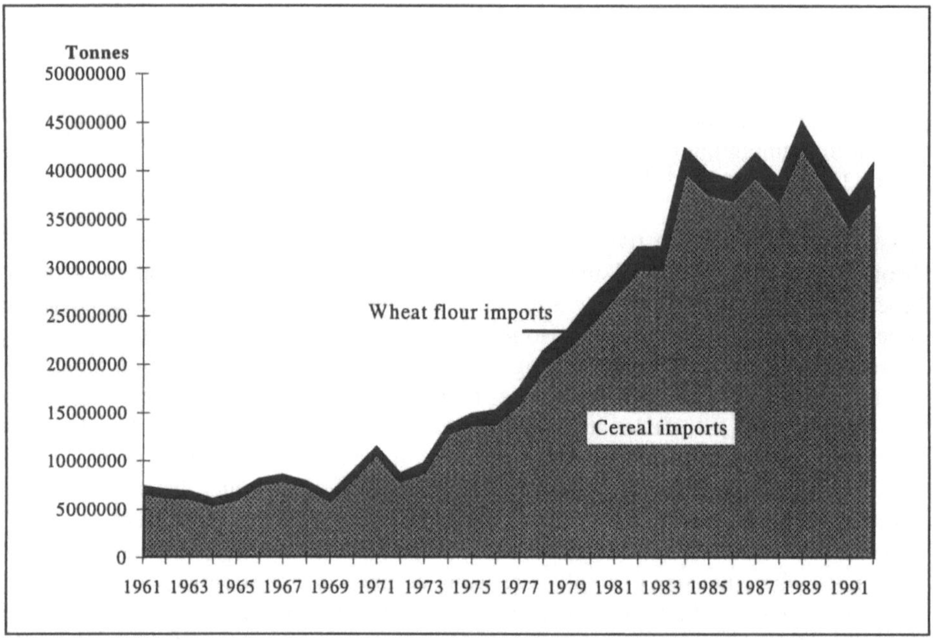

Figure 1: Middle East cereal and wheat flour imports (source: FAO, AGRISTAT).

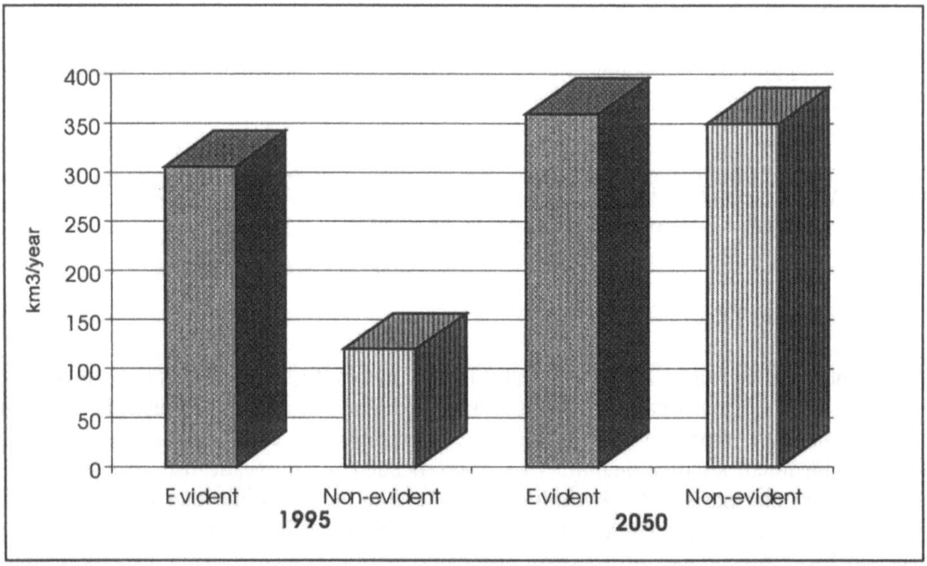

Figure 2: Evident and non-evident MENA water: 1995 and 2050 (source: author's estimates based on UN demographic data and estimates of water availability).

using sector such as irrigated agriculture can lead to severe stress for all users, including the urban domestic sector. The Libyan experience is a case in point (Allan, 1971). Spain is currently experiencing serious inter-sectoral competition for local waters (Del Moral, 1996). 'Virtual water' remedies national water deficits in a particularly effective way. The remedy is operationally effective. About 25 per cent of the water needs of the MENA region are accessed via trade in cereals – each tonne of imported wheat has 1000 tonnes of water 'embedded' in it. Figure 1 shows how the cereal imports rose quickly in the 1970s when the MENA region started to run out of water. By the mid-1980s these imported cereals enabled 40 billion cubic metres of water per year to be imported to the region to remedy the regional water deficit. Figure 2 provides an estimate of the future significance of virtual, 'non-evident', water in the MENA region.

Analysing the impacts of progressively deteriorating water supplies caused by demographically driven rising demands or the occasional single or multiple-year drought without taking into account the 'virtual water' remedy is a very partial and blind approach. It is rather like assuming that the only way for an individual to get to Milan from abroad is by road or rail, ignoring the airline option. Local water deficits can be accessed by hard pressed politicians by importing water from the soil profiles of very distant tracts. Farmers combine huge volumes of soil water with the other inputs to raise food extremely effectively in the industrialised and humid North. Northern governments have intervened for much of the twentieth century to enable such commodities with their embedded water to reach the world market at about half their production cost. The highly developed global grain trade readily conveys the grain with its embedded virtual water to deficit and drought affected regions. Politicians in water deficit economies get solutions for their problem in international 'problemsheds' not in local watersheds. The latter are a form of closed system. Economies and polities operate in open economic and political systems. Politicians and economies are impeded by closed hydrological systems but their options are not determined by them.

Other concepts of the economist are also important because they explain why economies can avoid the stress of a drought by ensuring that scarce water is effectively and efficiently allocated. Principles of water use efficiency provide foundation ideas for the supply and demand management approaches to water management (Brooks, 1994; Merrett, 1997; Winpenny, 1994). Water can be engineered so that higher returns to water are achieved. 'More crop per drop' (IIMI, 1996), the mission statement of the International Irrigation Management Institute, captures this principle. Doubling water use efficiency by improvements in the distribution of water and the reduction of leakage would go a long way to meeting the future needs of increasing populations. It is entirely possible that engineering measures will achieve such improvements as these via productively or technically efficient measures.

Much greater economic benefits than those achieved by technically efficient methods can be achieved by allocating water more effectively by economically or allocatively efficient measures. A hectare of wheat, requiring about 5000-6000 cubic metres of water, would provide a livelihood for a family in a poor country. It would provide only a fifth or less of a livelihood in a northern economy. The same area occupied by a service sector activity with the same water input could provide a thousand livelihoods and services turning over US$50 million per year. Allocatively effective water using activities can provide 10000 times the returns to water of agriculture. The greater livelihood providing potential of such activities enables communities and whole national economies to ensure their 'entitlement', to access to water intensive food commodities in trading systems. The notion of 'entitlement' transformed the analysis of famine in the early 1980s (Sen 1982, 1993, 1994). Above it was suggested that productive efficiency be captured in the phrase 'more crop per drop'. The concept of allocative efficiency is captured by the term 'more jobs per drop'.

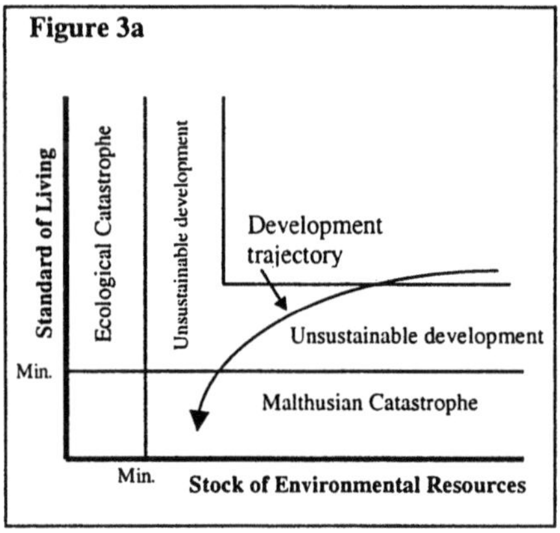

Eco-environmental space: consequences of low environmental capacity and 'over-use' of environmental capital

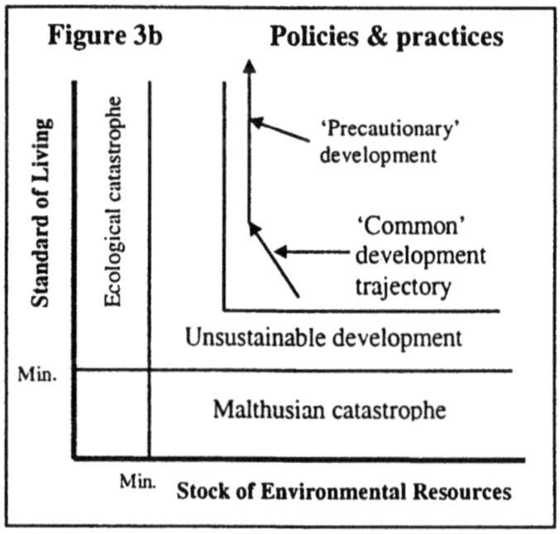

Common & 'precautionary' development trajectories

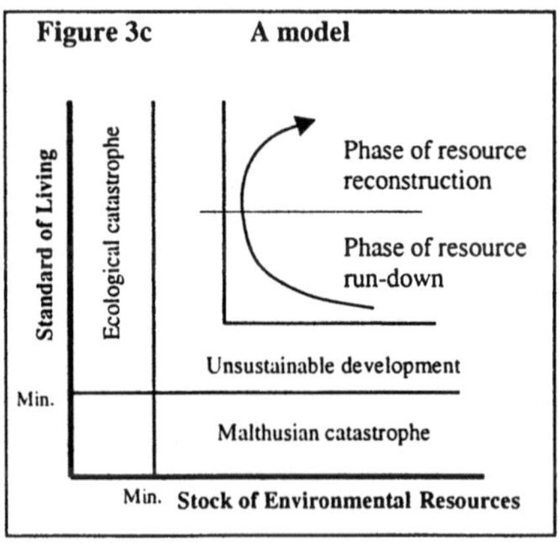

Diversifying economy trajectory

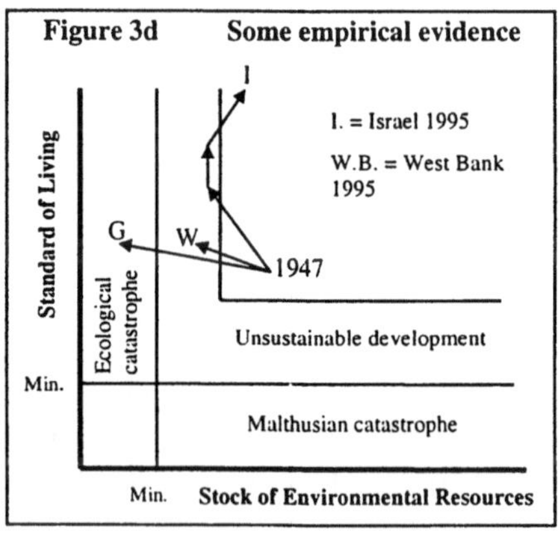

The cases of water in Israel, West Bank & Gaza. Trajectories 1947 - 1995

Figure 3: Environmental capital (water) and economic development.

The importance of analysing water stress in socio-economic contexts is further highlighted in a third group of economic concepts linking environmental resource use and socio-economic development. These models are very helpful in highlighting the socio-economic contexts that enable or impede water policy reform (Karshenas, 1994; Karshenas and Allan, 1996). In brief, the models attempt to show that a political economy goes through a number of predictable phases in its approach to managing its environmental capital, for example water. In an early phase environmental capital tends to be an input enabling economic development. As a result, the stock of environmental capital, for example water, tends to be degraded. When an economy reaches a phase of economic diversity and strength it can take different approaches and enjoy options not available when the economy was weak. One option is to curtail environmentally damaging policies. In a further phase, when the economy is even stronger, it is possible for governments and communities to rehabilitate the environment. The importance of the model is that it shows that remedies for drought impacts are not equally available. Concomitantly it must not be assumed that all political economies are equally afflicted by drought (Wilhite, 1993: 5). Social adaptive capacity varies and is transformed in the course of the socio-economic 'development' of a political economy. The case studies of Israel, the West Bank and Gaza in Figure 3 indicate the very differing predicaments of these equally drought-challenged political economies and their unequal capacities to respond to the challenges of implementing desirable policy reform.

3.2. Political and Social Theory

The first concept of political theory, which is useful in analysing drought and its consequences, is the notion of the 'interests' of the communities, bureaucracies and nations affected by drought events. These political entities will try to ensure that they gain any beneficial outcomes of a drought, or at least the most damaging outcomes are avoided. A number of analysts have shown how drought events in the 1970s (Sen, 1982) and the 1980s (De Waal, 1989: 30) have had severe impacts when economic and administrative infrastructures work to amplify the disruptions of a drought event rather than to alleviate them.

Secondly, discursive analysis (McHoul and Grace, 1993), an approach used by social scientists, helps analysts of the water sector and of drought events in particular. These analytical techniques show how those coping with the absence of water in water using and water policy making communities contend with those advocating policy reform advocated from outside drought affected regions. The process of contention and the achievement of 'consensus' take time and outcomes never accord exactly with the goals of any of the contending parties. A version of this explanation is that of 'sanctioned discourse' (Tripp, 1996; Figure 4) which is a process especially evident in political economies that have a very long tradition of water management, for example Egypt. In brief, communities and governments oppose unfamiliar innovation. There is a condition of embattled tradition. Governments are involved with traditional water using communities, usually in agriculture, in a discourse that is sanctioned. There is only one

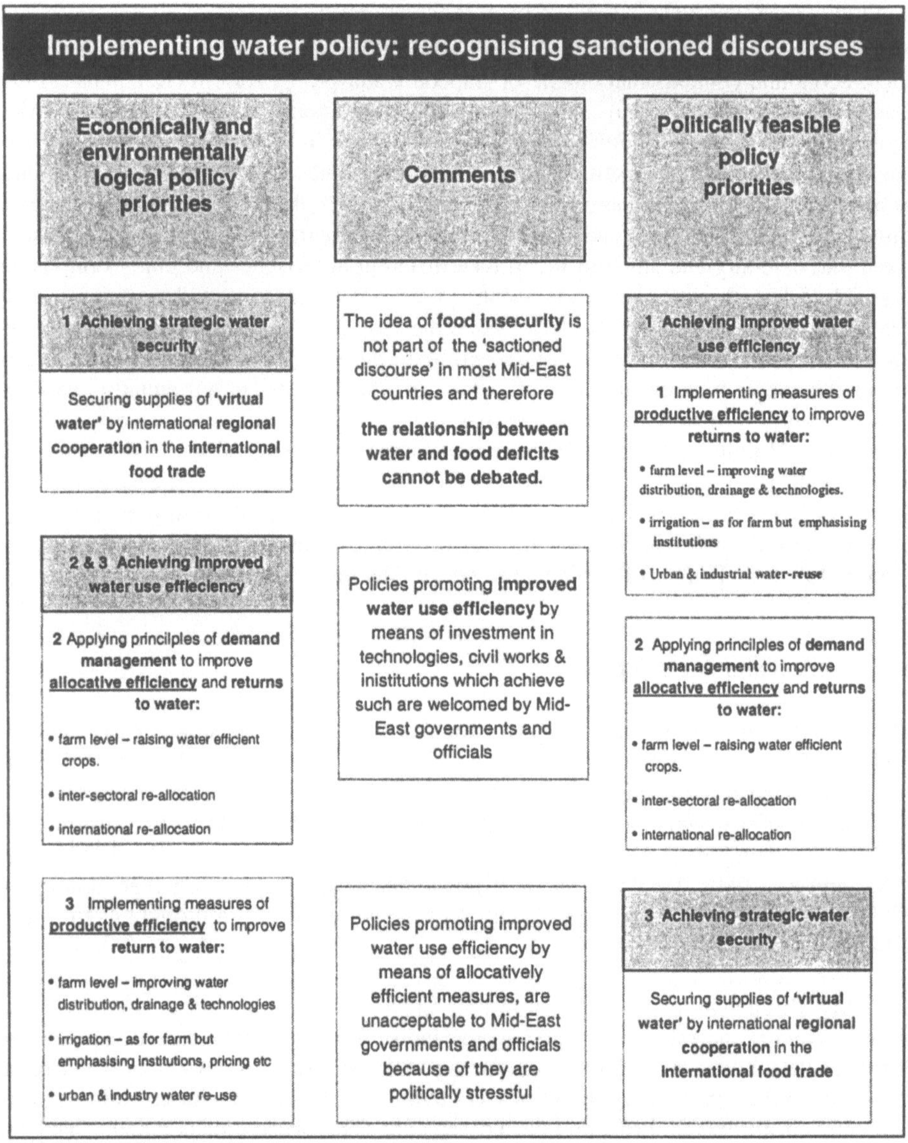

Figure 4: Implementing water policy: recognising sanctioned discourse.

agenda, the traditional agenda. Any deviation from this, to consider, for example, allocative efficiency measures would have such a high political price that it is not considered. Similarly the notion of recognising that 'virtual water' is the actual remedy to the increasing national water deficits and drought events carries an even higher political price and is also rejected.

Thirdly, the concept of strong societies and weak states (Migdal, 1988) reinforces the explanation derived from discursive analysis. In many political economies the size of rural water using communities is large and vested with such long traditions of water management so that their views prevail over those officials acting on behalf of the state. This situation is particularly true of long established irrigation societies. In weak states, in regions depending on rainfed agriculture, state institutions cannot ensure that there is institutional capacity to respond to drought events and ideas and actions at the local level are determining. Capacity to cope with major drought events is also absent at the local level. The impact of such events is exacerbated when populations have risen and there are impediments to traditional responses to drought such as migration, resulting from local political differences or international boundary regulations. In these circumstances the outcomes can be as devastating as they were in the droughts of the 1970s and the 1980s in Sahelian Africa.

Fourthly, the concept of ‘windows of opportunity’ (Kingdon, 1984) is especially relevant to water management in water scarce circumstances. The occurrence of a drought causes a short period of convergence in the ideas held by government officials and ministers (Feitelson, 1996), as well as by the public and the media (Allan, 1996a). Such convergence provides an opportunity for those wishing to introduce water policy reforms. The case of Israel is especially useful in illustrating this phenomenon (Allan, 1995). The droughts of 1986 and 1991-92 were not by themselves sufficient to bring about a change in national water allocation policy. But they proved to be the essential window of opportunity to swing the necessary public focus onto the issue of the sustainable utilisation of national water resources. Two very important additional political economy contexts were also important in bringing about the shift in policy.

Figure 5 illustrates the significance of drought. It also shows how Israel conformed to the Karshenas model until 1992, as have the water managing authorities of Australia (Chatterton 1995). Since then, other political forces brought on by the Middle East Peace Process have been overwhelming and Israeli water use has returned to its mid-1980s levels.

The first factor was domestic. It was the discourse that had been in train since the early 1960s about the need to adopt allocatively efficient (Palmer, 1962) as well as environmentally sustainable water policies. Activist scientists, Israeli officials and green pressure groups gained the attention of the media with talk of ‘red-lines’; these being dangerous drawdown thresholds with respect to lake levels in Lake Tiberias/Kinneret and in observation wells monitoring the West Bank aquifer. The second factor was international. In 1986 the US Reagan Administration embarked on a campaign to steer the Israeli economy. The US Government made the granting of a US $10 billion financial support dependent on major reforms in the Israeli economy. Water was one of the sectors targeted. Water reform was linked to drought, international pressure and internal discourse.

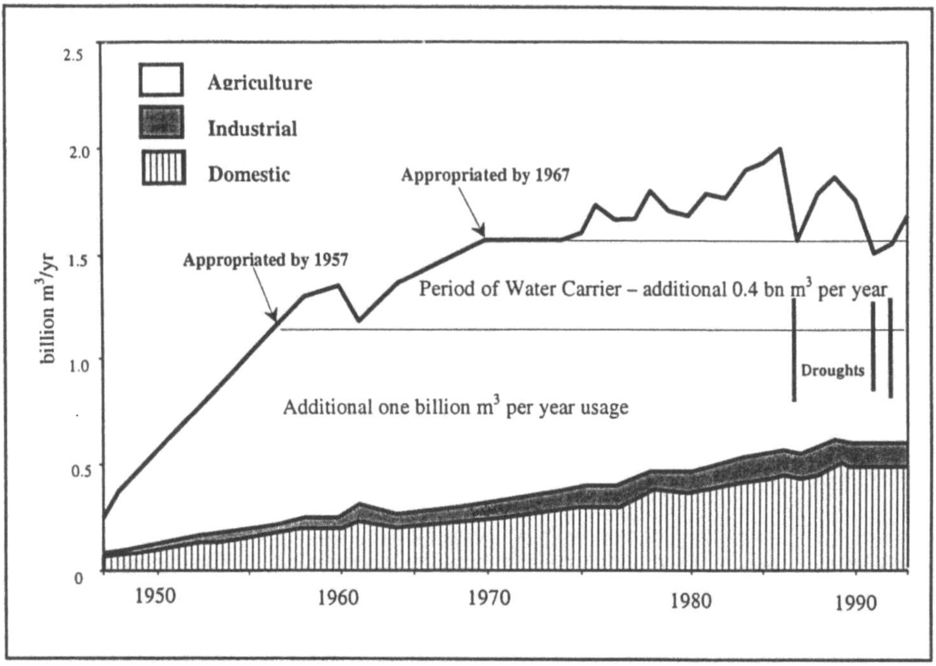

Figure 5: Israeli water consumption 1947 – 1993 by sector, showing the impact of the 1987 and 1992 droughts on sectoral water allocation.

The 1991-92 drought brought further shifts in water allocation out of agriculture, in accord with principles of economic efficiency. For example, an announcement was made that Israel proposed to reduce allocations to agriculture by 60 per cent from their mid-1980s levels. Half of this target had already been achieved (Voice of Israel, 1991). There has been a dramatic reversal in approach since 1992 by Israel. The period since 1992 has been an important phase in the Middle East Peace Process. Great progress was made by 1995; a peace was signed in September 1994 between Jordan and Israel. In September 1995 the Oslo Accord was signed between Israel and Palestine (Allan, 1996a: 211, 215). Throughout the 1992-99 period Israel has allowed its water consumption to rise to its 1986 level, a reversal of the 1986-1992 policy. The shift confirms again the dominance of politics in political economies and the perceived need by Israel to negotiate over water from a position of high levels of water consumption, 2 billion m^3/year, rather than the low 1991-92 level of 1.6 billion m^3/year.

Meanwhile the options available to strong political economies and to states in internationally significant theatres are revealed in the recent announcement by Ariel Sharon, the Israeli Infrastructure Minister, of a proposal to desalinate 400 million cubic metres salt and brackish waters annually for The West Bank, Gaza and Jordan. And a second phase of 800 million cubic meters annually for the 2010 needs of those three entities as well as those of Israel (Ha'aretz, 4 February 1999).

A fifth set of explanations, deriving from social theory, help explain why the ideas of alien hydrologists, drought specialists and economists have little impact on the national policies of the governments of water scarce political economies. The notions of 'belief systems', 'information gaps' and 'mutual knowledge' (Giddens, 1984: 334-343) provide profound insights into the processes which alien innovators and local politicians and communities encounter when new knowledge is being recommended by outside professionals and scientists. 'Belief systems' can reflect the experience of five millennia of occasional drought but not of systematic permanent deficit. Many political economies in the Middle East and the Mediterranean, including Spain, entered a period of progressively serious national water deficits – defined as insufficient water to meet all needs including that for self-sufficient food production. The idea of 'mutual knowledge' is useful. Gidden's notion captures very well the existence of 'mutual knowledge' in circumstances very common for the consulting community and scientists working on, for example, communicating the importance of water as an environmental resource. Ideas purveyed by alien scientists may come to be understood, that is become 'mutual knowledge' for part of the population, usually an educated elite. The phase of 'mutual knowledge' does not mean that water policy can easily be reformed. The process of reform is subject to the protracted community-wide discourses mentioned above which are subject to the interests of stakeholders other than those enjoying 'mutual knowledge'. The attenuation of the adoption of ideas and their incorporation into policy can take 30 years as in the Israeli case of the adoption of the concept of allocative efficiency. In less well-found economies without the adaptive social capacity (Ohlsson, 1998) the process could take much longer.

4. CONCLUSIONS

The analysis has shown that 'new knowledge', often very new indeed, on such issues as water resources and drought can take three decades to move from the existence of the new idea to its incorporation into national environmental and infrastructure polices in another country. Hydrological theory is not enough. In practice it can be misleading. Such theory tends be based on unsound assumptions that water resources determine economic, social and political outcomes.

The three main conclusions are that the impact of drought is very different on a diverse and strong political economy than on a weak political economy. Secondly, it is essential that those devising remedies to drought events take into account political and economic futures different from those that exist now. The future will frequently be a socially and economically developed future. The third main conclusion is that global economic systems are very important indeed with respect to local periodic drought. Global economic systems are especially important for those political economies which now, or will in future, face longer-term permanent water deficits. The global system also has the capacity to meet any local periodic droughts provided that economic and political systems to move supplies to the stressed locality are enabling rather than impeding.

ACKNOWLEDGEMENTS

The theoretical and case study material used in this paper results from decades of discussions with professionals, scientists and officials in Middle Eastern, North African and Mediterranean countries. Awareness of the new hydraulic mission – to use water in environmentally and economically sound ways – has resulted from attending many workshops and through involvement in studies of water policy for a number of international and bilateral agencies. To all the convenors and participants in these activities thanks are due for their diverse and improving insights.

REFERENCES

Allan, J.A. (1971) *Libya the Experience of Oil,* Croom Helm, London.

Allan, J.A. (1995) The Role of Drought in Determining the Reserve Water Sector in Israel, *Drought Network News,* International Drought Information Center, Vol 7, No. 3, October 1995, 21-23.

Allan, J.A. (1996a) Drought as a Concept and Drought as an Instrument of Policy, in M. Haddad and E. Feitelson (eds.), *Management of Shared Aquifers,* Third Workshop, Palestinian Consultancy Group and the Truman Institute for the Advancement of Peace, Jerusalem, 11-19.

Allan, J.A. (1996b) An Exceptional Return to Water, *MEWREW,* Number 7, SOAS, London, 1-2.

Allan, J.A. (1996c) The Political Economy of Water in the Jordan Basin, in J.A. Allan (ed.), *Water, Peace and the Middle East, Negotiating Water in the Jordan Basin,* Tauris Academic Studies, London, 71-116.

Chatterton, B. and Chatterton L. (1995) Closing a Water Resource; Experience in South Australia (SOAS Web Page - http://www.soas.ac.uk/geography/waterissues).

Brooks, D.B. (1994) Economy, Ecology and Equity: Lessons from the Energy Crisis in Managing Water Shared by Israelis and Palestinians, in J. Isaac and H. Shuval (eds.), *Water and Peace in the Middle East,* Elsevier, Amsterdam, 441-450.

Del Moral, L. (1996) The Debate on the Financial and Economic Regulation of Water in Contemporary Hydrological Planning in Spain, in J.A. Allan and L. Radwan (eds.), *Perceptions of the Values of Water and Water Environments,* Proceedings of the European Community Erasmus Seminar, Departments of Geography of SOAS and Middlesex Universities, London, 37-46 .

De Waal, A. (1989) *Famine that Kills: Darfur, Sudan 1984-85,* Clarendon Press, Oxford.

Feitelson, E. (1996) Economic and Political Dimensions in Changing Perceptions of Water in the Middle East, in J.A. Allan and L. Radwan (eds) *Perceptions of the Value of Water and Water Environments,* Proceedings of the European Community Erasmus Seminar, Departments of Geography of SOAS and Middlesex Universities, London, 17-22.

Ferguson, J. (1994) *The Anti-Politics Machine: "Development", Depoliticization and Bureaucratic Power in Lesotho,* University of Minnesota, Minneapolis.

Giddens, A. (1984) *The Constitution of Society,* Polity Press, Cambridge.

Ha'aretz, (1999) Sharon proposes major desalination projects, An Announcement by Ariel Sharon, Israel's Infrastructure Minister of a Proposal to Desalinate Water for The West Bank, Gaza, Jordan and Israel for 2010 Needs, Ha'aretz Newspaper, 4 February 1999, Jerusalem.

IIMI, (1996) *Mission Statement of the International Irrigation Management Institute,* IIMI, Colombo.

Karshenas, M. (1994) Environment, Technology and Employment, *Development and Change,* 25: 4, 723-757.

Karshenas, M. and Allan, J.A. (1996) Managing Environmental Capital, in J.A. Allan (ed.), *Water, Peace and the Middle East: Negotiating Resources in the Jordan Basin,* Tauris Academic Publications, London.

Kingdon, J. (1984) *Agendas, Alternatives and Public Policies,* Harper-Collins, New York.

Lonergran, S. and Brooks, D. (1993) *The Economic, Ecological and Geopolitical Dimensions of Water in Israel,* Centre for Sustainable Regional Development, University of Victoria, Victoria, British Colombia.

McCalla, A. (1997) Water, Food and Trade, Proceedings of the Mediterranean Development Forum convened by the World Bank in Marrakech, Morocco in March 1997, 33-42.

McHoul, A. and Grace, W. (1993) *A Foucault Primer: Discourse, Power and the Subject,* UCL Press, London.

Migdal, J. (1988) *Strong Societies and Weak States: State-Society Relations and State Capabilities in the Third World,* Princeton University Press, Princeton.

Merrett, S. (1997) *Introduction to the Economics of Water Resources,* UCL Press, London.

Ohlsson, L. (1998) *Environment, Scarcity and Conflict: a Study of Malthusian Concerns,* University of Göteborg, Department of Peace and Development Research, Göteborg.
Palmer, M. (1962) A memorandum by an Economic Secretary at the US Embassy in Tel Aviv in which the author pointed out that the future economic prosperity of Israel would lie in water economical industry and services and not in agriculture.
Sen, A. (1982) *Poverty and Famines: an Essay on Entitlement and Deprivation,* Oxford University Press, Oxford.
Sen, A. (1993) The Economics of Life and Death, *Scientific American* 268: 18-25.
Sen, A. (1994) Population and Reasoned Agency: Food, Fertility and Economic Development, in Lindhahl-Kiessling, W and Landberg, N (eds.), *Population, Economic Development, and the Environment,* Oxford University Press, Oxford.
Tripp, C.H. (1996) personal communication.
Voice of Israel (1991) An announcement quoting the Israeli Minister of Agriculture that water was to be cut again to agriculture. Jerusalem and IDF Radio, Tel Aviv, 2200 GMT, 27 January 1991. This comment was confirmed by the by then ex-Water Commissioner of Israel at a meeting in December 1992 in the University of Zurich when Israeli and Palestinian scientists and officials were meeting for the first time publicly. He indicated that water to agriculture would be cut by 60 per cent from mid-1980s levels; half had already been achieved.
Warren, A. and Agnew, C. (1988) *An Assessment of Desertification and Land Degradation in Arid and Semi-Arid Areas,* International Institute for Environment and Development, Drylands Programme, London.
Wilhite, D. (1993) The Enigma of Drought, in D. Wilhite (ed.), *Drought Assessment, Management and Planning: Theory and Case Studies,* Kluwer Academic Publishers, Boston.
Winpenny, J. (1994) *Managing Water as an Economic Resource,* Routledge and ODI, London.

DECISION PROCESSES WITH REGARD TO DROUGHT MONITORING AND MITIGATION

B. CORNÉLIS [1] and A.A. VIAU [2]

[1] *Département de géomatique, Université de Liège 4000 Liège, Belgium*

[2] *Centre de recherche en géomatique, Université Laval Ste Foy, Québec, G1K 7P4, Canada*

Abstract. This paper presents the decisional fountain - a generic model of decision processes. According to this model, decision taking is based on a decision analysis phase that relies on a documentation/information phase. It is then followed by the implementation and the evaluation phases. The foundation of the documentation/information step is composed by three interconnected aspects (data, issue and context). Based on this model, the study of the decision structures at the various decision levels of the EU is suggested in the frame of drought monitoring.

1. INTRODUCTION

The implementation of geographical information system (GIS) technologies often aims at supporting decision-making through data collection, data analysis, and simulation. While data quality and management have been studied for several years in the frame of GIS researches, little or no attention was given to decision itself. In that context, decision is usually seen as an outside event, based on the system outputs, resulting from the decision-makers' choice. When dealing with multiple objectives and/or multiple criteria, decision support systems (DSS) are used to suggest solutions. Nevertheless, so far, none of these have integrated information on the resulting human decision.

The work discussed in this paper aims at stressing, within a spatial decision support system (SDSS), the importance of human decision with its associated errors and uncertainties by proposing a way to manage it. Among other things, it looks at the nature of GIS and DSS, it proposes a typology of decisions, it includes perspectives on the spatial and temporal scales of application of decisions, and it underlines the different types of uncertainties associated to each decisional phase. The concepts developed are then discussed in the light of decision structures on drought related issues in the EU.

After presenting the theoretical background, the basic concepts of the decisional fountain will be presented. Then the methodology suggested for application to the drought phenomenon will be presented.

J.V. Vogt and F. Somma (eds.), Drought and Drought Mitigation in Europe, 279–290.

2. THEORETICAL BACKGROUND

2.1. The Need

For Heywood *et al.* (1994: 638), « *there appears to be insufficient research by the GIS community into how decision makers actually make decisions* ». Decision-making relies on the objective(s), on a set of information, on the knowledge and on the experience of the person(s) deciding. While some decisions depend upon little information and upon basic knowledge, others require huge amounts of data and complex models. The data sets used in a decision process have already been subject to some decisions in terms of acquisition, integration, processing and diffusion. The errors and uncertainties associated with each one of the processes should be clearly identified for the thorough documentation of the decision-maker on the reliability of the information he uses. Since spatial information has specific characteristics, « *more attention must be focused on understanding the role of spatial information in the decision making culture as a whole* » (Heywood *et al.*, 1994: 638). This research can be seen as an essay to fill the gap between decision-makers and the GIS community.

2.2. What is Going on Elsewhere ?

Numerous laboratories, institutes and companies worldwide have initiated research on Decision Support Systems (DSS). Among their research priorities for the next ten years, the University Consortium for Geographic Information Science (UCGIS) and the National Center for Geographic Information Analysis (NCGIA) identified « Uncertainty in Geographic Data and GIS-based Activities » (UCGIS, 1996). For Hunter (1998), this concern about uncertainty has grown in importance for the four following key reasons: (1) the requirements in some jurisdictions for mandatory data quality reports, (2) the need to protect personal and agencies reputations, (3) the need to safeguard against litigation, (4) and as part of the basic scientific quest for representing the reality as close as possible. More specifically in the cartographic field, Monmonier (1995: 299) stated « *future cartographic controversies will most certainly involve costly litigation over missing data and misinformation about data as well as errors in data* ».

2.3. Can GIS Support Decision ?

For Bonham-Carter (1994: 3), the « *ultimate purpose of GIS is to provide support for making decisions based on spatial data* ». In 1989, Aronoff is writing about « *using a GIS for decision-making under uncertainty* ». GIS definitions range from « *a powerful set of tools for collecting, storing, retrieving at will, transforming and displaying spatial data from the real world for a particular set of purposes* » (Burrough & McDonnell, 1998: 11) to a « *decision support system involving the integration of spatially referenced data in a problem solving environment* » (Cowen, 1988: 1553). However, for Densham (1991: 403), GIS « *do not adequately support decision making because they are lacking in analytical modelling capabilities and do not easily accommodate variations in either the context or the process of spatial decision making* ».

2.4. What are DSS then ?

For Watson *et al.* (1997: 263) a DSS is « *an interactive system that provides the user with easy access to decision models and data in order to support semistructured and unstructured decision-making tasks* ». For Sauter (1997: 13), a DSS is a « *computer-based system that supports choice by assisting the decision maker in organizing information and modeling outcomes* ». She also states (p. viii) that « *seemingly specialized forms of DSS, such as Geographic Information Systems or Purchasing Support Systems, actually follow the same principles as a "basic" DSS* ». Another definition by Turban and Aronson (1998: 856) states that DSS are « *computer-based information systems that combine models and data in an attempt to solve nonstructured problems with extensive user involvement* ».

3. THE DECISIONAL FOUNTAIN

Out of these flourishing research fields, the gap between researchers and decision-makers seemed to grow bigger and bigger over time. There was a need for global reflection on the various aspects of decision: actors, procedures, tools, outcomes, perception, and diffusion. The decisional fountain originated from this global reflection. It is a generic model that can be used for screening the decision processes. It apprehends decision-making through the following phases: Documentation/Information (DO), Decision Analysis (DA), Decision Taking (DT), Decision Implementation (DI) and Decision Evaluation (DE) (Figure 1).

3.1. Documentation/Information

In decisional processes, the documentation/information phase relies on three inter-linked bases: issue, context and data. The issue can be stated in terms of problem(s), of need(s) or of objective(s). These formulations are just ways to express the concern of the decision-maker(s). Each one of them can be reformulated as a problem, need or objective, which in turn might orient the way the issue will be dealt with. The context refers to the decisional philosophy, the experience and the knowledge. There are four basic philosophies: proportional, egalitarian, maximalist and fluctuating. They can be combined into decisional strategies. The experience is a personal or collective form of knowledge. The knowledge relates to the hard and soft scientific knowledge as well as to background information not necessarily stated or known.

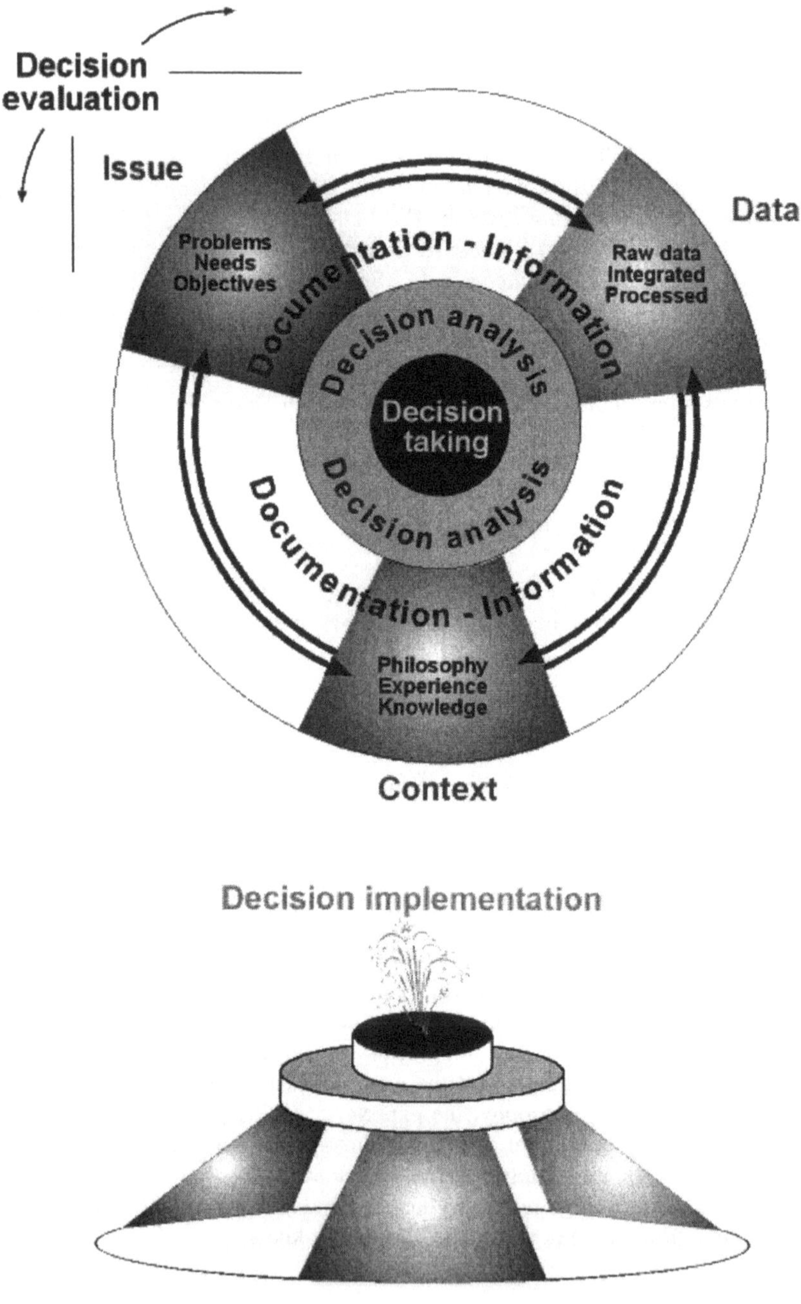

Figure 1: The decisional fountain. Zenithal view at the top and oblique view at the bottom.

The data can either be raw, integrated or processed. The processed data for one party might be considered as raw data for another party. The connections between issue, context and data are numerous. Sometime, the boundary between one or the other might appear fuzzy. Data can generate a problem through unexpected values. The other way around an issue will identify data to fulfil needs. If an objective is expressed in terms of maximisation of benefits, for example, it already includes philosophic elements. Knowledge is nowadays extracted from databases. Out of all this, the documentation /information is extracted for the decision analysis phase.

3.2. Decision Analysis

The fields dealing with decision science are numerous. Examples are operational research, mathematics, psychology, management, or political science. Each one of them developed some kind of decision culture, paying more attention to one aspect or another. For example, the methods involved in DSS either structure problems and decision-making through interviewing, multidimensional scaling, card sorting, and/or tend to imitate the human thinking. DSS are either data-driven or model-driven. Genetic algorithms can be used for evolving solutions, neural networks for simulating the brain while solving problems, rule-based systems for putting expert reasoning in a box, fuzzy logic for dealing with linguistic ambiguity, cased-based reasoning for solving problems by analogy, machine learning for deriving rules from data (Dhar & Stein, 1997). When working in a group or in the environmental domain, the decision support system used is often based either on the multiple criteria / multiple objectives theory (Colson & de Bruyn, 1989; Colson & Mareschal, 1994; Eastman, 1997), or on benefit / cost analysis (Dixon *et al.*, 1994). The decision analysis step is the phase that analyses and integrates the documentation/information. It comes right before the decision taking.

3.3. Decision Taking

The decision-taking phase is the moment at which the choice is done. It is characterised by the person(s) taking the decision, its (their) behaviour and by the formal procedure followed. Who is taking the decision? It can be one person, several persons with their own interests and objectives, a group of persons sharing the same interests and objectives, or several groups of persons with their own interests and objectives. The procedures followed can be despotic, negotiated, agreed, voted, freely taken, controlled, weighted, etc.

Figure 2 shows the behaviours decision-makers can have at the moment of taking decisions. The decision-makers have two basic decision behaviours: a positive one and a negative one. They can decide either to select (positive decision) or to eliminate (negative decision). Decisions such as binary, plural, ordered, null, self-conflicting, conditional or nuanced can be identified (Cornélis, 1998). A **binary** decision identifies a single element out of two elements of a set. In a binary positive decision, the resulting choice is the element selected. While in a binary negative decision, the resulting choice

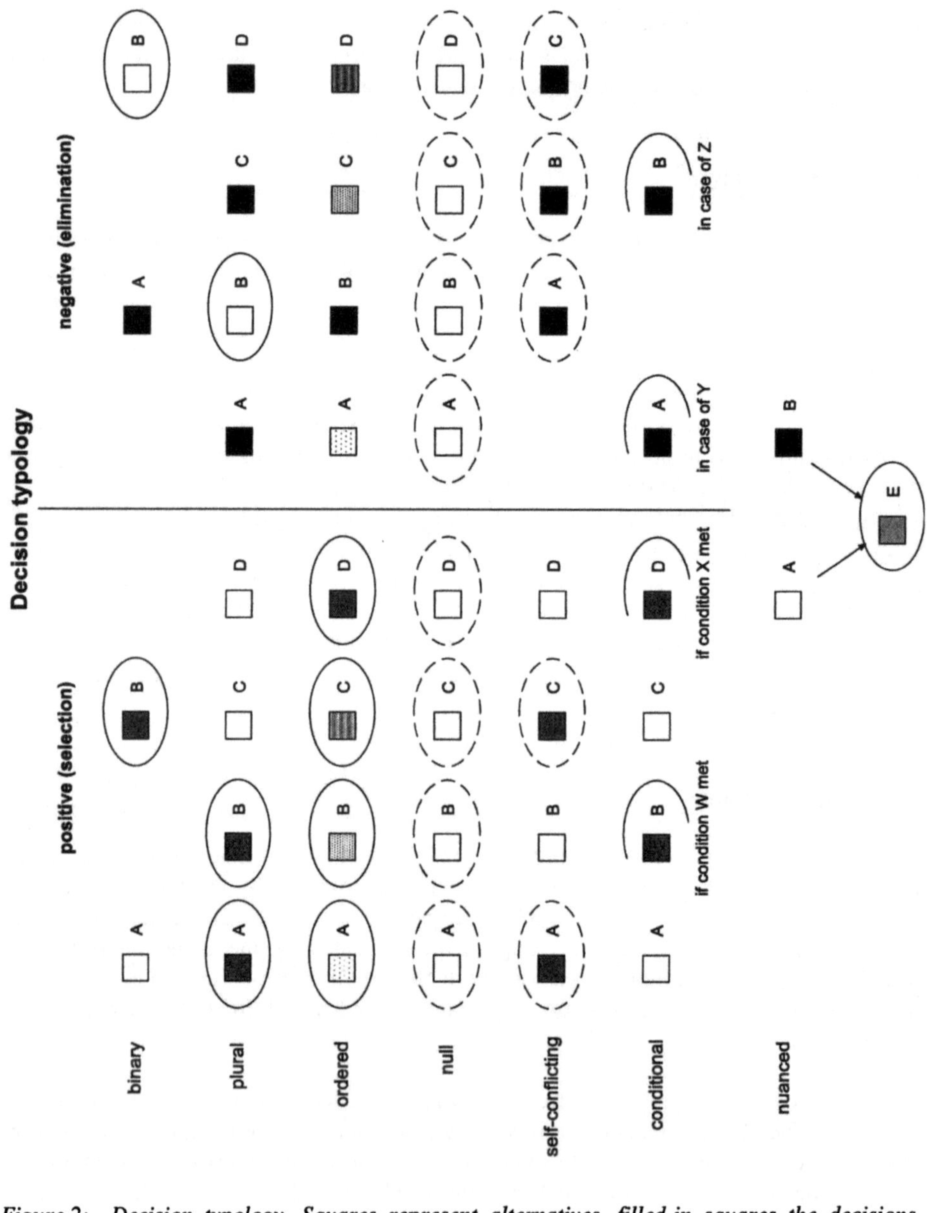

Figure 2: Decision typology. Squares represent alternatives, filled-in squares the decisions, ellipses the resulting choices, dotted ellipses the possible resulting choices.

is the element not eliminated. A **plural** decision is a decision that designates more than one element of a set. In a plural positive decision, the resulting choices correspond exactly to the selected elements. On the contrary, in a plural negative decision, the number of resulting choice varies according to the number of alternatives in the set minus the number of eliminations. So, the plural negative decision can lead to a single choice.

An **ordered** decision is a decision putting the elements of the set into a methodical arrangement based on the preference of the elements. An ordered positive decision will return a set of resulting choices directly dependent on the resources available, giving priority to the most preferred elements. In an ordered negative decision, it is the opposite. The elimination priority is given to the least preferred element. In this case, the number of resulting choices depends on the way the decision is to be taken. A **null** decision happens when a decision-maker decides not to decide. In a null positive decision, he decides not to choose. While in a null negative decision, the decision is not to eliminate. Both cases lead to no choice being made. A **self-conflicting** decision takes place when one is facing mutually exclusive desires or tendencies. In a self-conflicting positive decision, several elements of the set with the same preference level are pre-selected while fewer have to be selected. The decision-maker wonders then « which solution will be the most profitable keeping a satisfactory level as high as possible for a risk level given? ». For the self-conflicting negative decision, the elements of the set are all eliminated with the same level of dislike while at least one is to be chosen. The question turns to « which solution will be the least unpleasant? ». **Conditional** decisions are decisions taken in advance stating an (or several) event(s) to be met before the decision is valid. For conditional positive decision, one (or several) condition(s) is (are) to be met before the resulting choice is determined. For the conditional negative decision, the occurrence of one (or several) event(s) guides the elimination. When a decision-maker brings up a new solution resulting from parts of the elements of the set, he is taking a **nuanced** decision. This type of decision is neither positive nor negative since only the interesting bits of the elements remain. So no element is completely selected or eliminated. The extra solution is the resulting choice.

Combinations of these seven decision types are possible. For example, the decision-maker can select some elements and eliminate others within the same decision phase. Possible combinations include plural ordered positive decisions. A composed decision might result from a self-conflicting positive decision followed by a conditional negative decision or a null positive/negative decision if the condition(s) is (are) not met at the time the resulting choice is supposed to exist.

3.4. Decision Implementation

This phase succeeds to the decision taking. It can present implications to the different phases of the decisional fountain. It characterises the relation decision has with time: the stability in time (stable, evolving, definitive), the time of application, the prescription time, etc. The decision scale of application (e.g. regional government, municipality, farmland) characterises this step. This phase in other words turns the decision into action, policy, knowledge. The decision confidence level and the diffusion methods are typical of this level. The decision implementation can be a time for legitimisation and justification of the decision taken.

3.5. Decision Evaluation

Last but not least, the evaluation of decisions is in itself tricky, because it can not be complete and will be based only on one point of view. Just like when taking a picture of a landscape, the decision evaluation will look at the core of the fountain from one point of view. This point of view might present an emphasis on the issue, or on the data or on the context. But just like the picture of the landscape it will only be a partial view. Even a zenithal view would not be complete since the implementation would be diffused over the bases.

3.6. Uncertainties in the Decision Process

Uncertainties come along the entire decision process and can be found at the different stages. In the DO phase, uncertainty appears in data itself, or in its processing, in the knowledge used and in the values the decisions rely on. The uncertainties of reliance and of demarcation characterise the DA phase. At the moment of taking a decision, the uncertainty of legitimacy might arise. During the implementation phase, the uncertainty comes from the interpretation done on the decision taken. Belief and methodology uncertainties are met in the DE phase. The GIS community has only been busy with some types of uncertainties, which politicians and stakeholders would like to ignore having sufficiently uncertainties to worry about (Brunet and Cornélis, 1999).

4. METHODOLOGY SUGGESTED FOR ANALYSING DROUGHT RELATED DECISIONS IN THE EU

Drought related decisions are numerous and vary, among others, in terms of their effects on the impact of a drought event, in terms of the organisational structure they take place in, and in terms of the issue addressed. For example, at a farm level, the farmer is the only one deciding where to irrigate and which crop to grow in a certain field. He might be part of a co-operative democratically deciding for the community interests. For water resource allocation at the communal and regional scales, politicians have to take into account not only the farmers' interests but also the ones of the rest of the population. This might create conflicts between tourism and agricultural industries. At the national and European levels, favouring a crop to another one in order to follow the market rule might end up enhancing drought effects. To manage these different decisions for an improved drought awareness, the use of integrated SDSS is required.

In his book on Intelligent Spatial Decision Support Systems, Leung (1997: 6) states that « *the emphasis of the majority of present SDSS is placed more or less on software engineering involving integration of databases and models, rather than on knowledge engineering involving the use of human intelligence in decision making* ». For Heywood *et al.* (1994: 638), a « *complex problem is the need to fit GIS into the context of the whole decision making process rather than fit decision making into GIS* ». That's why the methodology should be needs/objectives-driven (Table 1). It should not only include the spatial and temporal scales of decision but also its confidence level. This approach should allow, for a given decision level, the identification of models and data to be used as well as their required precision and accuracy. This leads to the problematic of representation, visualisation, perception and interpretation. The management of decisions has also to be thought of in terms of responsibility, legacy, and diffusion. Not to mention the underlying philosophy. This management will get more and more important since according to Choffray (1992) the trend in decision-makers position allocation is to change every two to five years.

Table 1: Analogy and link between the data analysis and the decision process in a need-driven methodology for a European scale study.

Data	**Needs**
Data process	Decision making process
1. Data acquisition	1. Documentation-Information
2. Data integration	2. Decision analysis
3. Data processing	3. Decision taking
4. Data diffusion	4. Decision implementation
Spatial scale of data	Spatial scale of decision application
Point	UE
Inference	Country
Local	Regional
Region	Sub-regional
Country	Local
UE	Farm
Representation visualisation	Perception interpretation

4.1. Any Existing Methodology ?

The methodology for establishing such integrated SDSS could be based on the systemic approach and on its modified version (Figure 3) developed by Pantazis (1994). It might consist in an adaptation of the MECOSIG (*MÉthode de COnception de SIG*) method to SDSS, but adopting it could be sufficient.

In the modified version of the systemic approach (Pantazis and Donnay 1996), an organisation can be composed of up to five types of « systems ». One system would theoretically be sufficient. The « systems » are (Geographic) Information systems (G)IS, (Geographic) Operation systems (G)OS, (Geographic) Decision systems (G)DS, Complex systems (CXS) composed by several types of the previous systems, Auxiliary systems (AS) dealing with auxiliary and/or support activities of the organisation. These systems can all be found at the different functional levels of an organisation.

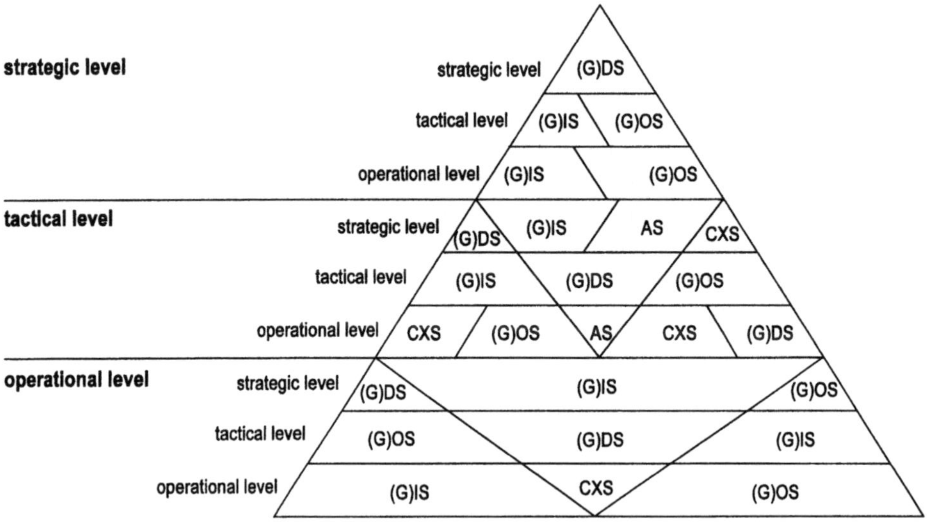

Figure 3: An organisation according to the modified version of the systemic approach (after Pantazis and Donnay, 1996).

The MECOSIG method distinguishes within an organisation the following functional levels: strategic, tactical and operational. « *In the strategic level, crucial information are needed to realise strategic actions and take strategic decisions (establishing policy guidelines, the long-term overall direction of the organisation, etc.). In the tactical level, decisions, information and actions are more specific and some of them directly apply to concrete cases: formulation and transformation of the policies into action plans. The operational level deals with the operational information, decisions and actions used and/or applied directly in every days activities of an organisation* » (Pantazis and Cornélis, 1997).

4.2. Practical Approach for the Analysis of Decision Processes

For the experimental part of such work, an extensive use of interview techniques should be applied. A good bit of time should be spent on questionnaires, meetings and inquiries of the different decision-makers at the different levels of decision making. In order to facilitate this part of data collection, it should be done within the framework of the *European Network on Drought Research and Mitigation* (ENDRM) as proposed by the Space Applications Institute of JRC, Italy and the Institute of Hydrology, UK. The set-up of an interactive inner-web-site and the interactions with the members of that network should give a direct feedback and allow the evolution of the original concepts to better fit the needs and the reality of the decision makers.

Such drought network could be seen as a common structure linking scientists, industries, politicians, public and administrations interested in the drought problematic. This structure should share tools and information in order to improve diffusion and communication at the different decisional level of the European Union, from farmland to the Commission.

5. CONCLUSION

While drought is a recurrent feature of climate occurring in virtually all climatic zones (National Drought Mitigation Center, 1995), human decisions can enhance or mitigate the phenomenon and its effects. Studying the physical aspects of drought is necessary but not sufficient for its management. The individual decision behaviour as well as the structure and organisation of the society have to be taken into account. With the development of computerised information and decision systems, linking the physical and human aspects is now feasible. Also, to set up a SDSS for drought mitigation in Europe, one should follow a multi-scalar approach and take into account the different facets of the decisional processes, especially since the decisions are inter-linked and influence each other.

REFERENCES

Aronoff, S. (1989) *Geographic Information Systems : A Management Perspective*, WDL Publications, Ottawa, Canada, 294 p.

Bonham-Carter, G.F. (1994) *Geographic Information Systems for Geoscientists: Modelling with GIS*, Pergamon, Kidlington, UK, 398 p.

Brunet, S. and Cornélis, B. (1999) Uncertainties in Spatial Decision: a Policy-maker Point of View, in W. Shi, M.F. Goodchild and P.F. Fisher (eds.), *Proceedings of the International Symposium on Spatial Data Quality*, 18-20 July, 1999, Hong-Kong, China, pp. 566-573.

Burrough, P.A. and McDonnell, R.A. (1998) *Principles of Geographical Information Systems*, Oxford University Press, Oxford, UK, 333 p.

Choffray, J.M. (1992) *Systèmes intelligents de management: diagnostic, analyse et assistance à la décision*, Collection Connaître et pratiquer la gestion, Éditions Nathan, Paris, France, 178 p.

Colson, G. and De Bruyn, C. (1989) Models and Methods in Multiple Objectives Decision Making, *Mathematical Computer Modelling*, 12(10/11): 1201-1211.

Colson, G. and Mareschal, B. (1994) JUDGES: A Descriptive Group Decision Support System for the Ranking of Items, *Decision Support Systems* 12: 391-404.

Cornélis, B. (1998) Managing Decision in Spatial Decision Support Systems, *Proceedings of the 1st AGILE Conference*, 23-25 April 1998, Enschede, NL, http://www.uniroma1.it/DICEA/Agile/Paper6.html.

Cowen, D.J. (1988) GIS versus CAD versus DBMS: What are the Differences?, *Photogrammetric Engineering and Remote Sensing* 54: 1551-1554.

Densham, P.J. (1991) Spatial Decision Support Systems, in D.J. Maguire, M.F. Goodchild and D.W. Rhind (eds.), *Geographical Information Systems: Principles and Applications* - Vol. 1, Longman Scientific & Technical, Essex, England, 403-412.

Dixon, J.A., Scura, L.F., Carpenter, R.A., and Sherman, P.B. (1994) *Economic Analysis of Environmental Impacts*, 2nd edition, Earthscan Publications Ltd, London, UK, 210 p.

Dhar, V. and Stein, R. (1997) *Intelligent Decision Support Methods : The Science of Knowledge Work*, Prentice Hall International, London, UK, 244 p.

Eastman, J.R. (1997) *Idrisi for Windows 2.0*, Clark University, Clark, USA.

Heywood, I., Oliver, J., and Tomlinson, S. (1994) Building an Exploratory Multi-criteria Modelling Environment for Spatial Decision Support, in J.J. Harts, H.F.L. Ottens and H.J. Scholten (eds.) *Proc. 5th European Conference and Exhibition on Geographical Information Systems,* 29 March – 1 April 1994, Paris, France, vol. 1, 632-641.

Hunter, G.J. (1998) Managing Uncertainty in GIS, *NCGIA Core Curriculum in GIS Science*, Unit 187, http://www.ncgia.ucbs.edu/giscc/units/u187/u187_f.html, posted February 3, 1998.

Leung, Y. (1997) *Intelligent Spatial Decision Support*, Advances in Spatial Science Series, Springer, Berlin, Germany, 470 p.

Monmonier, M. (1995) *Drawing the Line*, Owl Book Edition, New York, USA, 368 p.

National Drought Mitigation Center (1995) Understanding and Defining Drought, *Categories of Drought Definitions*, http://enso.unl.edu/ndmc/enigma/def2.htm.

Pantazis, D.N. (1994) Analyse méthodologique des phases de conception et de développement d'un système d'information géographique, unpublished Ph.D. Thesis, Université de Liège, 421 p.

Pantazis, D.N. and Donnay, J.P. (1996) *La conception de SIG : méthode et formalisme*, Hermes, Paris, 343 p.

Pantazis, D.N. and Cornélis, B. (1997) Designing and Implementing a GIS in an International Context, *Transaction in GIS* 1(4): 301-320.

Sauter, V.L. (1997) *Decision Support Systems : An Applied Managerial Approach*, John Wiley & Sons, New York, USA, 408 p.

Turban, E. and Aronson, J.E. (1998) *Decision Support Systems and Intelligent Systems*, 5th edition, Prentice Hall International, London, UK, 890 p.

UCGIS (1996) Research Priorities for Geographic Information Science, *Cartography and Geographic Information Systems* 23(3): 115-127.

Watson, H.J., Houdeshel, G., and Rainer, R.K. Jr (1997) *Building Executive Information Systems and Other Decision Support Applications*, John Wiley & Sons, New York, USA, 479 p.

Part 7
Conclusions and Recommendations

CONCLUSIONS AND RECOMMENDATIONS

J.V. VOGT[1] and F. SOMMA[1]

with contributions from

R.B. BRADFORD[2], J. BROMLEY[2], S. NIEMEYER[1], G. REES[2], A.A. VIAU[3] and all participants of the workshop

[1] *Space Applications Institute*
EC Joint Research Centre
21020 Ispra (Va), Italy

[2] *Institute of Hydrology*
Wallingford, Oxon, OX10 8BB, UK

[3] *Centre de recherche en géomatique, Université Laval,*
Ste-Foy, Québec, G1K 7P4, Canada

A lively and constructive discussion followed the presentations made in each of the five technical sessions of the workshop. A concluding session was also held, at the end of the meeting, during which previously raised issues were again discussed and a number of actions identified as being necessary for the development and set-up of a comprehensive Drought Watch System for Europe.

This chapter summarises the main issues raised in the presentations and the most pertinent comments made. It further outlines a series of proposed actions that were considered of high importance by the plenum. The participants of the workshop recognise that these are potential actions and that their accomplishment is subject to political and organisational constraints. Yet, the chapter offers an insight into viable ways to increase our odds in fighting drought, a natural hazard that should be accepted as an integral part of climate.

1. DROUGHT DEFINITION (Chairman: D. Wilhite, Rapporteur: S. Niemeyer)

Introduction

The scope of this session was to highlight the problems related to the definition of drought. The topic was discussed from a number of scientific and practical points of

J.V. Vogt and F. Somma (eds.), Drought and Drought Mitigation in Europe, 293–314.

view and the implications of the different definitions, especially with regard to drought monitoring, were further highlighted.

Summary of Presentations

The presentations gave an overview of various existing definitions of drought. Drought definitions have been discussed from climatological, hydrological, agro-meteorological and water resources management points of view. In general, these definitions go along with different drought indices that allow for a qualitative or quantitative assessment of water stress. At the same time, these indices are the basis for drought monitoring. From the hydrological point of view, groundwater drought has been discussed with particular emphasis on the consequences for water quality.

The difficulties related to the definition, detection and monitoring of agro-meteorological droughts have been highlighted. An agro-meteorological drought is directly related to the crop type, the current agricultural practice and other human influences.

Main Discussion Points

The discussion showed that any drought definition needs to be seen in the context of the application. It is impossible to come up with a unique definition of a drought event. However, as soon as a specific discipline is considered, a feasible and comprehensive definition can be given. This topic can be regarded as sufficiently covered by previous scientific studies. While an extensive overview of drought indices exists, there is a lack of inter-comparison and integration with regard to the needs of various applications.

A clear gap has been identified in the socio-economic field. Here no drought index ready for application exists. Further research is necessary in this field, in collaboration with the natural, agricultural and economic sciences.

An adequate regionalisation of drought indices was considered an important area of future research. The regions of Europe, not limited by national boundaries, were seen as most appropriate reference units for future activities.

The discussion also showed that the perception of drought is different throughout Europe. While ecological and water quality aspects are important features of a drought in north-western Europe, reduced irrigation and yield losses are the dominant concern in southern Europe. Because different tools for drought monitoring and detection exist, the information passed to the public and the political decision-makers must be tailored to the needs of the respective region and society.

Finally, the importance of a clearly identified customer was emphasised. It is possible to define a drought in its various aspects only with a specific user in mind. This definition can then serve as a basis for successful drought prevention and mitigation. Three levels of potential customers were identified:

(1) The regions of the European Community. They are closely related to individual groups directly affected by a drought (e.g. farmers, water authorities).

(2) The Member State authorities, their administrations and organisations related to civil protection and natural disaster mitigation programs.

(3) The European Commission, which has a high influence on regional development and agriculture within the EU. Here the concept of sustainable development that gained increasing importance over the last years is heavily touching on the problem of water stress.

Recommended Actions

(1) No need to define a single generic drought index (non-action).

(2) Compilation of an inventory of operationally used definitions and indices, with clarification of the relationship between (and relative merits of) different indicators.

(3) Development of socio-economic indicators for drought.

(4) Investigation and development of methods to describe the spatial extent of drought.

(5) Clarification should be sought as to who the potential user, or customer, is. That is, who requires the information at EU, national, regional and basin level and how will the information be used to influence policies?

(6) Applications of different indices in a variety of situations, with potential users canvassed on which are most appropriate.

(7) For each sector consider the adoption of a standard index (thus indices selection is sector specific).

2. DROUGHT RISK (Chairman: A.A. Viau, Rapporteur: J. Bromley)

Introduction

Drought affects everyone to a certain extent, but the severity and nature of the impact will vary from one group to another and from one society to another. A householder may be only slightly inconvenienced by the imposition of a hose-pipe ban while a farmer experiencing the same drought may face economic ruin. In poorer societies the drought risk is even more stark; it can result in starvation.

The aim of this session on Drought Risk was to provide a review of the methods and instruments of assessing the risk of droughts at a range of different scales, throughout Europe in particular. There were five presentations covering a range of topics: the need for drought risk assessment; techniques of analysis; drought policy; real-time, medium and long term prediction; and the importance of soil in the drought context.

Summary of Presentations

Assessment of drought risk

What do we mean by risk? This was a question posed during the session. Risk can be considered to be the result of imposing a hazard on something or someone that is vulnerable to that hazard. Following this reasoning the session adopted the definition of risk as:

Hazard x Vulnerability = Risk.

Presented in this way, the assessment of risk seems straightforward but in reality it is extremely difficult and complex. The hazard, which in this case is drought, is extremely variable in severity and extent in both space and time. Likewise those groups vulnerable to drought are also varied and the degree to which they are vulnerable depends very much on the social setting. Assessment of risk is particularly difficult in the light of changing climatic patterns and changing social conditions. Perceptions of drought are also changing. In Europe they are now beginning to be accepted as the 'norm' while in Africa improved management means that drought is no longer automatically assumed to mean food crisis.

To properly assess the risks posed by drought, improved methods of drought forecasting and a better understanding of the way in which society responds to droughts are required. This should be an important focus for future research.

Techniques of analysis

Droughts are not local phenomena. They are invariably caused by the development of persistent and widespread anticyclonic conditions. The Mediterranean droughts of 1988-89 and 1989-90 were particularly severe and linked to the development of large immobile anticyclones developed over Southern Europe. During the session the link between synoptic conditions and areas of severely reduced precipitation was demonstrated for a number of events.

The link between synoptic conditions and drought is clearly a prime area for research, and one that ultimately may provide a reliable means of predicting the onset of drought months before they occur.

Following on from the observation that droughts are not local but regional events, the session discussed the necessity for the analysis of river flows during these events to be undertaken at the regional and not site specific scales. It was shown that site-specific frequency analysis is based on a random sample from an unknown distribution. A regional approach on the other hand is much more robust, and through the spatial distribution of physical parameters can simulate the hydrological processes taking place.

From this investigation of frequency analysis two areas for future research were identified: firstly, the development of regional techniques of flow frequency analysis in which it is possible to specifically consider the hydrological processes involved; secondly, there should be more research to establish the relationship between synoptic conditions and stream flow in order to improve the capability to forecast droughts.

Drought policy and real-time, medium and long term forecasting

The work of the US National Drought Mitigation Centre (NDMC) based in Lincoln, Nebraska, has been discussed and it has been shown how the skills and services offered by this organisation have been used to plan for, monitor and mitigate droughts in the USA. All output from the centre is freely available on the World Wide Web and many world-wide links are offered to other sites which have drought-related themes. The Centre not only offers a real-time 'drought watch' for all the states, it also provides advice and expertise on policy and planning issues. For example, in 1982 only 3 states in the USA had drought plans in place, but with the aid of the NDMC a further 25 states have developed and implemented plans over the past 15 years. However, the measures in place still tend to be reactive, with too little emphasis on mitigation.

Following the severe drought of 1996 the National Drought Policy Act was introduced. This will set up a National Drought Policy Commission of which the NDMC will be an integral part. The objective of the commission will be to raise public awareness of the need for drought mitigation and prevention and to develop integrated local, state and national drought policies.

The lessons of the NDMC in the USA have demonstrated the value of having a drought mitigation centre and a network, which is able to provide:

- *(a) a world-wide clearinghouse for information (i.e. www site linked to other sites),*
- *(b) an up-to-the-minute 'drought watch' service,*
- *(c) advice to national and regional policy makers on drought planning and mitigation,*
- *(d) training workshops for planners,*
- *(e) research into drought-related issues.*

A similar organisation and network is needed to address the drought problems of Europe.

Integration of soil and climate data

The necessity to take into account the role of soil when estimating drought risk throughout Europe has been underlined. The impact of droughts on agriculture is manifested through a deficiency in soil water rather than in surface and ground water resources. It was demonstrated that maps showing the soil moisture status (the Soil Moisture Regime) could be produced by combining soil and climate data at the European scale. These types of maps can be produced to provide a real-time picture of soil moisture levels and an indication of the degree of stress being imposed on the vegetation.

The production of maps at a European scale to show real-time soil moisture status will be an essential component of any drought-watch network.

Main Discussion Points

(1) There is need to better understand how drought is perceived as a risk by different sectors in different regions. The assessment of risk is needed in social, economic and environmental terms, as well as in physical terms.

(2) The expression of drought risk in economic terms has greater resonance with decision-makers and is more likely to influence policy.

(3) Benefits could be gleaned by constructing a "knowledge base" of past drought events, the areas and sectors affected and the mitigation measures applied.

(4) The vulnerability of areas and (economic) sectors to drought risk should be assessed.

(5) There is need to raise the profile of Droughts and Drought Mitigation. It was noted that such 'knowledge construction' had been very successful in the fields of Climate Change and Biodiversity.

(6) The difficulty of reconciling political with scientific issues. Politicians work over a short time frame and will change their stance as the political situation changes. This is a major problem that needs to be addressed.

(7) The accurate forecasting of droughts will only be feasible when it is possible to predict changes, not only in the amount of rainfall, but also in its *distribution*. The spatial factor is of central importance.

(8) The impacts of climate change on crop production will be controlled by changes in the levels of CO_2 and temperature as much as by changes in rainfall.

(9) GIS systems are now being routinely used for spatial analysis, so that it is increasingly important that the existence of spatial data sets such as the European soil database should be publicised more extensively.

(10) Despite the importance of soil moisture it was observed that no European country has a national soil moisture monitoring network. In the absence of this type of data a possible way forward is to make more use of remotely sensed data.

(11) Geographical Information Systems' technology and the availability of improved spatial data-sets provide powerful tools whose potential have yet to be realised fully in drought research.

(12) Does Europe needed a Drought Mitigation Network or a Centre (or both)? The general feeling was that both a Centre and a Network are required but that success depends more on timing than the structure of the set-up. The United States National Drought Mitigation Centre was seen to be an excellent model.

Recommended Actions

(1) Undertake sectoral risk analysis in different regions.

(2) Construct a "knowledge base" of past drought events.

(3) Conduct research into:

- (a) trends and non-stationarity of data
- (b) synoptic weather patterns and teleconnections
- (c) regional hydrological analysis
- (d) continuous simulation, including the application of different climatic scenarios to assess the potential impact of drought
- (e) effect of climate change on drought vulnerability
- (f) application of remote sensing technology
- (g) drought risk in relation to other types of risk

3. DROUGHT MONITORING (Chairman: G. Rossi, Rapporteur: R.B. Bradford)

Introduction

An effective and comprehensive monitoring system is an essential part of a drought mitigation management plan. This needs to be based on a proper understanding of the complex characteristics of droughts. In addition, the spatial and temporal variability of drought events presents considerable problems in the generalisation of point measurements and in the evaluation of drought indices.

Drought severity is dependent not only on the duration, intensity and spatial extent of the drought, but also on the human and environmental demands on available water supplies. This will require precise, reliable and universally accepted definitions or indices expressed in terms that are meaningful to the different perspectives of policy makers, scientists, water planners, socio-economists and agriculturalists. A single index is unlikely to satisfy these different requirements.

The main objective of the session on Drought Monitoring was to review existing and potential methods of drought monitoring. Four presentations were given. These covered the status and trends in drought monitoring in the USA, a comparison of selected drought indices in Hungary, the application of satellite observations for drought monitoring, and a review of spatial and temporal scale issues in drought monitoring.

Summary of Presentations

Conventional methods for drought monitoring

In the USA, the National Drought Mitigation Center (NDMC) operates an Internet site that provides timely forecasts on potential drought conditions together with a wide range of information to assist decision makers at all levels to reduce the impact of drought events. Links with other sites provide information on, for example, snow pack, fire risk, river flows and groundwater levels. This enables users to rapidly assemble data and informational products relevant to their particular needs. An Integrated Climate Monitoring System (ICMS) is now being developed in collaboration with the Monitoring, Assessment and Prediction (MAP) Working Group of the Western Drought Co-ordination Council (WDCC). It is hoped that this will eventually extend nationally, but as yet a comprehensive, integrated nationwide climate monitoring or drought watch system has not been established in the USA.

Many countries have adopted some form of meteorological index or criteria for drought monitoring purposes, such as the decile approach in Australia and the Palmer Drought Severity Index (PDSI) in USA. As yet, however, there is no consistent approach across Europe. In the USA and elsewhere, the recently developed Standard Precipitation Index (SPI), which has a multiple time-scale capability and various other advantages compared to the PDSI, is now gaining wider acceptance for monitoring droughts, although further research is required to relate SPI to impacts in specific sectors.

PDSI and SPI values have been calculated using long-term data and data for the 1983 drought in Hungary to compare their suitability for monitoring purposes. The results suggest that SPI could be a more reliable drought indicator, although such indices need to be combined with real-time monitoring of streamflow, reservoir and groundwater levels and soil moisture deficits to provide a more truly effective drought monitoring system.

The preference for particular methods or indices for drought monitoring mainly depends on the particular application, availability of data and scale requirements. There is a need to improve the availability and standards of temporal and spatial monitoring data throughout Europe, particularly in the Mediterranean region, in order to develop suitable indicators and to provide timely information on drought and water stress. This could perhaps be achieved through the development of a Drought Monitoring System (DMS), which would benefit from a review of existing drought monitoring practices worldwide.

Potential methods for drought monitoring

The use of satellite data to monitor drought on regional to continental scales, either alone or in combination with other types of data, has increased over the past decade. The visible and near-infrared reflectances are used to examine the level of water stress by monitoring the vitality of the vegetation cover whilst the radiation emitted in the thermal infrared provides information on the evaporative flux over a larger surface.

A number of satellite-based drought indices have been developed, such as different Vegetation Indices (VIs), the Temperature Condition Index (TCI) and the Evaporative Fraction (EF). These can be calibrated with more conventional drought indices within a GIS environment to map the occurrence and severity of water stress. In this way, the high spatial information content of satellite data can be used to advantage. The potential application of remotely sensed data will be increased in the near future with the launch of new satellite platforms having enhanced capabilities and greater resolution.

Scale considerations are an integral part of the definition and objectives of drought monitoring and mitigation, and the level of detail required may vary depending on the approach. A particular challenge for drought monitoring is to identify the most appropriate scale (temporal, areal) for different management applications, appropriate historical perspectives and planning horizons.

A combination of satellite data, meteorological and land cover data in a GIS offers a promising and cost-effective method for the spatial and temporal monitoring of certain aspects of drought. Nonetheless, such information should be combined with more conventional indices into multi-scale indicators, in order to utilise the specific information provided by each approach.

Main Discussion Points

(1) There is need to improve drought monitoring in Europe, to provide timely information on drought and the general status of water resources in Europe; this may be achieved by an European "Drought Monitoring System" (DMS).

(2) A DMS could only be achieved by real-time data being more readily available than it is at present. Consequently, there is a need for improved access to environmental data, greater investment in hydro-meteorological networks and geological and hydrogeological mapping, and better standards of data collection. Water quantity issues need to have a higher profile within the broader context of water stress in Europe.

(3) The role of the European Environment Agency and its EuroWaternet implementation should be clarified in respect of a DMS.

(4) The requirements, benefits and users of a Drought Monitoring System (DMS) for Europe need to be clarified and any potential barriers to its implementation as well as its relation to other international databases and organisations identified.

(5) Earth observation systems (remote sensing) provide a new means of monitoring drought at a variety of scales.

(6) Research is required to identify the most appropriate multiple indices for drought monitoring. An integrated DMS based on a combination of such indices, GIS and remote sensing data may be an appropriate alternative to the more conventional approach for certain aspects of drought events.

(7) The most appropriate platform or pathway for disseminating the products of a DMS to a range of potential users needs to be investigated.

Recommended Actions

(1) Lobby EU, national governments and responsible agencies to improve access to environmental and real-time hydrometeorological data in Europe.

(2) Encourage public investment in hydrometric, meteorological, groundwater and soil moisture monitoring in Europe to arrest the present decline in observation networks (e.g. with INTERREG II support).

(3) Improve the mapping of geology, hydrogeology, soils and vegetation in Europe, with the information presented in digital form.

(4) Review drought monitoring practices around the world.

(5) Clarify the need for a drought monitoring system (DMS) for Europe; establish benefits of a DMS for different users; identify barriers to implementation; consider role of EEA/EuroWaternet, Eurostat, WHYCOS, etc.

(6) Establish a DMS, combining hydrometeorological data with Earth Observation data.

(7) Develop agricultural, social and economic indicators for drought monitoring and consider how these may be incorporated in a DMS.

(8) Determine dissemination platform/pathway for DMS products.

4. DROUGHT IMPACT (Chairman: G. Maracchi, Rapporteur: A.A. Viau)

Introduction

The main objective of the session was to review methods for the assessment of drought impact in various regions of Europe. Three presentations were given. These covered the socio-economic aspects of drought, the assessment of the vulnerability of river systems to drought and the key elements of a remote sensing based system for assessing drought impact on EU agriculture.

Summary of Presentations

Socio-economic aspects of droughts

Three objectives were addressed in this presentation: (1) the issue of drought and how it is perceived in Spain, highlighting the findings of the most recent literature; (2) a methodology to evaluate the socio-economic effects of drought on irrigated agriculture in southern Spain; and (3) the fact that drought has different impacts on various social groups belonging to the rural world.

Although the presentations scope was limited to irrigated agriculture, it looked beyond net revenue downfalls that result from water shortages. Intensive agriculture demands seasonal labour and variable inputs, and investments in irrigation equipment. Whereas net revenue measurements certainly provide a clear view of how droughts affect a regional economy, of equal importance are the trend of hired labour and the indirect effects on the rural economies.

Assessment of the vulnerability of a river system to drought

A method based on the regionalisation of recession parameters to estimate streamflow drought vulnerability along a river system was presented. Based on a multiple regression analysis with basin properties of 55 gauged catchments (ARIDE network), Recession parameters, initial discharges, QMS and threshold discharges were regionalised. Supported by GIS functions and network topology, aggregated recession curves were derived for all nodes along the river network and thus return periods could be estimated. A first test of the method was carried out in the study area of southwest Germany and the results and comparisons with earlier studies are satisfying. The visualised results show the degree of vulnerability of river stretches to droughts.

Key elements of remote sensing based systems for assessment of drought impact on EU agriculture

Based on drought related phenomena a retrospective learning process was suggested using satellite images either with low or high spatial resolution. Using low resolution satellite images the learning process refers to (1) the appearance of the crop vegetation cover in the image, (2) yield data of major crops from agrostatistics, (3) rotation patterns of cultivated crops, (4) phenological stages of cultivated crops, (5) local distribution of soil types, and (6) regional weather conditions. By combination of all these parameters over time it will be possible to reveal both periods and areas of high drought impact on crop growth. Moreover, at the end of a learning and calibration process a quantitative measure of drought impact can be related to yield.

Main Discussion Points

(1) It is important to balance long-term water demands with water supplies.

(2) The Common Agriculture Policy (CAP) influences the use of water resources in Europe.

(3) Policies and techniques to reduce water demands need to be further developed.

(4) Consistent criteria to assess a societal vulnerability to drought should be developed and tested.

(5) Drought encourages water trading and the control of water supplies by private companies.

(6) Hydrological criteria for expressing drought impacts are not necessarily consistent with socio-economic criteria.

(7) Socio-economic impacts of drought in the agricultural sector are important and farmer subsidies programmes are not well adapted to this situation.

(8) Subsidies often cushion the impacts of drought; there is little incentive for farmers to adapt and diversify.

(9) Impacts expressed in economic or monetary terms are often better understood by decision and policy makers.

(10) There is an increasing need to consider the ecological impacts of drought.

(11) The agricultural impact of a drought will vary according to the crop being cultivated and the timing, with respect to the growth stage of the crop.

(12) Not all impacts are negative (e.g. droughts can have a positive impact on tourism and, even, agriculture in some areas).

(13) There is a need for a European drought monitoring network.

(14) Early warning of drought may help farmers to cultivate crops more suitable for drought conditions.

(15) The establishment of a drought monitoring network for the EU needs the set-up of a thematic data base (GIS) including hydrological, meteorological, vegetation and soil information.

(16) Remote sensing offers great potential as a tool to assess drought impact on crop growth. Such an activity requires the availability of satellite data with different spatial resolution.

Recommended Actions

(1) Review previous examples of how drought impacts have been quantified and described by different users in different sectors.
(2) Establish best methods for expressing drought impacts.
(3) Identify methods for determining ecological impacts; what improvements can be made?
(4) Investigate how the CAP contributes to sustainable water use in Europe and how it may effect on the occurrence and impact of droughts.
(5) Conduct research into sustainable water use in agriculture and review how the use of new agro-technologies may reduce the impact of droughts.
(6) Consider the use of remote sensing, in conjunction with hydrological methods/data, to determine drought impacts.

5. DROUGHT MITIGATION (Chairman: J. Lomas, Rapporteur: F. Somma)

Introduction

The development of drought mitigation strategies and their implementation in political and organisational frameworks is a central issue for the management of Europes water resources. The regional and organisational differences existing over the extended territory of the European Union and the European continent as a whole need to be considered in these frameworks. Such a proactive approach to the problem of water stress should yield a much better basis for sustainable development in Europe

Summary of Presentations

Measures for mitigation of drought impacts include a very large number of actions, which can be grouped into three broad categories: supply-increase, demand reduction, drought impact minimisation. Within this schematisation, a distinction can be made between approaches. The *proactive* approach consists of measures conceived and prepared according to a planning strategy rather than improvised measures. The *reactive* approach consists of measures adopted once a drought occurs and impacts are perceived. It can be indicated as the “crisis management” approach because it is not based on plans prepared in advance. Although the reactive approach still represents the common response to drought events, there is an increasing awareness that such an approach implies hasty decisions and leads to expensive actions, which have often unsustainable environmental impacts and social costs.

The concept of drought itself and its perception within the community needs to be reassessed. For example, drought in Spain is a common event almost every year. This emphasises the need for a distinction between what could be classified as an exceptional event (drought) and what is perceived as such, but is instead a permanent or periodic condition of water shortage, stemming from unrealistic expectations placed on the water supply. In other words, shortage of water might be due to an excessive water demand or to a wasteful use of resources in a dry area, rather than to drought.

Economic impacts of drought events are difficult to estimate, as they go beyond the immediate, direct losses to agriculture, recreation, or power production, and include not readily quantifiable environmental losses. Social impacts are also difficult to evaluate, and involve health, water uses conflicts, rural migration, loss of jobs, reduced quality of life. Severe consequences of a drought event are sometimes reduced, thanks to international trade systems: large volumes of water are sometimes shifted from a country to another in the form of food commodities, thus diminishing the water requirement of the area receiving the commodity (concept of "virtual water", equivalent to the water required to produce the commodity itself).

Examples are presented in the session papers for some countries about past drought events, and the response to it at national level, showing how the most recent droughts have found both the institutions and the affected sectors largely unprepared. Awareness of a drought state only when the phenomenon has fully bloomed often prompts national and regional governments face the emergency mainly by increasing the water supply. This goal seems to be achieved mostly through structural measures (construction of new infrastructures for water storage and transport) and but also through the over-exploitation of existing supplies.

As pressure due to the emergency raises, such improvisation result in inappropriate use of public funds, corruption, and environmental impacts, stemming mostly from professional bias, lack of know-how and lack of preparedness. For example in Spain a valuable resource such as groundwater was not efficiently exploited during the last drought, even though surveys had shown that it could play a relevant role in mitigating the effects of drought in the agricultural and urban water supply. On the other hand the water transfer option, hastily planned, has lead to conflicts between different user groups and in some cases with the neighbouring country of Portugal. The environmental impacts of water transfers, aiming to achieve a balance between drier and wetter areas, need to be re-evaluated.

As such, the need arises for a change in policy that shifts the economic burden of drought mitigation measures (mostly of the structural type) from the national and regional budgets to those users groups benefiting from the measures themselves. This would stimulate a closer interest in the measures, and would assure a greater success.

The need is generally recognised for the development of a drought mitigation plan including a drought watch system. Very few countries are equipped with such plans/systems (e.g. USA, Australia). Attempts have been made in some of the most drought prone European countries (Italy, Spain, France) and general guidelines have been drafted in some cases. However, no comprehensive drought plan is operating in any European country at the moment.

Main Discussion Points

(1) A reactive approach to drought mitigation leads to questionable responses; a proactive approach is encouraged.

(2) Traditionally, mitigation measures have focused on increasing water supply rather than on reducing demand; greater emphasis on a policy of demand management is needed.

(3) Drought should be an established component of each water management plan.

(4) There is need for an EU policy on drought and an institutional framework for drought mitigation.

(5) There is need to improve public awareness of drought and the need to use water wisely.

Recommended Actions *(continued on next page)*

(1) Careful evaluation of water resources availability to improve water management. It is necessary to avoid neglecting precious resources (e.g. groundwater in Spain) and/or placing unreasonable expectation on resource availability, thus declaring a drought alert all too often (see examples in Spain again). Environmental impacts due to resource overexploitation need to be evaluated and minimised as well.

(2) Use water pricing as a tool to increase awareness and reduce water demand in the domestic sector. Higher prices will make consumers less prone to wasteful usage, while increasing the availability of funds necessary to improve the efficiency of the service provided.

(3) Apply the same concept to the agricultural sector, where it is suggested that the cost related to water management and resource augmentation (new infrastructures, maintenance of old ones) be born by the stakeholders, rather that being entirely supported by the national budget.

(4) Increase farmers' awareness to drought through training.

(5) Support farming management flexibility. Encourage the choice of less profitable, more drought resistant crops.

(6) Encourage the development of an adequate (subsidised, if necessary) farmers' insurance scheme.

(7) Establish an effective drought monitoring and early-warning system. Prepare a National Drought Policy. A comprehensive analysis of direct and indirect economic and environmental drought impacts should be included. Policy drafting and reform should take advantage of the "window of opportunity" (convergence of opinions and ideas between public, media and government) available during a drought emergency.

(8) Establish if there is a need for a specific EU policy on drought or if drought should be included in existing policies.

(9) Investigate the influence of fiscal mechanisms (e.g. insurance, compensation payments, tax relief, etc.) on drought mitigation.

(10) Assess the preparedness of countries to drought; review water management strategy, develop contingency actions (e.g. increase of water supply/ or reduction of demand) and Drought Management Plans.

(11) Review experiences of drought mitigation in Europe and identify best practices.

(12) Produce guidelines for the development of Drought Management Plans in Europe and establish a training mechanism (for stakeholders) for the implementation of the guidelines.

(13) Review the relative importance of politics and economics in the mitigation (or not) of past drought events.

(14) Consider the feasibility of an early warning system, how such a system would be used, by whom and at what cost.

6. GENERAL CONCLUSIONS AND PROPOSAL FOR A EUROPEAN NETWORK ON DROUGHT RESEARCH AND MITIGATION (ENDRM)

(Chairmen: J.V. Vogt and A. Gustard,
Rapporteurs: G. Rees, J.V. Vogt and A.A. Viau)

Introduction

The purpose of the last session was to draw the general conclusions from the foregoing workshop discussions and to arrive at some clear indications on where to prioritise actions in the future. In this framework the proposal to establish a European network with the purpose to stimulate and catalyse research and applications in the field of drought research and mitigation has been extensively discussed.

Two presentations where given in order to place the discussions in a broader context: The first presentation by Dr. P. Balabanis of DG RESEARCH-D4 gave an overview on European Commission funded activities related to drought and on future perspectives within the 5th Framework Programme of the EC. The second presentation by B. Cornélis highlighted the theoretical basis of decision mechanisms and proposed a study of current decision structures for drought management at various administrative levels in the European Union.

Summary of Presentations

Within the Environmental Research Programme of the 4th Framework Programme (1995-1998) the drought problematic was included under the headings of 'Climate Change', 'Climate Change Impacts on Agriculture', and 'Socio-Economic Aspects of Climate Change'. Drought related projects were mainly focused on the Mediterranean. The main action lines were the study of climate change and meteorological scenarios and the study of drought impacts. Examples are the MEDALUS (Mediterranean Desertification and Landuse), GRAPES (Groundwater and River Resources Action Programme on a European Scale) and ARIDE (Assessment of the Regional Impact of Droughts in Europe) projects.

The overall conclusions of the research performed under this programme highlight that:

(a) *Water resources management in Mediterranean countries is already a critical issue and current trends in demography will further increase the vulnerability of Mediterranean societies to drought.*

(b) *Better planning can significantly reduce the risk related to the high uncertainty in current projections of water availability on different time-scales. This needs an integrated long-term water management plan.*

(c) *Basic data for studying the drought phenomenon are not readily available.*

Based on past findings, the following research priorities have been defined for the future:

(a) *The historical perspective of drought needs to be investigated through long-term climatological studies.*

(b) *Aspects of ecology, landscape change and land degradation need to be integrated.*

(c) *The socio-economic factor needs to be seriously considered.*

(d) *Definitions of cost benefit need to be established.*

(e) *Specific case studies need to be performed.*

(f) *Advanced remote sensing techniques should be better exploited.*

Within the 5th Framework Programme (1999 – 2002) drought issues figure mainly under the thematic programme on 'Energy, Environment and Sustainable Development', more precisely under the key action on 'Sustainable Management and Quality of Water'. Targeted fields of research of this key action are, among others, 'Surveillance, Early Warning and Communication Systems', and 'Regulation of Stocks in Water-Deficit Regions'. Further relevance is found under the key action on 'Global Change, Climate and Biodiversity' and under the generic activities on 'The Fight against Major Natural and Technological Hazards' and on 'Development of Earth Observation Satellite Technologies'. Emphasis is put on the watershed as the basic reference unit. The link to the policy on regional development, the important role of socio-economic aspects and the need to develop decision support tools for a sustainable water management have been explicitly highlighted.

The presentation on decision processes highlighted the relations between the issue at hand (problem, needs, objectives), the data (raw, integrated, processed) and the context (organisation, experience, knowledge) in which a decision has to be made. The role of geographical information systems (GIS) and the problems in setting up a decision support system (DSS) integrating the necessary spatial data with modelling capabilities and the human decision process has been discussed. A study on the decision processes and the data and tools necessary for decisions regarding drought management and mitigation has finally been proposed. This study could be integrated in the proposed ENDRM, which should provide the link to both the research and application side and a test-bed for the DSS to be developed.

Main Discussion Points

The main conclusions of the individual sessions as presented by the rapporteurs have been discussed and amended by the plenum. Out of this discussion resulted a generally approved list of comments and possible actions structured according to the topics of the technical sessions. These comments and actions have been included in the reports of the various sessions (see above).

The participants of the workshop recognise that this list is a list of potential actions and that their accomplishment is restricted by political and organisational constraints. Taken this into consideration, the most important actions to be pursued with high priority can be summarised as follows:

Most Important Actions:

(a) Fast set-up of a first information server on the Internet (web server) as a focal point, including links to drought relevant projects.

(b) Identification of the various potential users of the information which can be provided. The different levels of users of such information should be considered (e.g. water authorities, regional, national, and European administrations).

(c) Seeking of financial support through network programmes, Concerted Action programmes or Shared Cost Action proposals.

(d) Development of some example applications for demonstration purposes.

Proposals for European Network on Drought Research and Mitigation (ENDRM)

Many of the identified gaps in data, research and actions are most likely to be filled through the set-up of a network of researchers, stakeholders and political actors in the field. The establishment of a European Network on Drought Research and Mitigation (ENDRM) was therefore proposed. The discussion of how such a network could be organised resulted in a variety of proposals regarding the organisational structure, the objectives, the components and the potential activities of such a network.

It was generally accepted that due to the very complex nature of the drought phenomenon a modular structure, implementing a limited number of working groups (or nodes) would be a practical solution. This structure could take advantage of already existing projects and/or networks in different disciplines such as hydrology and agrometeorology. The main question would then be how to effectively link these working groups. A structure with a regularly meeting steering committee and a permanent secretariat was proposed. The link between the working groups, including the exchange of data and dissemination of results could be established via the Internet. A schematic picture of such a network is given in figure 1.

It is important to note that the link to the customer as shown in this scheme is twofold. The customer should be able not only to retrieve information but also to communicate with the network by giving comments and feedback and by formulating specific requests according to his needs. It is further important to consider the link to similar activities outside the EU. Considerable experience has already been gained in the United States, Australia and Southern Africa. A regular exchange of information and experience with the relevant organisations in these countries will be a worthwhile effort for both sides.

Major objectives, components and activities of this network are summarised in the table one.

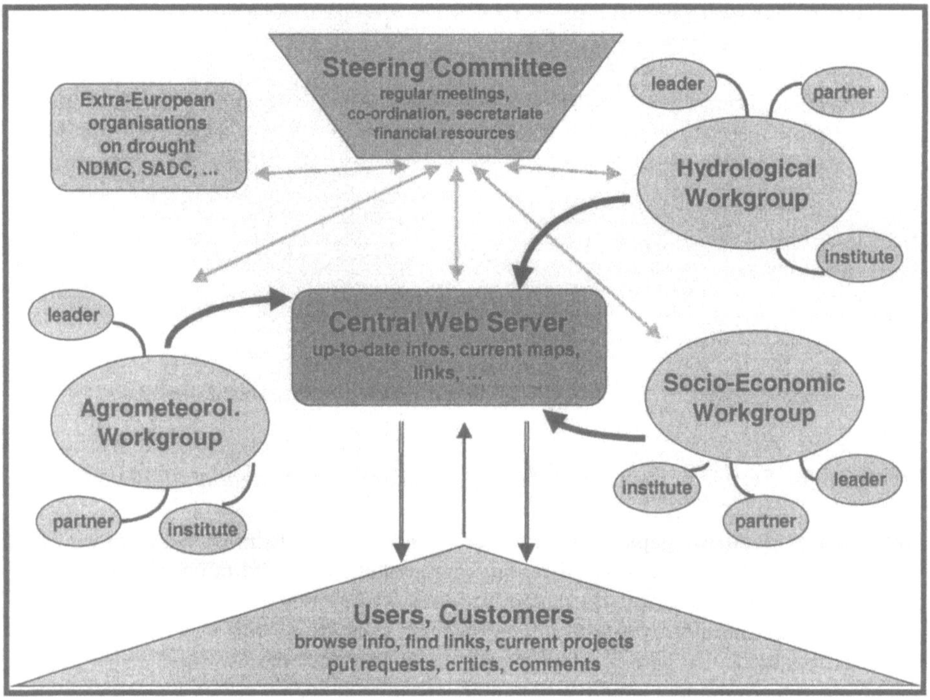

Figure 1: Proposed network structure.

Table 1: Proposed objectives, components and activities of a European Drought Research and Mitigation Network.

Objectives	Components	Activities
Promote the network objectives among policy makers and clients	Policy Administration Water Authorities	❑ Organise stakeholder workshops ❑ Raise awareness of the drought problematic ❑ Establish a central contact point ❑ Set up of web server (data and information)
Provide regional information on drought risk	Drought Monitoring Centre (DMC) Regional Working Groups (WGs)	❑ Develop web based monitoring products. ❑ Give past drought diagnostics ❑ Test real time risk assessment & forecast
Promote methodologies for drought monitoring and mitigation targeted towards specific sectors, regions and stakeholders	<u>Sectoral and Regional WGs</u> Climate — Mediterranean Hydrology — West Europe Agrometeorology — Northern Europe Socio-Economy — Central Europe Vulnerability — Eastern Europe Water Technology Health	❑ Contribute to DMC ❑ Develop on-line journal ❑ Disseminate research results ❑ Compile meta-data catalogue on existing data (e.g. meteorological, hydrological, land use), research networks, and projects.

Continued on next page

Table 1: Proposed objectives, components and activities of a European Drought Research and Mitigation Network (continued).

Objectives	Components	Activities
Stimulate effective drought contingency and mitigation plans	Development of Case Examples Include stakeholders, regional, and national administrations. Consider both integrated management and region specific issues.	❑ Document and disseminate innovation ❑ Demonstrate utility of successful planning ❑ Compile information on existing plans
Educate and train key audiences in drought monitoring and mitigation	Education Training Public Relation	❑ Generate education material ❑ Establish protocols for developing plans ❑ Organise application workshops
Develop decision support (DS) tools	Decision Makers at different levels (local, regional, national, European)	❑ Establish contacts and select test cases ❑ Demonstrate the use of DS tools

List of Contributors

Allan, J. Anthony
Water Issues Group, School of Oriental & African Studies, University of London, Thornhaugh Street, London, WC1H 0XG UK.

Bakker, Karen
Environmental Change Institute, University of Oxford, 1a Mansfield Road Oxford, OX1 3TB,UK.

Beaudin, Isabelle
Centre de recherche en géomatique, Université Laval, Ste Foy, Québec, G1K 7P4, Canada.

Bradford, Robert B.
Risks and Resources Division, Institute of Hydrology, Crowmarsh Gifford, Wallingford, Oxfordshire, OX10 8BB,UK.

Bromley, John
Risks and Resources Division, Institute of Hydrology, Crowmarsh Gifford, Wallingford, Oxfordshire, OX10 8BB,UK.

Cornélis, Bernard
Département de Géomatique, Université de Liège, Place du 20-Août 7, B-4000 Liège, Belgium.

Demuth, Siegfried
Institute of Hydrology, University of Freiburg, Fahnenbergplatz, 79098 Freiburg, Germany.

Downing, Thomas
Environmental Change Institute, University of Oxford, 1a Mansfield Road Oxford OX1 3TB Oxford,UK.

Garrido, Alberto
Departamento de Economía y Ciencias Sociales Agrarias, ETSI Agrónomos, Universidad Politécnica de Madrid, 28040 Madrid, Spain.

Gómez-Ramos, Almudena
Departamento de Economía y Ciencias Sociales Agrarias, ETSI Agrónomos, Universidad Politécnica de Madrid, 28040 Madrid, Spain.

Gustard, Alan
Head, Risks and Resources Division, Institute of Hydrology, Crowmarsh Gifford, Wallingford, Oxfordshire, OX10 8BB,UK.

Hayes, Michael J.
National Drought Mitigation Center, International Drought Information Center, University of Nebraska – Lincoln, 241 L. W. Chase Hall, Lincoln, Nebraska 68583-0749, USA

Jones, Robert R.J.A.
European Soil Bureau, Space Applications Institute, EC Joint Research Centre, TP 263, 21020 Ispra (Varese), Italy.

Now with: Soil Survey and Land Research Centre, Cranfield University at Silsoe, Silsoe, Bedfordshire, MK45 4DT UK

Kühbauch, Wolfgang
Lehrstuhl für Allgemeinen Pflanzenbau, Universität Bonn, Katzenburgweg 5, 53115 Bonn, Germany.

Lanen, Henny van
Sub-department of Water Resources, Wageningen Agricultural University, Nieuwe Kanaal 11, 6709 PA Wageningen, The Netherlands

Lehner, Bernhard
Institute of Hydrology, University of Freiburg, Fahnenbergplatz, 79098 Freiburg, Germany.

Llamas, M. Ramón
Departamento de Geodinamica, Facultad de Ciencias Geologicas, Universidad Complutense de Madrid (UCM), 28040 Madrid, Spain.

Lomas, Jakob
Director, WMO Regional Meteorological Training Centre for Postgraduate Training in Applied Meteorology, Meteorological Service, P.O. Box 25, 50250 Bet Dagan, Israel.

Maheras, Panagiotis Ch.
Department of Meteorology and Climatology, University of Thessaloniki, 54006 Thessaloniki, Greece.

Maracchi, Giampiero
Director, Istituto per l'Agrometeorologia e l'Analisi Ambientale applicata all'Agricoltura (IATA), P.le delle Cascine 18, 50144 Firenze, Italy, and Department of Climatology and Agrometeorology, University of Florence, 50144 Florence, Italy

Montanarella, Luca
European Soil Bureau, Space Applications Institute, EC Joint Research Centre, TP 263, 21020 Ispra (Varese), Italy.

Niemeyer, Stefan
Agriculture and Regional Information Systems Unit, Space Applications Institute, EC Joint Research Centre, 21020 Ispra (Varese), Italy.

Now with: Swiss Federal Institute for Snow and Avalanche Research, Flüelastr. 11, 7260 Davos, Switzerland

Peters, Elisabeth
Sub-department of Water Resources, Wageningen Agricultural University, Nieuwe Kanaal 11, 6709 PA Wageningen, The Netherlands

Rademacher, Ingo
Lehrstuhl für Allgemeinen Pflanzenbau, Universität Bonn, Katzenburgweg 5, 53115 Bonn, Germany.

Rees, Gwyn
Risks and Resources Division, Institute of Hydrology, Crowmarsh Gifford, Wallingford, Oxfordshire, OX10 8BB,UK.

Rossi, Giuseppe
Istituto di Idraulica, Idrologia e Gestione delle Acque, Università di Catania, Viale A. Doria, 6, 95125 Catania, Italy.

Somma, Francesca
Space Applications Institute, EC Joint Research Centre, 21020 Ispra (Varese), Italy.

Now with: Department of Agricultural Engineering, University of Catania, Via Valdisavoia, 5, 95123 Catania, Italy.

Stahl, Kerstin
Institute of Hydrology, University of Freiburg, Fahnenbergplatz, 79098 Freiburg, Germany.

Svoboda, Mark D.
National Drought Mitigation Center, International Drought Information Center, University of Nebraska – Lincoln, 241 L. W. Chase Hall, Lincoln, Nebraska 68583-0749, USA

Szalai, Sandor
Climate Research and Development Division, Hungarian Meteorological Service, 1024 Budapest, Hungary.

Szinell, Csaba
Climate Research and Development Division, Hungarian Meteorological Service, 1024 Budapest, Hungary.

Tallaksen, Lena
Department of Geophysics, University of Oslo, P.O. Box 1022, Blindern, 0315 Oslo, Norway

Tate, Emma
Risks and Resources Division, Institute of Hydrology, Crowmarsh Gifford, Wallingford, Oxfordshire, OX10 8BB,UK.

Viau, Alain
Centre de recherche en géomatique, Université Laval, Ste Foy, Québec, G1K 7P4, Canada.

Vogt, Jürgen
Environment and Geo-Information Unit, Space Applications Institute, EC Joint Research Centre, TP 441, 21020 Ispra (Va), Italy.

Wilhite, Donald A.
Director, National Drought Mitigation Center, International Drought Information Center, University of Nebraska – Lincoln, 241 L. W. Chase Hall, Lincoln, Nebraska 68583-0749, USA

Zdruli, Pandi
European Soil Bureau, Space Applications Institute, EC Joint Research Centre, TP 263, 21020 Ispra (Varese), Italy.

Now with: International Centre for Advanced Mediterranean Agronomic Studies, Mediterranean Agronomic Institute of Bari, 70010 Valenzano, Bari, Italy

Subject Index

E

Advances in Natural and Technological Hazards Research

Series Editor: Prof. Dr. Mohammed I. El-Sabh, *Département d'Océanographie, Université du Québec à Rimouski, 310 Allée des Ursulines, Rimouski, Québec, Canada G5L 3A1*

Publications

1. S. Tinti (ed.): *Tsunamis in the World.* Fifteenth International Tsunami Symposium (1991). 1993 ISBN 0-7923-2316-5
2. J. Nemec, J.M. Nigg and F. Siccardi (eds.): *Prediction and Perception of Natural Hazards.* Symposium Perugia, Italy (1990). 1993 ISBN 0-7923-2355-6
3. M.I. El-Sabh, T.S. Murty, S. Venkatesh, F. Siccardi and K. Andah (eds.): *Recent Studies in Geophysical Hazards.* 1994 ISBN 0-7923-2972-4
4. Y. Tsuchiya and N. Shuto (eds.): *Tsunami: Progress in Prediction, Disaster Prevention and Warning.* 1995 ISBN 0-7923-3483-3
5. A. Carrara and F. Guzzetti (eds.): *Geographical Information Systems in Assessing Natural Hazards.* 1995 ISBN 0-7923-3502-3
6. V. Schenk (ed.): *Earthquake Hazard and Risk.* 1996 ISBN 0-7923-4008-6
7. M.I. El-Sabh, S. Venkatesh, H. Denis and T.S. Murty (eds.): *Land-based and Marine Hazards.* Scientific and Management Issues. 1996 ISBN 0-7923-4064-7
8. J.M. Gutteling and O. Wiegman: *Exploring Risk Communication.* 1996 ISBN 0-7923-4065-5
9. G. Hebenstreit (ed.): *Perspectives on Tsunami Hazard Reduction.* Observations, Theory and Planning. 1997 ISBN 0-7923-4811-7
10. C. Emdad Haque: *Hazards in a Fickle Environment: Bangladesh.* 1998 ISBN 0-7923-4869-9
11. F. Wenzel, D. Lungu and O. Novak (eds.): *Vrancea Earthquakes: Tectonics, Hazard and Risk Mitigation.* 1999 ISBN 0-7923-5283-1
12. S. Balassanian, A. Cisternas and M. Melkumyan (eds.): *Earthquake Hazard and Seismic Risk Reduction.* 2000 ISBN 0-7923-6390-6
13. S.L. Soloviev, O.N. Solovieva, C.N. Go, K.S. Sim and N.A. Shchetnikov: *Tsunamis in the Mediterranean Sea 2000 B.C. – 2000 A.D.* 2000 ISBN 0-7923-6548-8

Advances in Natural and Technological Hazards Research

14. J.V. Vogt and F. Somma (eds.): *Drought and Drought Mitigation in Europe.* 2000
ISBN 0-7923-6589-5

Kluwer Academic Publishers – Dordrecht / Boston / London

FSC
www.fsc.org
MIX
Papier aus verantwortungsvollen Quellen
Paper from responsible sources
FSC® C105338

If you have any concerns about our products,
you can contact us on
ProductSafety@springernature.com

In case Publisher is established outside the EU,
the EU authorized representative is:
Springer Nature Customer Service Center GmbH
Europaplatz 3, 69115 Heidelberg, Germany

Printed by Libri Plureos GmbH
in Hamburg, Germany